Springer Complexity

Springer Complexity is an interdisciplinary program publishing the best research and academic-level teaching on both fundamental and applied aspects of complex systems – cutting across all traditional disciplines of the natural and life sciences, engineering, economics, medicine, neuroscience, social and computer science.

Complex Systems are systems that comprise many interacting parts with the ability to generate a new quality of macroscopic collective behavior the manifestations of which are the spontaneous formation of distinctive temporal, spatial or functional structures. Models of such systems can be successfully mapped onto quite diverse "real-life" situations like the climate, the coherent emission of light from lasers, chemical reaction-diffusion systems, biological cellular networks, the dynamics of stock markets and of the internet, earthquake statistics and prediction, freeway traffic, the human brain, or the formation of opinions in social systems, to name just some of the popular applications.

Although their scope and methodologies overlap somewhat, one can distinguish the following main concepts and tools: self-organization, nonlinear dynamics, synergetics, turbulence, dynamical systems, catastrophes, instabilities, stochastic processes, chaos, graphs and networks, cellular automata, adaptive systems, genetic algorithms and computational intelligence.

The two major book publication platforms of the Springer Complexity program are the monograph series "Understanding Complex Systems" focusing on the various applications of complexity, and the "Springer Series in Synergetics", which is devoted to the quantitative theoretical and methodological foundations. In addition to the books in these two core series, the program also incorporates individual titles ranging from textbooks to major reference works.

Editorial and Programme Advisory Board

Understanding Complex Systems

Founding Editor: J.A. Scott Kelso

Future scientific and technological developments in many fields will necessarily depend upon coming to grips with complex systems. Such systems are complex in both their composition – typically many different kinds of components interacting simultaneously and nonlinearly with each other and their environments on multiple levels – and in the rich diversity of behavior of which they are capable.

The Springer Series in Understanding Complex Systems series (UCS) promotes new strategies and paradigms for understanding and realizing applications of complex systems research in a wide variety of fields and endeavors. UCS is explicitly transdisciplinary. It has three main goals: First, to elaborate the concepts, methods and tools of complex systems at all levels of description and in all scientific fields, especially newly emerging areas within the life, social, behavioral, economic, neuro- and cognitive sciences (and derivatives thereof); second, to encourage novel applications of these ideas in various fields of engineering and computation such as robotics, nano-technology and informatics; third, to provide a single forum within which commonalities and differences in the workings of complex systems may be discerned, hence leading to deeper insight and understanding.

UCS will publish monographs, lecture notes and selected edited contributions aimed at communicating new findings to a large multidisciplinary audience.

Klaus Lucas · Peter Roosen
Editors

Emergence, Analysis and Evolution of Structures

Concepts and Strategies Across Disciplines

Springer

Editors
Prof. Dr. Klaus Lucas
RWTH Aachen
Lehrstuhl für Technische
Thermodynamik
Schinkelstr. 8
52062 Aachen
Germany
lucas@ltt.rwth-aachen.de

Dr. Peter Roosen
Gesellschaft zur Optimierung
von Energiesystemen und
thermochemischen
Prozessen (g.o.e.the)
Höhenweg 50
52074 Aachen
Germany
roosen@vollmer-roosen.de

berlin-brandenburgische
AKADEMIE DER WISSENSCHAFTEN

The publication is the result of the interdisciplinary study group *Structure Evolution and Innovation* of the Berlin-Brandenburg Academy of Sciences and Humanities. It has been made possible with the kind assistance of the Senate Department for Education, Science and Research, Berlin, and the Ministry of Science, Research and Culture, Brandenburg.

ISBN 978-3-642-00869-6 e-ISBN 978-3-642-00870-2
DOI 10.1007/978-3-642-00870-2
Springer Heidelberg Dordrecht London New York

Library of Congress Control Number: 2009927026

Cover design: WMXDesign GmbH, Heidelberg

Printed on acid-free paper

Springer is part of Springer Science+Business Media (www.springer.com)

Preface

In May 2002 a number of about 20 scientists from various disciplines were invited by the Berlin-Brandenburg Academy of Sciences and Humanities to participate in an interdisciplinary workshop on structures and structure generating processes. The site was the beautiful little castle of Blankensee, south of Berlin. The disciplines represented ranged from mathematics and information theory, over various fields of engineering, biochemistry and biology, to the economic and social sciences. All participants presented talks explaining the nature of structures considered in their fields and the associated procedures of analysis.

It soon became evident that the study of structures is indeed a common concern of virtually all disciplines. The motivation as well as the methods of analysis, however, differ considerably. In engineering, the generation of artifacts, such as infrastructures or technological processes, are of primary interest. Frequently, the analysis aims there at defining a simplified mathematical model for the optimization of the structures and the structure generating processes. Mathematical or heuristic methods are applied, the latter preferably of the type of biology based evolutionary algorithms. On the other hand, setting up complex technical structures is not possible by such simplified model calculations but requires a different and less model but rather knowledge-based type of approach, using empirical rules rather than formal equations. In biochemistry, interest is frequently focussed on the structures of molecules, such as proteins or ribonucleic acids. Again, optimal structures can usually be defined. They are synthesized by elaborate experimental methods, but also by computer based evolutionary approaches, similar in general but different in detail from those used in engineering. Quite different is the study and explanation of existing structures and structure generating processes in biology and in the social sciences. Here a mathematical model can, as a rule, not be formulated, because the internal interactions are too complex and are strongly governed by large statistical uncertainties or even are of a non-causal nature. Analysis therefore requires, at least partially, approaches of analysis which are rather different from those of the physical, chemical or the technical sciences. Thought and experimental observation, rather than mathematical analysis, are the basis of analysing and documenting structures and structure generating processes in these fields. The economic sciences play a kind of intermediate role as both types of analysis are found. Finally, the question of optimality does not present itself in an unambiguous manner in these fields.

Generally, biological and social structures, although developing in an evolutionary way, cannot be considered as proceeding towards an a-priori definable goal of optimality.

In view of the many fundamental analogies of structures and structure generating processes in the various fields, and also of the equally fundamental differences, the workshop resulted in a broad conviction that a systematic interdisciplinary investigation of the topic would be of considerable scientific interest. On this basis an interdisciplinary study group was inaugurated with the beginning of 2003 by the Berlin-Brandenburg Academy of Sciences and Humanities. It was instituted for a period of three years, i.e. over the years 2003–2005, with the dedication of studying the emergence, analysis and evolution of structures in the various disciplines. This book documents the research results of this group, working out the transdisciplinary fundamentals and analysing their operation in a number of case studies. From the results it can be speculated that a new transdisciplinary research field of structure analysis may emerge, condensing the various strategies to a common approach towards a better understanding of structures and structure generating processes in all fields.

It first became necessary to define the characteristic features of structures and to find a useful classification for them. Here, the fundamental notions of static and dynamic structures, originally proposed in thermodynamics, turned out to provide a largely common basis for understanding and analysis. Static structures, isolated from their environments are determined by internal interactions of their constituting components only, and evolve, as time proceeds, towards a predetermined or at least implicitly definable goal. They remain unchanged over time when this goal has been reached. On the other side, dynamic structures, additionally influenced by interactions with the environment, develop in an essentially unpredictable way and without approaching a predetermined goal. They require steady exchange of material, energy and information with the environment for their persistence. Many structures, notably the artifacts in the technical and economic sciences, appear to be static at first sight. However, a closer look reveals that their design and operation depends to a large extent on the environment, which makes them assume typical features of dynamical structures. When looking at their evolution over a long time it can be seen that their interactions with the environment in terms of energy, material and information frequently change in a basically unpredictable way, sometimes abruptly, and with them the related structures, too. In this general view it appears that the common properties of structures in a general sense are those found in the dynamic structures, such as dependence on external interactions, abrupt changes and intrinsic unpredictability.

The basis of analysis of structures in all sciences are models. So, the modeling of reality as executed in the various fields had to be investigated. It turns out that the major approaches of modeling are rather different in the various disciplines. Still, there is also much in common. As an example, any model aiming at representing reality has to cope with various types of uncertainty. Methods to cope with uncertainty are available, ranging from error propagation and stochastic approaches in the technical model simulations to the approach of knowledge based decision making in complex applications of engineering and sociology. In cases in which a model maps the input parameters mathematically into a performance function, an algorithm has to be used to work out this relationship, either provided by strict mathematics or by some kind of heuristic. Models of real structures and structure generating processes in some fields are analysed in terms of optimality, at least partially, i.e. restricted in space and time. However, the assessment of optimality even if justified is not straightforward, since various conflicting goals contribute to its definition in most applications. While these goals may be objective or subjective in themselves, their relative weights for a global optimality assessment are in most cases subjective. Also sometimes these weights change over time due to subjective esteem. The complex aspects of qualitative and quantitative modeling, and the goal-oriented and statistical approaches to understanding the meliorisation efforts in the various fields are discussed, with the aim to help readers familiarize themselves with methods hitherto unusual in their own fields.

These transdisciplinary fundamentals can be seen at work in a large variety of case studies from various disciplines, which constitute the main body of the text. They are taken from the fields of natural and technical sciences such as engineering, biology, medicine and chemistry as well as from social sciences such as sociology, economics and psychology. On the basis of these case studies the general purpose of this book becomes evident: the exploration of the status of knowledge about structures and structure generating processes and of the tools available for analysing them in various disciplines. It is expected that information is provided that may help the reader to cross the border of the disciplines, with the result of being inspired by new ideas profitable in his or her own field. Also the fundamentals as well as the case studies considered may serve to point out a frequently perceived resemblance of rather different scientific fields.

The contributors to this book are listed at the end. While the authors of the case studies have been explicitly identified, the introduction, the chapter on the transdisciplinary foundations and the chapter on the transdisciplinary perspective have been written by the editors based on the discussions within group meetings unless stated otherwise. The final typing and layout of the manuscript was realized by I. Wallraven.

On behalf of all participants of the study group, the editors wish to express their gratitude to the Berlin-Brandenburg Academy of Sciences and Humanities for generous support. The publication has been made possible with the kind assistance of the Senatsverwaltung für Bildung, Wissenschaft und Forschung des Landes Berlin and the Ministerium für Wissenschaft, Forschung und Kultur des Landes Brandenburg.

Aachen, Germany — Klaus Lucas
Summer 2009 — Peter Roosen

Contents

Chapter 1
Introduction

We live in a world of structures. In the form of plants and animals, as well as of landscapes and clouds we are surrounded by the beautiful structures of nature (Fig. 1.1). Human civilization has added artifacts of a different kind of beauty and geometry, from the egyptian pyramids to the structures of modern urban settlements (Fig. 1.2). Modern industrial development has brought to appearance artificial structures in the form of engines and products of various types, such as airplanes, automobiles, power stations, chemical production sites and many more (Fig. 1.3). We start learning to design molecular structures, such as proteins or nucleic acids, for medical applications by experimental methods. We organize our social and economical life in societies, companies and logistical decisions. We invent medical treatment procedures, mathematical algorithms and theories explaining the phenomena of the world and we create structures of art in music, poetry, sculptures and painting (Fig. 1.4). Structure generation, be it of a material or an immaterial nature, is a process of permanent development, accompanied by decay and resurrection, and, in particular, characterized by a trend to progress towards complexity.

What do we know about the principles of the processes creating structures? Since Darwin we believe that the structures of the living nature emerge from ongoing stochastic mutation and selection, leading to an increasing survival and reproduction of the fittest. The same principles can frequently also be observed to operate in processes leading to our artificial structures, no matter at which area we look, i.e. technical structures, structures in the social sciences as well as structures of art and intellect. This evolutionary procedure is, however, occasionally interrupted by a revolutionary invention or a new intellectual concept, apparently without any evolutionary roots, which strikes and alters the world and its further evolutionary development abruptly.

Given the fact that structure generating processes are studied in almost any scientific field the question arises to what extent they can be reduced to common foundations. In particular, it appears fruitful to investigate whether progress can be made in analyzing and, where applicable, optimizing them by transferring the knowledge accumulated in the various disciplines across the frontiers.

As can be expected, there are differences between the fields with respect to approach, analysis, and solution strategies, but there is much more in common than

K. Lucas, P. Roosen (eds.), *Emergence, Analysis and Evolution of Structures*, Understanding Complex Systems, DOI 10.1007/978-3-642-00870-2_1,

Fig. 1.1 Structures in nature: Plants and animals

scientists focussing on disciplinary work normally anticipate. Therefore in Chap. 2 some transdisciplinary fundamentals are discussed. We start from a first look at the structures and structure building processes in simple systems. The principles are abstracted and applied to the structures of the real world. Then we proceed to the problem of modeling reality. In this context the question of optimality is discussed, and general optimization methodologies are scrutinized.

In Chap. 3, which represents the main body of the text, a selection of examples illustrates the issues involved. Problems and solution procedures are presented that are typical for the disciplines contributing to this book, ranging from engineering, over natural sciences up to the social and economic sciences. They are not written as in-depth treatments for the specialist. Rather they are presented with the aim to explain area-specific questions and methods of scientific analysis in a way digestible for non-specialists.

Fig. 1.2 Artificial structures: Complex technical designs

Fig. 1.3 Artificial structures: Radio telescope with complex support construction to ensure the correct surface radius of curvature for all directions of operation

Fig. 1.4 Artificial structures: Works of art.

Finally, in Chap. 4, we summarize the results and further challenges of this interdisciplinary approach to understanding structures and illustrate the perspectives of a new research field.

Chapter 2
Transdisciplinary Fundamentals

The analysis of structures and structure generating processes in the various disciplines rests on the basis of some transdisciplinary fundamentals, more or less common to all fields of application. These fundamentals will be discussed in the present chapter. First we define our understanding of structures and their properties as treated in this book. After that we discuss the various aspects of modeling structures in terms of artificial mental concepts, referred to as models. Here we recognize the necessity of introducing simplifications to the real world structures, in order to make them accessible to analysis. We shall then discuss the fundamentals of model analysis, where we reflect on the question of optimality, of constraints in optimum seeking, of conflicting goals, and of objective and subjective quality. We conclude the chapter by discussing mathematical and heuristic methods of search for the optimum.

2.1 Theories, Models, Structures and Interactions

Four notions will frequently appear throughout the book: theory, model, structure and interaction. They are usually utilized in a wide range of contexts. None of the words has a universally accepted definition. Their meaning depends on the disciplinary environment in which they are used and on the educational, sociological, or scientific background of the persons employing them. This book cannot and will not make an attempt to provide precise definitions, but rather will explore the meaning of these notions in different contexts, explain their usage by examples, point at differences, and work out common features with the aim to help improve interdisciplinary communication.

Theories. To scientists, a phrase "... theory" signals a particularly well-tested set of scientific ideas for which there is some range of phenomena where the theory gives correct predictions every time it is applied. A theory represents the well-established understanding of a system of objects, mechanisms, or processes. In the technical and natural sciences, it is often formulated in terms of mathematical formulas. These make the intellectual concepts precise that have led to formulating a theory. A theory can never be proven to be complete and final. New discoveries

K. Lucas, P. Roosen (eds.), *Emergence, Analysis and Evolution of Structures*, Understanding Complex Systems, DOI 10.1007/978-3-642-00870-2_2,

may result in generalizations, such as the special theory of relativity or quantum mechanics extending Newton's laws of mechanics to treat processes with velocities approaching the light velocity and phenomena on an atomistic scale, respectively. Whether a system of ideas can be called a theory depends on the history and the development of a scientific field. While nobody doubts that there is a mature theory of electromagnetism, some may question whether psychoanalysis has a similar scientific status that warrants its denomination as a theory.

Models. Models are employed to describe certain aspects of objects or phenomena without claiming to represent reality. Typical physical models are those in the building sector. An architect would build a model of an opera house in a manner quite different from a sound engineer. While the architect would concentrate on the physical appearance a sound engineer would focus on the acoustically relevant aspects of the interior, such that experiments can be made to simulate the sound propagation.

In the technical and natural sciences, a conceptual or abstract model is often phrased in mathematical terms using variables, equations, inequalities, or other logical and quantitative relations. The aim of such mathematical models is to enable a quantitative analysis of the modeled object or process. One and the same object can be investigated with different goals by employing different mathematical models. Each of these mathematical models will ignore certain aspects and focus on others. The reason for using a multitude of models is that we are frequently unable to understand the process or object on the whole. Instead, we try to capture certain aspects by partial models. At present, for instance, nobody is able to set up a mathematical model of an opera house representing all aspects important for the construction of such a building at the same time. One mathematical model may reflect the statics of the building, another air conditioning and heating, a third sound propagation, etc. All these models depend on each other, but to reduce complexity to a presently manageable level, special features are singled out while others are ignored.

The relation between model and theory is not clear-cut. The word theory is usually employed for "general aspects". Quantum theory and thermodynamics in physics, algebra and probability theory in mathematics, and equilibrium and game theory in economics are well established theories addressing broad ranges of phenomena. One would, however, not speak of a theory of motor vehicles. Cars are modeled, and there are many of such models, such as mathematical and physical models for the aerodynamics, models for stability and crash simulations, models for the engine, for the catalytic converter, etc. Many theories are required to model a vehicle properly.

Theories and models are both judged by their ability to predict consequences. Gravitational theory can be employed to compute the trajectory of a space craft flying to Mars very exactly, and various mathematical models for cars have significantly helped to make vehicles safer, to reduce energy consumption, and to produce cars more efficiently.

Structures. Although a precise and universal definition is difficult, it seems that we have an intuitive feeling or maybe even an inherent understanding of what structure is. We encounter structures in many different appearances. They may be material ones, like ordered matter, or immaterial ones, like social interrelations or concepts of thought. Scientific problem solving reveals structures that may be characterized as distillates arising from data analysis, general problem knowledge, experience, experiments, contemplation, etc. Which structures are recognized and considered important for the specific niche of the real world dealt with and which structures finally emerge as key objects of study strongly depends, however, on the scientific and cultural background of the investigators and on the viewpoints chosen.

Material objects, i.e. structures, as well as concepts and ideas, are interlinked by more or less complex networks. We refer to this phenomenon of a structure being part of a network generally as its interaction with an environment. To illustrate this view, let us consider a simple object, e.g. a window in a house. A window has a certain internal structure as it consists of translucent elements, a frame, an operating handle, and some inner mechanics. On the other hand the window is embedded in the house and, together with the house, in the neighboring environment that may define varying interactions with it. When these interactions are taken into account other properties of the window become important, such as its direction with respect to sunshine, its relative position in the room, its size, its operability etc.. Last but not least, in a sociological or psychological context, the window in the house may be seen as a purely abstract functionality connecting the adjacent sides for visible information while at the same time separating them materially. This last aspect reflects the window as a purely ideational structure. Each view defines a particular net of interactions, that makes the window analyzable and optimizable in a systematic way. So structure analysis is a fundamental part of any problem-solving process.

We may even assume that the way our brain operates on its sensory input strongly determines the generation of (abstract) structures. Originating from the ancient impulse of making the material environment manageable by ordering and introducing causality, we extend this method to purely theoretical constructs as well. Only this ordering process makes them accessive to understanding and, sometimes, melioration. There is history, tradition, experience; there is progress in general: We do not determine structures anew in every act of problem-solving. Embedded in our collective memory, in our language, our rules of conduct, our social or technical norms, or our scientific knowledge and last but not least in our artefacts there are results of former acts of structuring, resulting from problem-solving in other, sometimes similar contexts. We usually start working on the basis of these already available mental structurations. In our everyday practices working with given structures prevails compared to inventing new structures. Therefore we tend to assume that these given structures are part of "the nature of objects". However, this essentialist view reveals itself as insufficient as soon as we come across new problems as it is often the case in sciences and humanities. Solving these new problems requires us to establish new views on the case and its context.

Interactions. The internal interactions of a structure as well as those with its environment have to be carefully considered in any analysis. Usually, we tend to isolate an object from the surrounding world for analytical and abstracting purposes in the sense of defining a system, i.e. we separate the object under consideration from the environment. Without this isolation an abundance of influences may have to be taken into account when an object is to be modeled and investigated. Let us return to the above window example and study its optimization with respect to its heat transmission. Then we limit our consideration to the thermal properties, taking all effects beyond heat transmission as belonging to a non-interacting environment. It is quite obvious that under such optimization restrictions the window will turn into something small (because the surrounding walls are better isolated), thick (as more material reduces the overall heat conduction), not very translucent (since heat-reflecting panes tend to be dim), and directed southward (in order to accumulate energy gains by direct sunlight). Clearly the "optimized" object is not the technical solution we had in mind as human beings when we started out optimizing it. Had we, instead, isolated our perspective to just one of the other aspects we would have come to other, but similarly unsatisfying results. So, any one concerned with the assessment of optimality of a window in a house, such as the carpenter, the psychologist, the police and, finally, the housewife cleaning it, will arive at different results. We conclude that an "optimal" window can thus only be achieved if the model includes the relevant interactions with its environment. Recognition of these interactions will by itself lead to a multitude of mostly conflicting and subjective property demands as will be treated later in this book. The impossibility of treating the window independent of its interactions with the environment explains the common observation that there is a wide variety of window concepts in buildings. They arise from the rather individual interactions with the environment along with a plentitude of conflicting and subjective preference settings. As can be seen, an analysis of structures and structure generating processes requires understanding of the objects under consideration, in particular their internal interactions as well as the interactions with their environment.

2.2 A Simple Example: Structures in Fluids

In order to arrive at fundamental insight and classifications, a first look on structures and structure generating processes is reasonably focused to highly simplified systems, where the objects and the interactions between them are well understood. An example of such systems are fluids. The objects are the molecules, the internal interactions are those between the molecules, and the external interactions those between the fluid and its environment. Due to the simplicity of these interactions important and formal insights can be attained that will be generalized to much more complex, macroscopic structures in other fields of interest. The science providing a universal theory in such systems is thermodynamics. Thermodynamics tells us that there are two fundamentally different types of structures, referred to as static and dynamic.

2.2.1 Static Structures

Let us first look at static structures. A prototype of a static structure generated by nature in a fluid is a snowflake (Fig. 2.1). Under certain circumstances, such static structures are more specifically referred to as structures in equilibrium. Here, equilibrium is that static state of a system attained after isolation from its environment when all relaxation processes have come to an end. In particular, the equilibrium state does not depend on the process conditions leading to it. In any case, for static structures in general and equilibrium structures in particular, all interactions with the environment are eliminated and in equilibrium we have a final, i.e. not any more changing state.

The global principles leading to "static structures" in fluids are well understood. In an isolated system, the second law of thermodynamics requires that the entropy of the system under consideration must increase in any real process. The state of maximum entropy, consistent with the constraints of the isolated system, is the equilibrium state and as such is the final state of an evolution towards a predetermined goal. In that state the rate of entropy production is zero because all processes have come to an end. Under suitable constraints (but not always) maximum entropy will require the generation of macroscopic equilibrium structures, such as that of a snowflake. When we follow the structure generating process, e.g. the freezing process, by looking at the modeled interactions of the molecules on a computer over time, we see that the evolutive relaxation process towards maximum entropy indeed

Fig. 2.1 The always varying structures of snowflake crystals. Although every crystal differs from all others there are common structuring principles, like the six-fold geometry.

leads to the emergence of a static structure, when the constraints of the system are appropriate. The particular and complex shapes of the snowflakes are determined by the particular interactions between the molecules, but also depend on additional parameters such as the rate of cooling. As such, they are not really equilibrium structures but rather static structures frozen on the way to equilibrium. An example for a true equilibrium structure, however, is the vapor-liquid equilibrium of a pure fluid. When we cool a pure gas, say water vapor, we observe that the molecules will at a certain temperature generate a structure, in particular a spatial separation of two phases, gas and liquid. This structure, much simpler than that of a snowflake, will persist after isolation of the system, i.e. after eliminating all interactions with the environment. It is thus a particularily simple example of a static equilibrium structure.

More complicated static equilibrium structures are known to appear in fluid mixtures as a result of interactions between the molecules by cooling processes from the disordered gaseous state. Well-known examples are the phase and reaction equilibria exploited in an industrial scale in chemical processes. The particular appearance of the structures depends on the particular intermolecular interactions. The large variety of internal interactions between the objects, i.e. the molecules, is the reason for the enormous plentitude of static structures in fluids, that we observe. The entropy maximum principle involves that the existence of a structure is subject to particular constraints. It will disappear when the constraints are violated, e.g. at lower and higher temperatures and pressures, since the entropy maximum will then require a structureless state. We note that for purely mechanical systems, i.e. systems at zero temperature and entropy, the maximum entropy definition for isolated systems transforms into one of minimum energy. So, the geometry of a simple molecule is calculated today in a standard way from quantum mechanics by looking for the minimum energy. Similar remarks apply to that of a folded protein, as studied in biochemistry. In general, such molecular geometrics can be considered as static equilibrium structures (Fig. 2.2). In this application the free energy has to replace the energy of purely mechanical systems, when entropy effects are not negligible. We note that conditions can occur in which the minimum of (free) energy is not attained by a system on its way to equilibrium since the required activation energy is not available. We then find static structures associated with frozen states, depending e.g. on the speed of change like in the case of snow flakes. However, when the hindrance is eliminated, they invariably move to the state of minimum free energy with the emergence of the associated equilibrium structures.

We summarize that static equilibrium structures in simple fluids originate from the particular interactions between their molecules after eliminating any interactions with the environment. They are generated by relaxation processes towards a predetermined goal, which here is maximum entropy, unless hindered by particular constraints. These relaxation processes can be reproduced on a computer as an evolutive process, although the final result can in such simple cases more easily and reliably be calculated in one rational step by a direct mathematical optimization routine. Static equilibrium structures are predictable and reproducible. So, given the full definition of the molecular interactions in a thermodynamic system, the

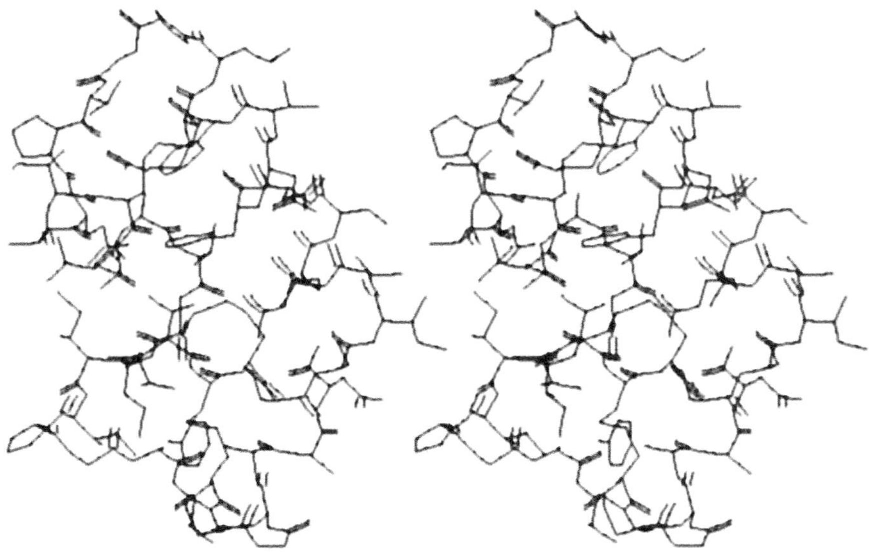

Fig. 2.2 Folded three-dimensional structure of a protein molecule, representing the state of minimal free energy. The structure is displayed in a stereographic representation. If observed with a squinting view the spatial structure appears after a while of eye adaption.

equilibrium state can be formally predicted, at least in principle, and the same static structure will be reached independently from the starting condition and from the process of the development. Clearly, static structures in fluids are no more than a simple example for much more complicated static structures placed into the society as technological artifacts, such as engines, buildings etc. We shall see, however, that those artifacts share important fundamental properties with those of the static structures of fluids.

2.2.2 Dynamic Structures

Static structures, i.e. structures originating from relaxation processes towards a predetermined goal and remaining constant after eliminating all interactions with the environment, are only a small part of the structures in the world. Evidently there are other types of stable structures in fluids which are generated and kept in existence by interactions with the environment through a continuous transfer of energy and matter in open systems. This input of energy and matter stabilizes a state away from equlibrium, and so, these structures are referred to as nonequilibrium or dynamical structures. Two simple and well-known examples may serve to illustrate the point and will then be generalized.

One is the Bénard-instability, studied in hydrodynamics (Fig. 2.3). A horizontal fluid layer between two plates of different temperatures in a constant gravitational

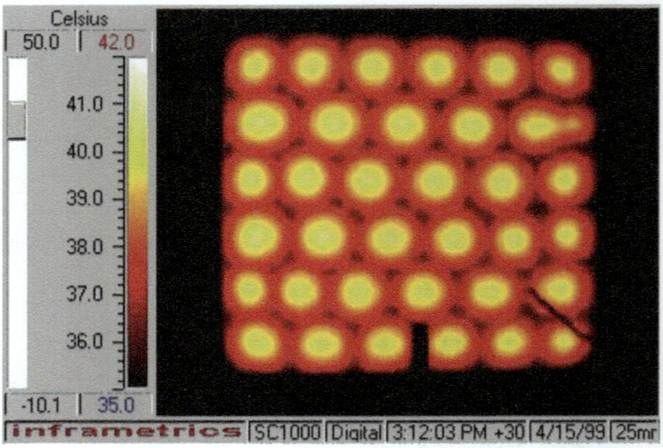

Fig. 2.3 Infrared camera view of the free surface temperature field of a well-established Marangoni-Bénard instability in a 5 mm layer of silicone oil heated from below. The pattern is hexagonal, apart from the influence of lateral walls. The size of the surface is 10 cm^2. (Curtesy of P. Colinet, Université Libre de Bruxelles, Department Transfers, Interfaces and Processes, Fluid Physics Unit)

field will show a structureless heat conduction phenomenon at small temperature differences. However, at a sufficiently large temperature difference the structureless state becomes unstable. Convection occurs, and the heat flow and along with it the entropy production increases. Although accompanied by a production of entropy, this increase of entropy does not produce chaos but rather generates regular hydrodynamic patterns, a stable dynamical structure. So, although the entropy evidently is increased as a whole there are areas in the system which are obviously characterized by a local reduction of entropy, the structured Bénard cells. A non-zero rate of entropy production, kept up by an input of energy, is the basic source of this structure generating process.

The second example refers to dynamical structures generated by a certain kind of again well-known chemical reactions, the Belousov-Zhabotinskii reactions [Win72]. When specified chemicals are transferred to a reaction apparatus under suitable circumstances, a regular change of colours and also a migrating spatially coloured pattern can be produced. A similar moving pattern structured by the same principles of an oszillating biochemical process, although with much lower frequency, has been observed in a special mutant form of mice (Fig. 2.4). Again, there is a considerable rate of entropy production associated with such a reaction process. And again this rate of entropy production is a source of structure generation.

So, in both of the above nonequilibrium structure generating processes, we find that a non zero entropy production rate is associated with the structure generation and preservation. This entropy production is made possible by specific interactions with the environment. The particular type of structure depends on both, the interactions with the environment as well as the internal interactions of the molecules. When looking at the processes of dynamic structure generating in such comparably simple systems, i.e. fluids, in detail some interesting features become visible that can be generalized to more complex situations. So, we observe that in the course of the still rather simple-structured nearest-neighbor interactions between the molecules long-range, macroscopic structures are created. In the case of the Bénard effect, i.e. in a hydrodynamic system far from equilibrium, small fluctuations, which are

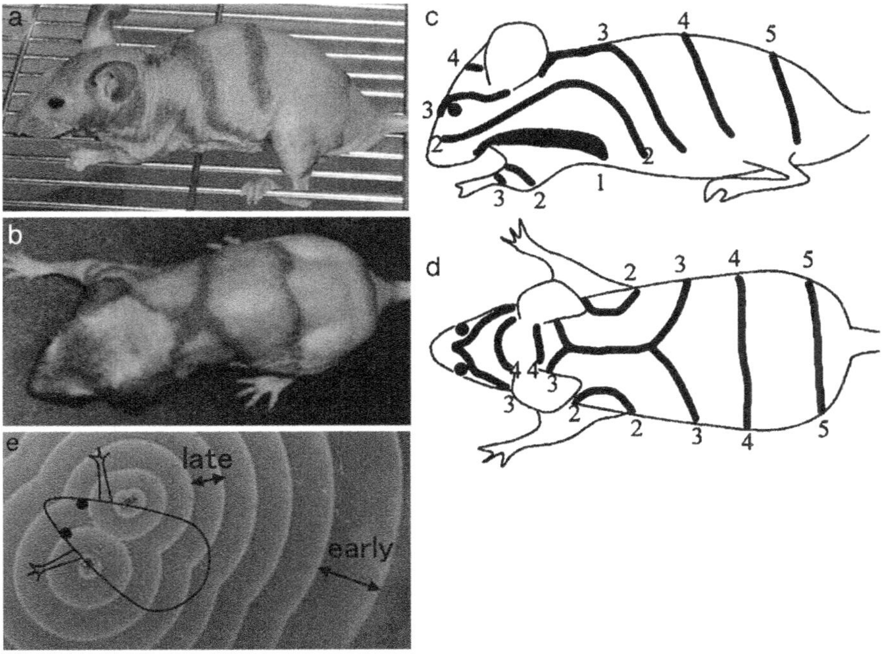

Fig. 2.4 Traveling stripes on the skin of a mutant Foxn1^{tw} mouse, with (**a** and **b**) showing *side* and *top* views of an adult mouse, respectively. (**c** and **d**): Schematic drawing of a typical wave movement in an adult mouse. The numbers by each *line* represent the time course of a wave. The time interval between the numbers is $\approx$30 days. (**e**): Traveling waves formed by a Belousov-Zhabotinskii (BZ) reaction. The shape of a mouse is superimposed. The outermost wave is the first wave. Intervals between waves are relatively wider for the first few waves, then become shorter and constant. Experimental BZ waves are made by the standard method described in [Win72]. (Courtesy of Noboru Suzuki, Masashi Hirata and Shigeru Kondo)

damped close to equilibrium, are augmented as a coherent effect and create a new type of macroscopic structure, stabilized by the exchange of energy with the environment. In the Bénard cells the global temperature difference between the lower and upper interface plate is so great that the *local* environment of the interacting molecules is modified in such a way that in a first step of structure evolution small-scale deviations from the global temperature distribution arise by stochastic effects. They modify the environment and the associated interactions. These cumulated effects, taking place in the immediate surrounding, add up to promote and extend the self-organizing and stabilizing of long-range structures. Similar effects become apparent in the chemical or biological structures referred to above.

As a common feature of dynamic structures in fluids, we find an unexpected element of randomness, non-reproducibility and non-predictability. The precise and detailed form of the Bénard cells is not predictable and not reproducible. They may have different sizes and rotation directions. The same is true for all dynamical structures of thermodynamic systems, notably those arising in chemical reactions.

The theoretical foundation for this non-predictability is the fact that the appearance of dynamical structures is a stability problem. Increasing the deviation from the equilibrium state leads to a point where this state becomes instable. The point of instability is referred to as a bifurcation point, at which fluctuations of an intrinsic nondeterministic character lead to an impredictable new state of the system, which is usually chosen out of various possibilities. Once this state is attained the system will further develop in a deterministic way on increasing its distance from equilibrium until a new bifurcation occurs. This new bifurcation is, of course, predetermined by the outcome of the earlier bifurcation, i.e. there is an element of history in structure generating processes far from equilibrium. They thus obtain an evolutionary character. There is no predetermined goal for the structure being obtained in detail. The particular appearance of the structure is coarsely classified by the particular interactions effective within the system and with its environment, but in detail it is the result of chance.

It should be noted, that increasing the driving forces beyond strenghts sufficient for (long-range) structure creation will usually lead to increasingly finer substructures by further bifurcations until at some point deterministic chaos production sets in: The dynamically created structures change so fast that they are no more identifiable as such. Accordingly, a certain local predictability can be maintained only for short periods, with decreasing exactness in structuring forecast due to accumulating nascence of new structure nuclei. The ordering principle leading to such states remains the same, however.

We note the fundamental difference of the dynamic vs. static structures in simple fluids. The latter are characterized both by a maximum of entropy in an isolated system (when equilibrium has been attained) and a zero entropy production, while the former are stabilized by a non-zero entropy production, accompanied by an inflow and outflow of mass and energy. Also, static structures are determined by the interactions between the molecules only, while dynamic structures are also influenced by the interactions with the environment. So, contrary to static structures, dynamic structures cannot persist after isolation. They need the interactions with the environment. As soon as the energy inflow and outflow, along with the resulting entropy production, are reduced under a critical value, the dynamic structures will collapse to a structureless or staticly structured state.

2.3 Real-World Structures and Structure-Generating Processes

Let us now generalize the fundamental properties of structures and structure generating processes, as elucidated by a consideration of simple fluids, to the material and immaterial structures of our society. They appear to us in the beauty of living nature, as well as in form of processes, social networks and in practically all operational appearances of our man-made artifacts (Fig. 2.5). Again the study of fluids, on the basis of interactions between simple molecules, is a convenient basis for getting insight into the generating and sustaining processes of such dynamical structures.

Fig. 2.5 A bridge as representative of a man-made dynamic structure. Even if it appears static at first glance, the continuous care in terms of inspection and maintenance (interaction with the environment) should be kept in mind that is needed to retain its functionality over a long period of time.

Are such material structures as an automobile or a bridge static or dynamic structures? The answer evidently depends on the type of analysis. An artificial structure may well be classified as static, when we consider its production from a constrained viewpoint. Although the internal objects interacting may be quite specific for each object under consideration, they will be identifiable. Let us consider an automobile as an example. Here we have the wheels, the engine, the body and a number of electronic and mechanical devices as interacting objects. The predetermined goal for generating the static structure of an automobile may be minimum cost, minimum fuel consumption or others. Indeed, designing processes may in principle be set up which lead to an automobile of predetermined quality, based on the selection and composition of its interacting components. However, if we broaden our viewpoint and consider the production of an automobile with reference to its performance on the market, we have to consider it as a dynamical structure. It is determined by the interactions with the environment such as, e.g., acceptance by the customer. Further broadening the view to operation we realize that without maintenance the automobile will eventually be destroyed and turn into structureless heap of rust and plastic. Without considering the interactions with the environment, no successful automobile will emerge from a design process and can be kept in operation. The external interactions with the customers make it depend on a plentitude of objectives and the plentitude of automobiles on the market reflects such interactions with the environment.

The structure generating processes leading to technical structures can be considered as evolutive relaxation processes towards a short-time predetermined goal, which, however, changes over long time periods by interactions with the environment. So, they have elements of static as well as dynamic structures. Value assessment for automobiles has altered the predetermined goals over time such as design, safety, fuel consumption, all being summarized as success on the market. A similar example is a power plant process with a varying number of feedwater preheaters. Subjected to an optimization towards maximum efficiency, such a system will eventually approach an idealized Carnot process by adapting an infinite number of feedwater preheaters. However, when minimization of investment and fuel costs is considered simultaneously as predetermined goals, quite different power plant processes will arise as static structures from the optimization process. When external interactions with markets are considered additionally the power plant structure generating process attains features of a dynamical structure. Then, when analyzed over a long time period, there will not be a predeterminable goal for a power plant, since assessments and values change in the society in an unpredictable way. The actual power plant processes will then adapt themselves to the varying short time goals in an evolutionary process. So, we see, not only the intrinsic interactions of the objects but also external interactions in terms of objective functions and value assessments are responsible for the plentitude of structures that we observe.

While structures of the technical world frequently lend themselves to manageability and more or less stringent optimization efforts, those of the social and biological domain do not. In the social sciences a most common structural issue of vast importance is what we call social stratification, i.e. the unequal distribution of life chances among the members of a society. Income, education, job opportunities, talents, status attributed on the basis of recognized social capital – all this adds up to a view of a structured society. Social structures are results of dynamical structure generating processes, which can be hardly be managed, and even less optimized. While social stratification is a type of macro structure effected only indirectly by individual acts, we have likewise to deal with meso-level and micro-level structures where the actors' influence on structure-building processes are easier to see, although the microstructures of interaction are most often invisible to the actors. Examples are work flow and hierarchies in organizations, sets of relations in peer groups and the like.

Structures of knowledge, as human beings derive them from observation, change with them, but not necessarily to growing objective truth or complexity. If we think of the Inuit and their traditional detailed and differentiated knowledge about snow and ice, we can easily see that the ability to differentiate varies with the tasks to perform and that established fine grained structurations of knowledge can get lost as soon as they lose their relevance for problem-solving. For driving a car on a road in winter it is pointless to distinguish 30 or 40 different types of it, as the Inuit do. Instead, we just differentiate by a binary distinction: snow or no snow, frozen or not frozen. In domains where the properties of "snow" are more important, like in winter sports, at least some additional differentiations (powder snow, crusted snow, ...) are maintained even in our modern civilization. Clearly, the Inuit, living in and on snow, depend on a much finer distinction between the various types of it. It may be deducted that the thoroughness of structuring is, to a certain extent, economized depending on the subjective practical value it has for the performing human being.

We conclude that generally the real world's material and immaterial structures and structure generating processes that we wish to study in this book belong to the class of dynamical structures. All structure generating processes, technical, sociological, biological, are associated with interactions with the environment. Besides matter and energy information is another prominent type of external interaction in most systems. As a consequence the processes are accompanied by entropy production. Based on simple balances of energy and entropy, this requires an input of energy with a low entropy content and the capability of the system to export the entropy produced by the structure generating process to its environment. This can easily be verified for the earth as a whole. Here, we profit from the influx of solar energy as the driving force for structure generation, including the production of food. Solar radiation has a low entropy content due to the high temperature of the sun. The necessary entropy export is realized by the low temperature heat radiation from the earth to the surrounding space, which carries all the entropy with it. Looking at technical systems confirms the general conclusions. Low entropy energy such as fuel or electricity, must be transferred to the system, to generate and preserve structures such as a running engine or an illumination during night time. For living organisms the low entropy input is contained in the food. High entropy energy must be exported in the form of waste heat or waste material.

The details of structure generation depend on the specific intrinsic interactions of the system as well as on the interactions with the environment. Generally we have to consider rather complicated interactions in analyzing the structures of our civilized world. They are in many cases not really understood and cannot be fully expressed in terms of mathematical functions. For example, generating the structure of a bridge or a power plant requires the consideration of interactions between material properties, technological components, natural laws, and, in particular, man made values of efficiency and desirability. Sociological structures are determined by interactions between human beings, which makes their analysis even more complex. We shall return to the analysis of interactions in Chap. 3, when we discuss structures and structure generating processes in various disciplines. No matter how complicated the considered structure is: As soon as the inflow of energy and the export of entropy are blocked, all structures will collapse. A bridge without maintenance, i.e. after isolation from all artificial energy and mass transfer, will eventually rust away in a structure destroying process. A structure in logistics will cease to exist when nobody will obey it and keep up its existence. An infrastructure will eventually disappear when it is not continuously used. A structure of thought will be destroyed as soon as the human beings adhering to it will stop thinking it. The analogous fate is associated with any natural and artificial structure, including human life.

On this basis, it is clear that all structure generating processes in principle share the basic properties of dynamic structures, such as non-reproducibility, non-predictability in detail and lack of a predetermined goal. This is immediately evident, e.g. in sociological structures, in those of diseases and in those associated with weather and climate. But it also holds for the structures studied in the technical and natural sciences, when full consideration is given to the external interactions associated with them. So, in general, structure generating processes are difficult to control and impossible to predict in detail. However, the laws of their generation can be analyzed by considering the relevant intrinsic interactions and fruitful generalizations as well as average predictions can frequently be made by statistical methods. In a restricted view we will see that even dynamic structures *can* be predicted and optimized when their complex nature is eliminated by representing them by simplified, quasi-static models. This is a reasonable strategy as long as one is aware of its limitations. The intrinsic interactions in such system models as well as those with the environment are to be carefully modeled, and the answers obtained from such an analysis must be carefully assessed in full recognition of their limited relation to reality.

2.4 Modeling Reality

It is an established fact that the vast majority of systems or processes in the real world are so complicated that there is no hope and even no sense in trying to analyze them in full detail. Instead, scientific analysis has to be liberated from the confusing plentitude of phenomena in order to reduce the complexity of seemingly unsurveyable problems to something that is amenable to analysis. The method of analysis may be mathematics in the technology and natural science oriented applications, or observation and thought along with creating notions and their operational interactions in the social sciences and in the humanities. The very process of modeling even a small part of reality is naturally accompanied by a loss of information, in the sense that some aspects are deliberately eliminated from further consideration. So, there is always the danger of an unacceptable disparity between model and reality which has to be taken into account when conclusions about model behavior are transferred to conclusions about reality. Models cannot be justified or evaluated within the categories right or wrong. Rather, they are either useful or not useful. The judgement about this classification is subject to a comparison with reality.

2.4.1 State Models

We wish to illustrate the problem associated with modeling reality first with relation to state models. Under "state models" we summarize models representing a well-defined state of a system. We start with a simple example, once again chosen from the thermodynamics of fluids. We shall treat varying levels of complexity, each level being adequate to describe certain physical phenomena in different state domains of a substance. We know that almost any material can be encountered in a liquid or a gaeous state, both being summarized as the fluid state. How can the properties of this technically very important state adequately be modeled, e.g., in order to derive information about its behaviour in certain artificial boundary conditions, such as chemical equipment?

Let us first consider a gas of monatomic molecules at normal temperatures and pressures. An adequate model for this reality is that of small billiard balls. In the billiard ball model of a fluid the molecules are hard balls with a negligible volume, and the nature of the interactions is such that, apart from occasional elastic bounces, there is no interaction at all. So, the billiard balls move independently of each other and their motion is described by the laws of classical mechanics. On the

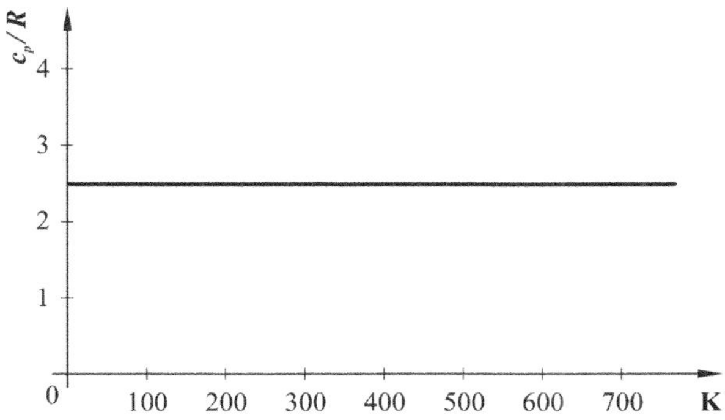

Fig. 2.6 Fluid phase behaviour of Argon, shown by its molar isobaric heat capacity divided by the universal gas constant against temperature, for a limited range of pressure and temperature.

basis of this model one finds formal expressions for the energy, the temperature, the pressure and the entropy of such a gas which are in perfect agreement with the properties of a real gas like Argon over a significant temperature range at normal pressures. To illustrate, we plot the molar isobaric heat capacity of Argon divided by the universal gas constant against temperature in Fig. 2.6. It is a constant of value 5/2, in perfect agreement with precise measurements. This quantity determines the temperature dependence of thermodynamic behavior and plays a significant role in various engineering applications.

It is remarkable that such an ultrasimplified model can be useful in describing the properties of a fluid. It claims that the molecules of Argon do not have any type of interaction with each other, neither that of attraction nor that or repulsion. Quite evidently this is wrong since it is in conflict with our basic knowledge of the interactions between the molecules in fluids. The contradiction is only apparent, however, and its resolution is that the simple billiard ball model is only probed in a very limited region of the phase diagram, i.e. only at normal temperatures and pressures. In this region, indeed, the interactions between the molecules, although basically effective, are not essential for the macroscopic properties of the fluid, due to the relatively large average distance between them. Events, in which two molecules approach each other so closely that the interactions between them become noticeable are simply too rare to be statistically significant and therefore can be neglected in calculations of the macroscopic properties. So, while the billiard ball model is a useful model for the properties of a real monatomic gas at normal temperatures and pressures, it does definitely not represent the general situation in an arbitrary fluid.

When the properties of a fluid like Argon are to be analyzed over a larger region of states, it becomes necessary to change the model. Going to higher temperatures, say 10,000 K, real Argon starts to ionize, an effect which cannot properly be described by the billiard ball model since it does not allow for a disintegration of the balls into nuclei and electrons as happens in reality. Going to higher pressures and lower temperatures the phase behavior of Argon becomes entirely different from the prediction of the billiard ball model. Figure 2.7 shows the fluid phase behavior of Argon over a large region of states. The region adequately described by the billiard

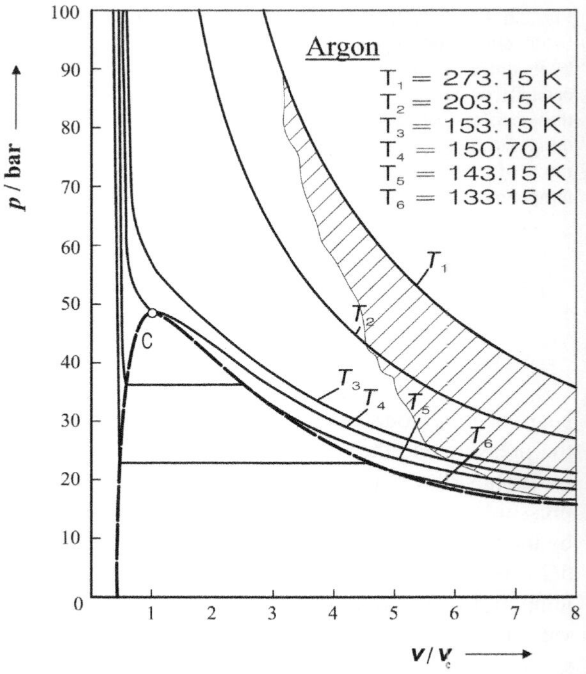

Fig. 2.7 Fluid phase behaviour of Argon over a large region of states [Luc07].

ball model is shown as a shaded region in the phase diagram. The rest more or less deviates from the simple model. A most significant deviation from the prediction of the billiard ball model is the effect of condensation. It becomes noticeable in Fig. 2.7 by the unsteady appearance of the isotherms T_5 and T_6. This is a simple example for a structure generating process in a fluid. Increasing the pressure at a sufficiently low temperature from the low pressure billiard ball region eventually leads to a situation where the average distance between the molecules decreases so strongly that their interactions cannot be overlooked any more. Condensation takes place, and in a gravitational field the liquid and the gaseous region are separated by a meniscus, clearly pointing to the appearance of a spatial structure. Hence, a model for a fluid over the full region of states requires the interactions between the molecules to be taken into account. These can be modeled on the basis of quantum mechanics. With these interactions the properties of fluid Argon can then be predicted over a large region of states by the methods of statistical mechanics along with mathematical algorithms. We note that even this more demanding problem of predicting the fluid phase behavior of Argon over a large region of states, including the structure of the vapor-liquid equilibrium, is usually based on a model which significantly reduces the real complexity.

The latter particularly holds for more complicated molecules, where the interactions depend on their internal structures, and so, on many coordinates. The models for the intermolecular interactions currently used for predicting structures of fluid

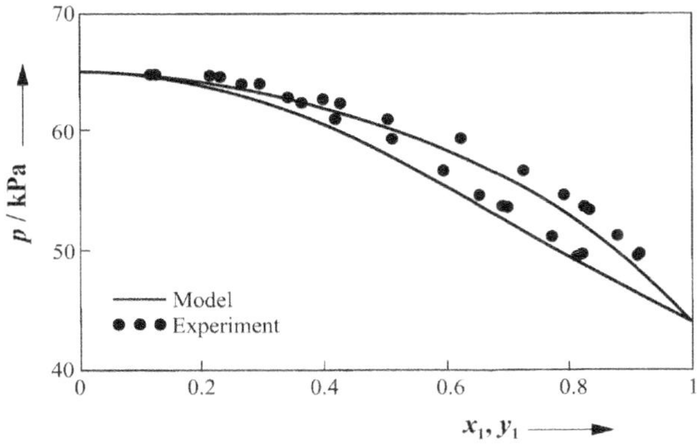

Fig. 2.8 Vapor-liquid equilibrium modeling prediction compared to measured data for Benzene-Hexane, showing good conformance [Luc07].

phase behavior are strong simplifications of the real world. However, when used in combination with an adequate model for the connection between the macroscopic fluid phase behavior and the intermolecular interactions we are able to make predictions of great practical value. This is shown in Fig. 2.8 where a model prediction of the vapor-liquid equilibrium in the system Benzene-Hexane is compared to experimental data. The relative enrichment of the vapor phase with hexane can clearly be seen in reality as well as in the model. Evidently, the model, while being far away from the true situation in the fluid, is a useful representation of reality in the context of vapor-liquid equilibrium prediction. It is by no means guaranteed, however, that this model predicts equally well other aspects of fluid phase behavior of the same system, as can indeed be verified.

Various aspects of state models, which hold quite generally in all branches of science, can be studied from the simple example of fluid phase behavior. First, a model has to be tailored to the problem to be discussed, e.g. predicting the structure of the vapor-liquid equilibrium in a fluid mixture. There is no sense in extending the model to more and more complexity if only a segment of reality is supposed to be analyzed. So, when the low pressure gas phase properties of a fluid such as Argon are to be analyzed, there is no sense in setting up a model for the intermolecular interactions between its molecules. Further, when only the vapor-liquid equilibrium structure in a fluid mixture is to be considered, a satisfying model can be based on a rather crude model for the intermolecular interactions. The demands on model complexity become more and more complex as further aspects of fluid phase behavior are to be studied, such as liquid-liquid equilibrium structures and heat effects. A model representing a large segment of reality will necessarily be much more complex and thus require more effort on applying it than a simple model for only a small portion of the real system.

Models for fluids are simple examples for static modeling. Many aspects of them can be generalized to more complicated static structures. Generally we shall, with increasing demands of reflecting reality, need an increasing detailedness of the model. As an illustration, let us turn to a different application, taken from energy engineering, and have a closer look on the energetic supply of a residential area with district heating and electrical power supply. We treat the system as isolated from the environment. The energy supply system thus represents a static structure. In a first order of rather crude detailedness the added-up peak needs in heating and electrical power for the intended set of houses can be taken as a limiting value. If the existing electrical power generators and heating stations provide these respective maximum values the demand will be satisfied. If the district is to expand by building additional houses and a sensible statement is required on how many of them could be supplied with the same system the power requirement model has to be refined. The coincidence that all customers are requesting their peak power demand at the same time will never occur. Hence it would be much more realistic to model the requirements by a time-resolved demand structure on known peak demand days separately for electrical energy and heat. Finally an efficient energy system for the residential area could be based on cogeneration technology which requires a correlation of electrical power and heat demands, opening up the possibility to utilize low-temperature waste heat of the electrical power generation for domestic heating purposes. Here the coincidence of respective demands of the two types of energy will have to be mapped, requiring an even more complicated model.

2.4.2 Process Models

In the technical and natural sciences usually process models with a certain self-organizing intelligence are used. Such a model is particularily fruitful in an interdisciplinary context since in many applications it can be implemented in the form of a computer code with an ab-initio overwhelmingly large abundance of possible system reactions upon changes of boundary conditions. Such models are usually called simulation systems, as they try to mimick the causalities of the real world, and can serve as models for structure generating processes.

In the social sciences a somewhat different notion of process modeling prevails, as the primary goal of those disciplines is mostly analysis and description, not targeted change. Here a descriptive social network is one of the primary ways of modeling, contrary to the simulation of physical causalities. Through the use of formal representations – such as directed graphs showing how each unit is linked to certain other units in a network – the model-based analysis of social networks has the aim of uncovering structural patterns and studying their emergence and evolution. Various techniques are employed to develop and analyze models using mathematical, statistical, and computational methods (e.g. simulation of network processes). Archives of network data and related computer programs to analyze them are typical resources available to social network analysts. In this way, the area of social network modeling relates to the general goal of understanding how social processes generate, maintain and change social structures, and how social structures both serve to constrain and enable social action. Typical modeling target issues are structured social inequalities, including social demography, socio-historical research, discourse analysis, ethnographies, and policy analysis.

In the causality-emphasizing models a given set of starting conditions is fed into the implemented simulating rule set of mutual dependencies, and the computer calculates the evolutive development of related output or monitoring variables. Even if only a limited number of causal dependencies is implemented in the computer code the variation span of possible outcomes can be vast, due to the effect of the so-called combinatorial explosion. This is due to the fact, that possible outcomes of modeling variable state changes tend to be multiplicative. A nice episode in the musical domain, taken from [Ste99], pinpoints this phenomenon:

> *As a boy, John Stuart Mill was alarmed to deduce that the finite number of musical notes, together with the maximum practical length of a musical piece, meant that the world would soon run out of melodies. At the time he sank into this melancholy, Brahms, Tchaikovsky, and Rachmaninoff had not yet been born, to say nothing of the entire genres of ragtime, jazz, Broadway musicals, blues, country and western, rock and roll, samba, reggae, and punk. We are unlikely to have a melody shortage anytime soon because music is a combinatorial system. If each note of a melody can be selected from, say, eight notes on average, there are 64 pairs of notes, 512 motifs of three notes, 4,096 phrases of four notes, and so on, multiplying out to trillions and trillions of musical pieces.*

So, even if the amounts of modeled variations in an evolutive simulation system are few and easily manageable, the potentially reachable variation space needs not necessarily be so, because: Simple rules can lead to complex and seemingly purposeful behavior.

Proof of this fact can be given and supported by more or less abstract formalisms, but an illustration may be more convincing. For this purpose we choose a two-dimensional cellular automaton, known as the Game of Life, running on a rectangular grid of infinite or finite size. Grid nodes (mostly depicted as cells) are occupied or empty, and there are only four rules determining the pattern in the next generation:

i. Occupied cells with no or one occupied cell in the neighborhood become empty,
ii. occupied cells with two or three occupied cells in the neighborhood stay occupied,
iii. occupied cells with four or more occupied cells in the neighborhood are emptied, and
iv. empty cells with three occupied neighboring cells become occupied.

In this concept the "neighborhood" is defined as the set of eight cells around the one in question.

Given these rules the unfolding of the cellular automation in time is completely determined by its initial conditions, but nevertheless, an incredible richness of different dynamical behaviors results from the "Game of Life" rules. Even very slight changes in the pattern of initially populated cells can change the development of later generations dramatically. Some of these patterns are even suggestive of purposeful design. For example, from one neither random nor fully ordered initial pattern a forever living periodical structure develops that emits small gliding motifs in one precisely defined direction. Figure 2.9 shows the recurring sequence of populated cells' distribution that is reached after some time of more or less irregular

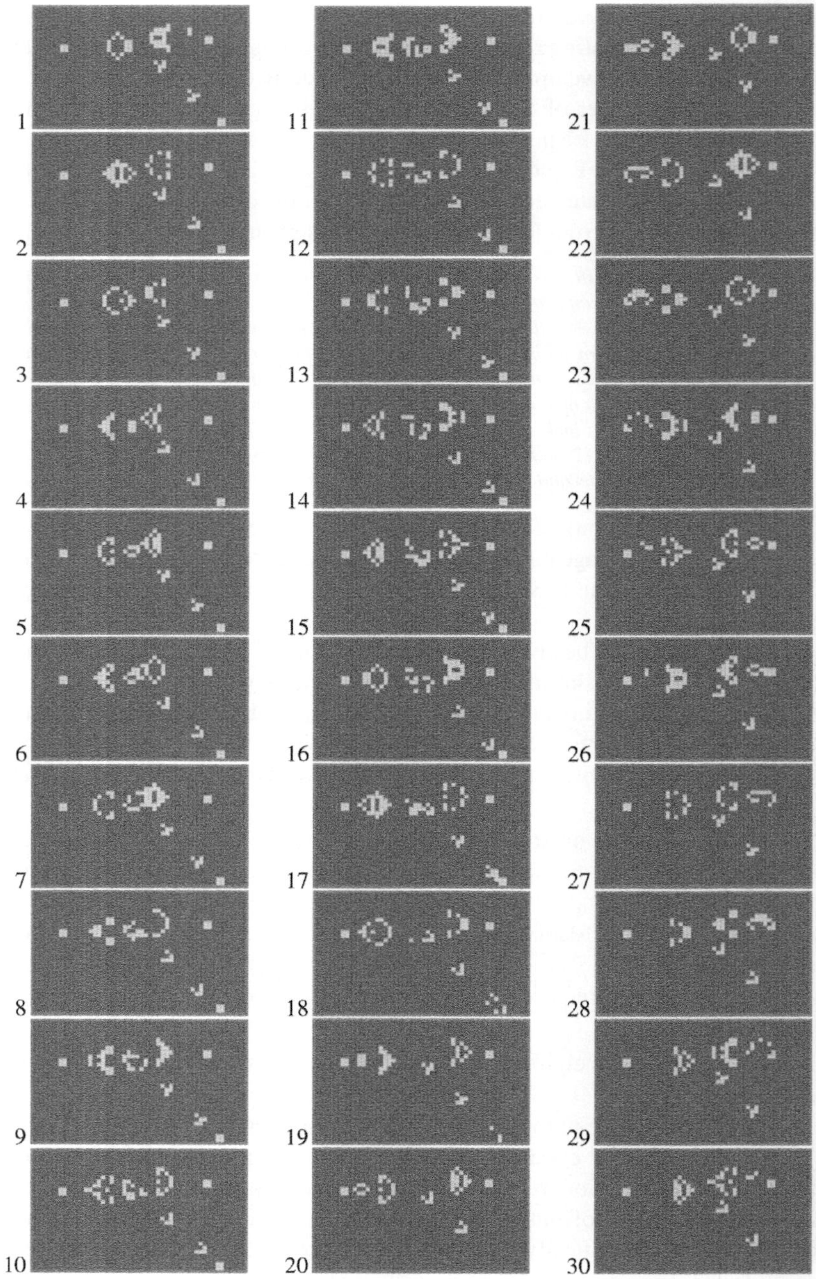

Fig. 2.9 Example sequence of a Game of Life configuration, repeating its core structure with a period of 30 iterations and continuously producing "gliders", being emitted to the lower right corner. Images are sorted column-wise.

approach to it. As anticipated, larger structures, although adhering to the same simple rules, show even much more diverse and astonishing features up to the point of reacting to "environmental" influences in almost intelligent manners [Hen06].

Contrary to the Game of Life, characterized by a very small rule set, a well-known evolutive simulation engine, the computer game "Civilization" [Wika] (Fig. 2.10), may serve as an example that combines a large rule set and a mixture of continuous and discrete variables. It is a model of society evolution. In a vast network of mutual dependencies human activities such as exploration of unknown territory or technology, war and diplomacy are simulated in their effect on an endeavoring society. The player, thought as a ruler of his people, has to make decisions about which improvements or units to build in each city, where to build new cities, and how to transform the land surrounding the cities for maximum benefit. The number of possible simulation engine reactions is so large that it can be interpreted as almost limitless, even though there are boundary conditions for every variable. The actual state of the game is represented by a larger number of system variable states. Apart from the variety of purely deterministic simulation answers there is an additional inclusion of stochastic elements, such as a modeling of natural

Fig. 2.10 Screen shot of the society simulation game "Civilization", taken from [Wika]. Some of the state-determining variables (such as amount of food for the people, *top left*, or treasury of state, *bottom right*) are depicted or listed in the corners of the screen. Others, such as the number of roads and hence the facilities of transportation, are represented by respective parts of the drawing.

catastrophies. This set of features causes each game to develop different from any preceding, even if the player tries to repeat a former scenario.

Similar simulation systems, although less popular, are used in scientific modeling of development processes, e.g. to understand the development of social groups depending on political boundary condition settings over time. As soon as the behaviour of human beings is modeled the simulations tend to become unrealistic, though: Representing the twists of the human mind in its reaction to more or less comprehensible boundary conditions is still a wide and mostly not yet understood field of future simulation development.

A final remark appears adequate in relation to such process simulation in systems. Mathematical analysis strongly favors steady and differentiable dependencies, since they allow the application of rigorous and well-assessed optimization algorithms for system melioration. But this frequently represents an undue simplification. Instead, typical systems also contain large numbers of internal decision variables of a discrete nature to reflect real-world conditions. If, for example, the real features of a hot water piping in a house, or the set of interconnections of chemical engineering apparatuses, is to be realistically represented, the discrete diameters of tubes available for construction purposes must be respected. Any calculation in the respective model is dependent on tube diameters and its outcome will accordingly "jump". To represent tube diameters in such systems by a continuous variation usually impedes realistic modeling.

2.4.3 The Economy of Modeling

It is essential to reflect on the effort that should go into the conception of a model from a practical point of view. Even if more detailedness would be desirable in many cases with respect to a noticeable reproduction fidelity improvement, it may prove rather impractical to proceed in this direction. These main reasons to restrict modeling depth exist: (i) the time requirements for evaluating a complex model, (ii) the required labour to set up the complex model, compared to the expected improvement the model promises to cause in an analyzing project and (iii) incomplete knowledge on the system under consideration.

An example for the first case is the weather forecast. If there is a model to describe the complex phenomena of atmospheric processes, but the time needed to solve the modeling equations is too large, the real phenomena that should have been forecasted are on a faster timescale than the forecast itself. The results of the model are then useless. This problem of excessive computing power requirements may in most cases be a transient one, caused by temporary non-availabilities of adequate calculational power. Similar effects, however, show up if rather fast phenomena, such as real-time control units and similar applications, are to be scrutinized. The operation of an unmanned vehicle, striving to cover a given distance in a sensible amount of time, is such an application. Equipped with a set of sensors, like stereographic cameras, tire slip and bearing controls etc. the steering control unit must interpret many signals in combination, as no single input channel provides satisfying information on its own. Each channel input is ambiguous to a certain extent, so the interpretation of all available inputs needs to be balanced. If the model underlying this interpretation is too complex and thus too slow the vehicle is stuck in the ditch before the control unit gives the signal to circumvent the obstacle. Here a timely circumvention of a non-existing ditch, assumed by a crude but conservative interpretation of the acquired sensors data, is better than a too late one of a correctly recognized ditch.

The design of a chemical production plant may serve as an example for the second case. A general task in this field is the anticipatory performance evaluation of a coupling of several basic types of equipment into a so-called flowsheet (Fig. 2.11). The boundary conditions the real plant has to obey are manifold, being mostly of a thermodynamic nature. First of all, conservation laws for matter and energy have to be fulfilled. Second, the properties of the reacting substances, such as separation volatilities in distillation columns, impose limits on what a single apparatus can achieve. Third, at several points in the flowsheet there are decision points on the principal layout of the future plant. In the flow sheet this is represented by alternative paths the materials can take, leading to different product strategies, (human) reaction possibilities upon changes in market prices, apparatus developments etc. Fourth, there are up to two recycle branches, recirculating some intermediately created

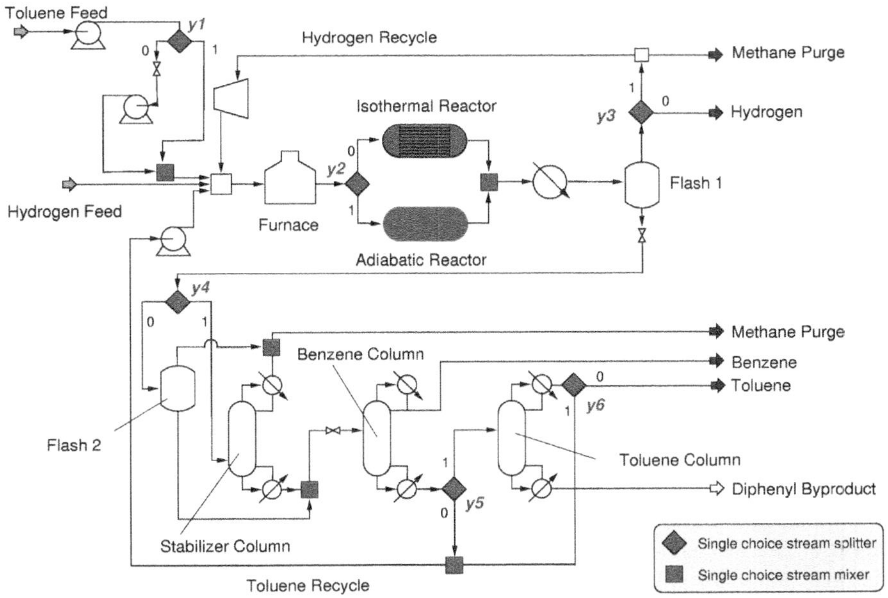

Fig. 2.11 Flowsheet of a benzene production plant, taken from [GR99]. Two educts are fed into a (mostly) serial connection of several basic types of equipment (reactors, pumps, distillation columns, ...), producing the desired output of benzene, but possibly also an additional by-product (diphenyl), depending on the concrete choice of the chosen path.

substances back to the plant inlet. These potential recycle streams must be considered in the balances.

Basically there are two differing modeling approaches to obtain an optimal path choice and settings of respective operative values – the equation-based and the sequential modular approach. In an equation based approach all simultaneously valid equations and inequality conditions are formulated for the complete system, creating quite a vast mathematically formulated collection of dependencies. The problem class may generally be defined as a mixed-integer non-linear (MINLP) problem. Note that in this approach values and settings expressing conditions in the educt section of the plant are frequently connected to values at the end or in the center of the system. Accordingly it is a very complex and time-consuming task to model a chemical reaction system on this basis. Generally adequate algorithms are not available. If a particular solution procedure has been tailored to a given problem, one is rewarded by a "closed model" that can be subjected to a purely mathematical, rigorous and usually very effective and fast treatment. If small parts of the setup are changed, however, a lot of reorganizing work has to be put into the model, due to the strong interconnections even of spatially far separated units. There is a caveat in the depicted flowsheet, though. It contains six decision points where a binary selection of one path or the other has to be set. Depending on the chosen path, the mathematical system of equations and inequalities changes significantly, barring a combined modeling of the complete plant. Instead, separate models must be created for every sensible combination of binary decisions. Even if some of the $2^5 = 32$ purely combinatorically defined ones can be dismissed by engineering reasoning quite a number will have to be elaborated to cover the interesting range. Accordingly such a modeling treatment is very expensive with respect to the involved human effort. To summarize, the setup of an equation-based simulation mode requires a considerable amount of human effort that will be paid off both by fast optimization times and, even more important, stable system answers in case of mathematically difficult conditions. Whether this effort is warranted in a practical sense depends on the expected return relative to an existing design.

An alternative method is the setup of a sequentially modular model where mathematical representations of the individual apparatuses are coupled in an manner strongly resembling their physical interconnection: Substance and energy streams are coupled mostly on a local basis, splitting up the very complex interrelationships (e.g.: the sum of discrete mass flows summarized over every parallel stream of the flow sheet must remain constant) into local subsystems called unit operations. These unit operations, being internally represented by smaller sets of equations and inequalities, may relatively simply be switched on and off for decision variable resetting, requiring rather few changes in global balance adaptations. The modeling effort on the whole is less demanding than the equation-based one, but produces, by its very nature, a computational problem. Due to the localized, hierarchical nature of the modeled interactions the unit operations need to converge in themselves before respective stream redefinitions are to be forwarded to the hierarchically higher level of the unit-combining equation sets. Especially in the case of recycling streams such models have a tendency to oscillate as changed output streams of one unit influence the next in row, but eventually affect their own inputs again due to the circular loops. Such computational problems *may* be overcome by some increased calculational efforts, such as decreased time steps, or upper bounds on the changes of important variables with auto-adapted time step decrement. In both approaches it is clear that reducing modeling complexity will lead to a faster obtainment of results but is accompanied by the threat of an ill-defined model response.

The third case, i.e. incomplete knowledge of the system is, in fact, the most fundamental but also realistic situation. In the application of designing a chemical production process typical examples of incomplete knowledge may be uncertain prediction of future sales, uncertain physical properties or an uncertain by-product spectrum in the reaction process and the like. In view of the nearly infinite number of potential flowsheet options one then needs a methodology tool box consisting of a set of rules based on experience with earlier projects. Detailed modeling on the basis of balance equations and physic-chemical property profiles are not adequate on this level, but rather will follow once a knowledge-based decision has been made on the process structure. Similar ideas have been developed in the social sciences where decisions are made on the basis of bounded rationality: essentially sets of heuristic rules tailored to the problem under consideration. In the typical applications of these rules a mathematical model cannot even be formulated, not to speak of its complexity and the many sources of uncertainty.

2.4.4 Explicit Modeling of Uncertainty

In many cases the real world shows numerous imponderabilities which pinpoint different categories of uncertainty. Embodying these uncertainties into engineering models without turning to rule-based approaches provides ways and means to introduce more reliability into the computation of real world phenomena computationally. In harmony to the definition of uncertainty introduced by BOTHE [BOTHE93], uncertainty can be understood as a gradual assessment of the truth content of a proposition, related to the appearance of a specified event.

The following diagram (Fig. 2.12) demonstrates the potential subcategories of uncertainty and assigns them to corresponding theories that have been elaborated in the last two decades. Three different subcategories can be distinguished:

- stochastic uncertainty, which describes the random results of multiply repeated experiments where the boundary conditions have always to remain unaltered,

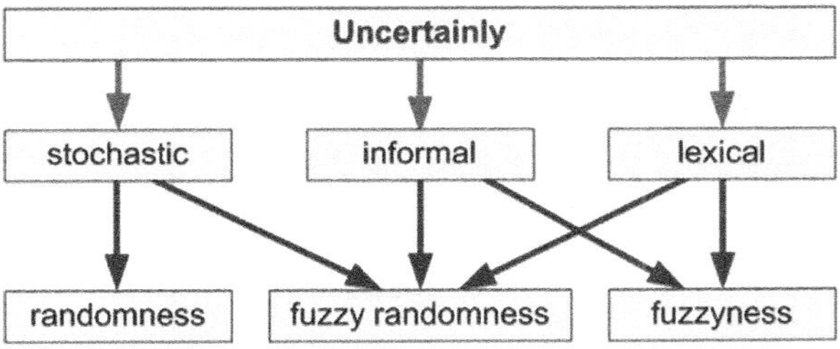

Fig. 2.12 Subcategories of uncertainty

- informal uncertainty, which describes information deficits due to the limited information sources and only small numbers of observations,
- lexical information, which quantifies relevant real world facts in terms of linguistic variables by ranking the memberships of these facts to defined uncertain sets.

There is a rather traditional treatment framework to calculate the effect of uncertain input or measurement values on the outcome of a functionally dependent output value: the error propagation method. This kind of uncertainty effect is almost omnipresent in the modeling of real world systems. It poses no major problem on the interpretation of results as long as the model is not a self-evolutive one, meaning that the modeled responses of the system are directly depending on a set of input parameters. As an illustration, let us consider a very simple model for determining the required power output P of a vehicle engine, driving at constant speed on a motor way. The modeling is done with a simple formula:

$$P = \frac{\rho}{2}(v + v_0)^2 A v c_w$$

(with ρ = density of the air, v = velocity of the vehicle, v_0 = headwind velocity, c_w air resistance coefficient, and A = head face of the vehicle)

The rule of error propagation then directly tells us how uncertainties in a given variable, say, the air resistance coefficient of the vehicle c_w, lead to a respective uncertainty in the calculated required locomotion power:

$$\Delta P = \frac{\rho}{2}(v + v_0)^2 A v \cdot \Delta c_w$$

Or, with other words, the *relative* uncertainty of required power, $\Delta P/P$, is identical to the relative uncertainty of the vehicle's air resistance coefficient, and the *absolute* uncertainty in P is constant for an assumed constant uncertainty in c_w. This is a dependency which common sense usually expects. On the contrary, if we apply calculus to the modeling equation to elucidate the effect of uncertainty in the velocity value, we find for the absolute error in P

$$\Delta P = \frac{\rho}{2} c_w A(3v^2 + 4vv_0 + v_0^2) \cdot \Delta v$$

or, if we again look at relative uncertainties,

$$\frac{\Delta P}{P} = \frac{3v + v_0}{v + v_0} \cdot \frac{\Delta v}{v}$$

which gives us a definitely more complicated dependence to reflect on. If we experience no headwind ($v_0 = 0$) the relative uncertainty in P is three times as large as that in our observed variable v. This dependency can change dramatically, though, if we have a stronger tail wind ($v_0 < 0$), e.g. in the order of half the speed over ground. In that case, a miss on our measured velocity will raise the propagation of relative uncertainty to a factor of five, far away from what common sense would expect!

Although this effect of error propagation may come unexpected for a person not regularly involved in technical calculations, things tend to grow much worse in case of self-evolutive models. In these models, frequently applied to simulate the temporal development of a system, the calculative result of a value at a certain time interval Δt_i is needed to determine its value at the next interval Δt_{i+1} and later stages. In such cases simple error propagation techniques will soon lead to accumulated errors well beyond the actual parameter mean values, thus rendering their calculation worthless. In such cases, more elaborated techniques, based on in-depth probability and density function considerations, must be put to work.

Well known for many years, but still advancing, is the probabilistic approach or the concept of randomness which captures uncertainty through random variables and/or random processes in an objective fashion, excluding subjective views on the problem. Thus, based on long-lasting observations or experiments, an effective probabilistic assessment of the quantities, patterns, actions and structural responses within systems and system processes can be made available. For stochastic variables, the key ingredients are expected values, mean values, variances, and also moments of higher order, quantile values, probability density functions (pdf) and cumulative density functions (cdf). For the wide variety of time-variant stochastic processes, functions for the average behavior, auto-correlations, auto-covariances, covariances and auto-correlations as well as cross-correlations are of interest. All of them can appropriately map the stochastic phenomena of uncertainty. Two problems, however, have to be realized if conventional randomness is applied: (i) a sufficiently large sample size is required if the pdf and cdf of a random variable are to be accurately defined; (ii) often, it is not an easy task to find the characteristic description of a real world process, with respect, e.g., to stationarity, ergodicity, spectral moments, etc. Consequently, if unfounded assumptions are made in a stochastic simulation model of a technical system or process, additional uncertainties are introduced on top of the attempted incorporations of uncertainty into a model (i.e. uncertainty of the uncertainty).

In the past years, fuzziness has tremendously increased in prosperity as a new paradigm to computerize uncertainty. Introduced already by ZADEH [ZADE65] around 1965 despite numerous hostilities, the fuzzy-based optimal decision finding and fuzzy control have been extremely successful in many applications. In particular, this applies to highly complex problems of automatic control engineering where the solution of non-linear partial differential equations represented the regular modus operandi for a long time. As a matter of fact, many highly non-linear or near-chaotic problems, which could not appropriately be solved through numerical simulation (e.g. by means of non-linear finite element models), proved manageable by applying the fuzziness paradigm. The nucleus for fuzziness is the human attitude, even if no viable numerical representation of a problem is possible, to express solution behaviors in terms of fuzzy rules which connect uncertain input with uncertain output quantities. Both, input and output quantities are modeling informal and lexical uncertainties, i.e. non-statistical properties, where particularly subjective aspects are entered into the consideration. To this end, specified crisp input and output quantities of a basic set are transformed by fuzzification into fuzzy sets using so-called membership functions of different functional shapes (e.g. triangular, trapezoidal or curved). Based on the above triad "fuzzification", "fuzzy rules" and "defuzzification" and based on the α-discretization and α-level optimization in structural analysis (see the book published by MOELLER [MOELL04]), data as well as model uncertainty of systems and processes can be handled. It should be mentioned that fuzziness can analogously be expanded from elementary quantities to functions leading to fuzzy functions, fuzzy processes and fuzzy fields.

Finally, fuzzy randomness has to be introduced if neither randomness nor fuzziness is sufficient to describe the "crude reality". This happens when the rigorous preconditions and laws of randomness can not be matched, e.g., because a stochastic quantity is affected by informal and/or lexical uncertainties. Typical for this are (i) only small numbers of samples such that the type of the pdf or cdf has to be approximated with considerable uncertainty and (ii) the violation of a fundamental principal of probability, i.e. the constant reproduction condition, by which it must be guaranteed that all samples of the taken universe obey identical boundary conditions. In practice, both premises are often violated. The universe may contain only a few samples, since in many cases the boundary conditions are varying with respect to time and location. Also, the assumption of statistically independent stochastic variables and/or stochastic processes, used in many engineering applications, does frequently not correspond to real world scenarios. The effect of how a random quantity is fuzzificated can be seen in (Fig. 2.13) where a cdf is superposed with fuzziness.

In view of these unavoidable uncertainties all model predictions necessarily assume the character of a range of values, depending on the sensitivity of the parameter uncertainties to the macroscopic predictions. Any rational application of a model, including forecasts of the optimal target function values and settings of parameters, must therefore take such uncertainties and their consequences on the conclusions into account. This observation affects in particular one typical method

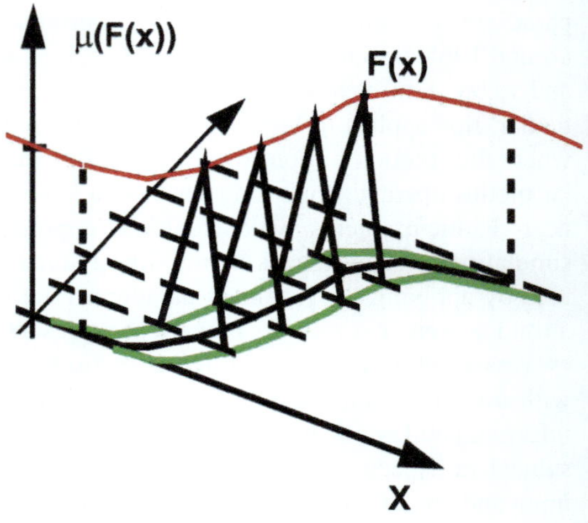

Fig. 2.13 Fuzzification of a stochastic

of model enhancement: the inclusion of increasing numbers of system describing parameters. On the first glance, the addition of more details will improve the detailedness of the model providing more adjusting screws to fit it to the observed reality. But this holds true only as long as the additional parameters can be associated with sufficient sharpness. If parameters tend to be unsharp, however, additional parameters will only result in an obfuscation of the simulated causal dependencies.

2.5 General Aspects of Optimization

Optimization assumes that a goal can be defined. This almost trivial prerequisite, taken for granted in most cases, does in fact require further consideration if we leave the usual textbook examples. We will show several practically important cases where the definition of the "correct" target function is by no means stringent. There are even systems to which such a thing as a target function in the common understanding cannot be attributed in principle.

2.5.1 A Reflection on Optimality

What is an optimal society, or an optimal public transportation system? Even if we leave out any subjectivity-related optimality definition, the problematic nature of this question becomes clear by the similar one: What is an optimally adapted plant or animal?

If we ponder on those questions we will notice some properties of systems that make them hard to treat in any optimizing framework. The predominant problem with the mentioned systems is their multitude of principally sensible quality measures that all seem more or less arbitrary, both by definition and relative weighting. If we look at biology, the concept of an optimally adapted animal has been ever-changing over time, space and context, as the phylogenetic tree [WA] and the diversity of the living world tell us quite impressively. Nevertheless there seems to be a kind of directional evolution in the development of a certain species along its historical course. This observation in turn tempts to assume some kind of teleological positivism [Wikc] by interpreting these developments as directed towards a final state – implying that a targeted melioration or even optimization is feasible. There are objective observations, however, that suggest this assumption not to be realistic.

If we compare different answers of biological evolution to very similar environmental conditions in spatially distant corners of the world we find quite disparate concepts that obviously have proven as well fitting: The rather large number of different marsupial[1] species on the Australian archipelago inhabit habitats very similar to those presently dominated by ordinary (placental) mammals in the rest of

---

[1] Marsupials are mammals that fawn premature offspring and feed it during the first part of its further development in a skin pouch outside the womb. The young remain there for a period

the world. There is an interesting *a posteriori* interpretation of the differing developmental traits of those concepts, though [Wikb]: "The early birth of marsupials removes the developing young much sooner than in placental mammals, and marsupials did not need to develop a complex placenta to protect the young from its mother's immune system. Early birth places the tiny new-born marsupial at greater risk, but significantly reduces the risks associated with pregnancy, as there is no need to carry a large fetus to full-term in bad seasons. Because a newborn marsupial must climb up to its mother's nipples, the otherwise minimally developed newborn has front limbs that are much better developed than the rest of its body. This requirement is responsible for the more limited range of locomotory adaptations in marsupials than placentals; marsupials must retain a grasping forepaw and cannot develop it into a hoof, wing, or flipper as some groups of placental mammals have done."

It may be estimated that many systems that a human optimizer would like to meliorate fall into a similar category: Interpreting a historical development in terms of an identified melioration effect succeeds quite well, but a prediction of further development, or even the derivation of an optimality point is not viable. Many ways of potential development will lead to similar qualitative improvement with respect to singled-out target functions, but with unforeseeable side-effects on further "total quality defining" target aspects.

Even if a multitude of potentially important targets can be defined in an optimization effort, they may change their relative weights or even appear or disappear during the process of ongoing melioration, when effects neglected at the start of the consideration become decisive in later stages, or limiting boundary conditions eventually dissolve. To illustrate, let us inspect the century-long development process of a typical technical system – the evolution of power plants.

When the vast expansion of electrical power usage started in the earlier decades of the last century the foremost quality of plants was the cycle efficiency while the total power output of a given design and its adaptability to changing load situations were considered to be secondary aspects. Due to still lacking large-scale transportation facilities usually near-by primary energy sources, as, e.g., local coal mines, played an important role. If exhausts were regarded at all consideration was mostly limited to the pollution by particulates in the nearest neighborhood of the power plant, usually resulting in quite high chimneys to enforce better mixing of exhausts with clean air. With growing size of the energy market the cost efficiency became more and more important as the electrical energy entered a steadily growing range of end consumer products (like household aids, electrically operated industrial machines etc.). The growing demand created dependencies on far-away suppliers of primary energy sources that led eventually to the first world-wide "oil crisis" in the seventies of the last century – a new competing goal of power station (or at least: power provision) valuation sprang into existence: the provision security.

---

corresponding to the late stages of fetal development of a placental mammal. The best-known representatives of these creatures are the Australian kangaroos.

A relief from this situation was sought by the technology of nuclear energy plants, lessening the dependence of the industrial countries on fossile energy imports. Only regarded from a technical point of view this emerging technology promised large amounts of cheap electrical energy. But in various countries the opposition to nuclear power production was under-estimated.[2] The additional expenditures for shielding and securing the required infrastructure against opposing groups, as well as fulfilling politically imposed requirements of strongly raised security assessments, made the price of electrical energy production explode relative to earlier estimates. Had the social and political side effects been foreseen they might have been counteracted on a definitely lower impact level. But as history tells this chance had been missed.

As a short-term effect the return of the fossile energy burning power plants, meanwhile very cleanly operating with respect to particulate emissions, collided with the rather recently emerging concerns of volatile exhausts leading to the emergence of a strong renewable energy movement. While on the long run this method of consumable energy production on the basis of renewables will be the only one proving practicable, it will lead to raising costs in the immediate future, potentially leading to undesired side effects on economic and social development.

Summarizing this historical development of striving for the best (i.e. cheapest, most productive, stable-operating) power plants we obtain some insights into the emerging optimization process peculiarities. The persecution of a central main objective for certain periods leads to the upcoming of additional goals, not having been regarded before, that grow to play limiting and counteracting roles in the whole process. It turns out that such a process is unpredictable in detail, changing in an abrupt manner, and thereby has the properties of dynamical structure generation. Even if later-on appearing goals had been taken into account at the start of the long-term evolution they would not have contributed to a more pointed and less crisis-driven development, as the additional targets would have led to *an inferior outcome at an earlier point in time.* Alternatively proposed solutions, anticipating later objectives, would not have been competitive at earlier points in time. This, finally, leads to the conclusion that the system "effective electrical energy preparation" is one without an objectively definable long term target.

There are many more systems and optimization objectives that are comparable in this respect. Here, just one other shall be discussed that is surely familiar to (almost) any reader: the development of easy and efficient individual locomotion over larger distances.

Earth-bound individual locomotion is strongly coupled to the improvement of passenger cars. Due to the technical success of the respective automobile development and the present exorbitant use of them (at least compared to the times of early development) traffic jams, too little parking space, fuel costs, and air pollution have become severe limiting factors, which can only be counteracted by additional enforced technical and logistic developments. Those changing boundary conditions

---

[2] This description is mainly abstracted from the historical development in Germany. Other countries, like France, exhibited significantly different developments.

in turn modify the development paths modern automotive technology is taking. The structure generation process of individual mobility thus assumes the properties of a dynamical structure. While in earlier times maximum velocity, spacyness and comfort were anticipated future development goals, public attention and customer focus has strongly shifted towards economical operation, crash stability, unobtrusiveness and endurance, with an upcoming perspective on navigation support and automatized, instantaneous avoidance of traffic jam conditions.

While this description primarily holds for passenger cars, an almost opposite direction of development is observed for the two-wheeled versions of automotive devices, the motor bikes. In earlier times they were mainly seen and used as a simple means of individual transportation, with mostly cost-effectiveness, solidity and ease of maintenance being the predominant goals. Contemporary aspects strongly deviate from this view, bringing individuality, enforced demonstration of power etc. to front. Extrapolating views of the past, a typical recent motor bike would have been quite probably attributed as too heavy, too difficult to service, and too expensive in former times.

We may interpret these views in terms of static vs. dynamical structure generation and sustainance. We may regard the goals of a technical system, eventually in combination with an (averaged) subjective view of a potential evaluator, as static as long as we deliberately blind out longer-range perspectives. A power station or a car *may be* optimized with respect to a certain goal, or even a set of goals, if its or their apprehension is regarded as constant. Changes of such apprehension frequently manifest themselves in their distribution within a society, quite often with a gradient on individual age, leading to a certain type of ‘generation conflicts. So, in an unrestricted view, neither sociological nor technical structures lend themselves to an analysis of optimality. However, technical structures do so in a restricted view. In a limited time horizon as well as in a limited location technical structures can be optimized. We shall discuss the constraints of the optimality algorithms in the next section.

2.5.2 Constraints in Optimum Seeking

In the partial view discussed above, many technical structures and structure generating processes lend themselves to melioration and even optimization endeavors. Then actual executions of these procedures are determined by a number of constraints, also frequently referred to as boundary conditions.

2.5.2.1 Parameter Ranges as Boundary Conditions

Preparing a system for (numerical) optimization, we first model the causal dependencies of one or more target function values upon a set of configuration parameters. These dependencies need not necessarily be direct, sharp, or simple. On the contrary in most practical cases complex multiple, non-linear and unsharp dependencies are quite common. Configuration parameters are typically *bounded*, i.e. they cannot assume arbitrary values: Modeling the uplift of an airship we have to lower-bound the weight of the lift gas to that of helium, as there is no one lighter and at the same time safer gas than that. Modeling the performance of a gas turbine power station (depending on the properties of its operative parts, such as turbines, pumps, heaters etc.) we have to limit the maximum temperatures of the underlying thermodynamic process to the maximum values that the material contact faces of the structural elements can tolerate.

These constraints usually limit the attainable values of the target qualities. In many cases it is not obvious, though, what extremal target function values can be obtained. Here the underlying model must be queried, either by direct mathematical evaluation of dependencies, or by numerous samplings of target values depending on respective sets of input variable settings.

The notion of limitations or bounds in the model definition, implying limits in attainable target values, seems rather trivial at first sight. However, as soon as the potential ranges of configuration variables depend on each other, this grows into a sometimes very complex problem. Revisiting the power station example, we may observe that the maximum surface temperature on the gas turbine blades and thus the efficiency depends on the amount of additional cooling gas input through fine air ducts inside the blades. Additional cooling gas reduces the maximum output of the turbine, however. Furtheron the available cross section for such air ducts is limited by the required material stability for the very fast spinning blades, thus limiting even more design variables. While this description is by no means exhaustive it should suffice to demonstrate the complex dependencies of a target function value on the configuration factors in the presence of constraints. This makes it rather difficult to estimate the influence of them on available configuration parameter spans. It directly brings us to the problem of potentially excluding interesting regions of (unbounded) configuration parameter space without the ability to grasp it.

Bounds in the sense of forbidden regions are not necessarily maximum or minimum limiting conditions. They may as well appear in the interior of an otherwise already limited configuration parameter range. As an example, let us consider another thermodynamic system: water under environmental pressure at it's boiling point. In that state we will either observe it in the rather dense form of liquid water (density approx. 960 kg/m3) or in its diluted state of water vapor (density 0.6 kg/m3) . There is no in-between, and system configurations requiring a homogeneous form of water with an intermediate density value cannot be realized. If one tries to approach such a state from a formerly homogeneous material (either liquid water or vapor) the system separates into two phases with strikingly different mechanical and thermodynamical properties. This process may be highly fatal for appropriate layouts for turbomachinery, compressors and even tubing systems.

2.5.2.2 Superstructures

As soon as structural variants can be defined for the realization of a system we enter the domain of discrete optimization. For the matter of argument we restrict our discussion here to a problem class containing integral as well as continuous configuration variables. A typical representative of such a problem is the chemical engineering process optimization discussed in Sect. 2.4.3. Besides several continuous variables there are structural alternatives, each of them defining their individual subset of dependent parameters.

A frequently adopted method to manage the space of potential structural alternatives is the definition of a so-called superstructure.[3] It contains all allowed structural variants as binary switching options, expressed in terms of allowed parameter values of 0 and 1 respectively. As with continuous variables, simple decision variants are readily intelligible. But as soon as intermingled and mutually dependent switch settings are required the system may become highly complicated. Again, we note the problem of ascertaining that every desired parameter setting, here with respect to structural alternatives, should be able to be reached during the optimization process. If this is not the case we may unknowingly exclude important ranges of the "true" (i.e. pragmatically feasible) configuration space from systematic optimization.

In general, such superstructures occur in combination with continuous variables, thus creating mixed-integer linear or non-linear optimization problems. Especially the latter are feared for their algorithmic complexity with respect to a rigorous and affirmed location of their respective global optima. They continue to be subject of current research and development.

One of the major problems in such systems is the fact that configuration parameters are related to each other in a hierarchic way: The continuous parameters defining the properties of a chemical engineering plant in some switchable sub-tree (such as pressures, temperatures, sizing, material selections) only *exist* if the sub-tree is switched on. Otherwise they are completely meaningless. In simulative practice, usually the complete decision space with all potentially needed continuous parameters is set up. When a respective branch of the flowsheet is switched off they are just considered existing, but meaningless with respect to the target function value(s).

2.5.2.3 Open-Ended Structures

A boundary condition of a system's structure can also be defined implicitly and self-adaptingly. In that case the system's higher-level semantic definition must automatically be interpreted and translated into a functional model amenable for computational evaluation. As this is a highly complex task there are not too many reported examples for this methodology. Therefore we would like to mention this kind of approach more as a desirable future goal in system modeling than a broadly applied technique.

As an example we may re-inspect the chemical engineering flowsheet as shown in Fig. 2.11 on page 27. The conventional method to model such a system is the superstructure approach discussed in the last paragraph. Here connections of pre-determined apparatuses are fixated, each being defined by its respective set of

[3] The concept of superstructures is much more encompassing than the pure interconnections of structural alternatives if we use it in a more transdisciplinary context, but for the reason of argument we only understand its meaning here in the narrower sense usually attributed to it in the area of chemical engineering.

design variables. Mathematical algorithms may then be applied to identify a best possible setting of the variables, leading to the respective value(s) of target function(s). An alternative semantic definition of the flowsheet may be given by just describing the effect that is desired from a concatenation of various kinds of generalized unit operations. We might say: "The lower-boiling fraction separated in 'Flash 1' is to be fed into a distillation column (representing y4 $\longrightarrow$ 1)" and so forth, of course favourably in a more formalized language structure, but with the same level of (im)preciseness. Layout changes can be expressed relatively easy in such an abstracting language, as on this level of generalization it is, by definition, not necessary to define precise configuration parameter settings in details. Instead we rely on computational intelligence for re-adapting them at least to a certain measure of quality. If we restrict ourselves to such system definitions there is ample room for modifying the layout of the interconnection structure, just by changing some elements of our description. We may introduce additional items or take items or whole branches out.

Compared to the more mathematically oriented superstructure approach it is somewhat simpler to assure the definability of all sensible *structural* variants. But as the optimizing effort needs the assistance of computerized hierarchically underlying continuous parameter definition intelligence, we in turn cannot be sure that for all structural variants the complete scope of continuous parameter settings is realized, so the general problem of potentially excluding interesting configuration space ranges remains.

It is very difficult to subject this kind of semantics-based structural redefinition to a rigorous, algorithmic system optimization. In the rather few documented cases this approach was put to work evolutionary concepts were applied, e.g., in [Sci96] for the optimization of a chemical plant. A kind of intermediate method is the application of 5th generation programming languages, such as Prolog [Wik06b]: Here the human programmer can concentrate on the logical structure of the simulation model, without having to bother about the inner looping and variables control. These inner structures are organized by the Prolog interpreter automatically as soon as a simulation is run. But for realistically complex simulation models this claim of complete procedural automation has to be significantly limited, almost withdrawn, as the high-level Prolog program has no notion of practical relevance or additional, mostly unsharp contextual limitations that would impose respective evidence limits on the governing resolution methodology of "back-tracking". These issues have rather to be dealt with by manually introducing so-called "cuts" that manifest lower-level reasoning based on individually occurring parameter settings.

2.5.3 The Chore of Choice: Trading Off Conflicting Goals

In all practical applications structures to be generated underlie a set of different, frequently conflicting measures of value. The associated problem of multiobjective optimization is not new. Already in 1871 Vilfredo Pareto investigated in his work the problems of multi objectiveness and suggested the mathematical concept of best compromise solutions [Par71]. Many, if not almost all, melioration problems of the real world exhibit more than one target function. Cars are not only valued for their maximum speed but also for a low fuel consumption value, high reliability, low maintenance costs etc. A delivery service not only needs to deliver the goods in good quality but also is expected to handle them carefully, to be competitive in its costs, or to obey time windows for pick-up and delivery as close as possible. Usually such sets of target aspects cannot be satisfied optimally for each criterion at the same time. Let us, in this respect, reconsider the passenger car example: Satisfying the desire for high maximum speed will require large and complex engines, as well as elaborated braking equipment, which in turn leads to higher investment and maintenance costs. A part of the raised costs may be caught by applying less solid constructive layouts but will in turn lead to lower reliability. Almost any scenario of constructive change will stress different aspects of the arbitration effort of "optimizing everything at the same time".

Accordingly, the systematic, objective optimization of a system will eventually reach a bound it cannot transgress, and the domain of the subjective choice sets in: If two or more conflicting goals cannot be resolved at the same time, the individual valuation is the last resort to choose *the one* solution suggestion that is to be realized in actual utilization. Nevertheless there is a *set of best solutions*, frequently called the "Pareto Set" of the given problem (Fig. 2.14) in honour to Vilfredo Pareto. It represents those solution suggestions where one target aspect can be improved only by impairing at least one other. Mathematically, it creates a mapping

$$\vec{x} \varepsilon D \implies \vec{f}(\vec{x})$$

with $\vec{x}$ representing the vector of system design variables and $\vec{f}(\vec{x})$ the vector of dependent individual target function values. This mapping can be discrete (in the case that only distinct configuration parameter settings are feasible) or continuous

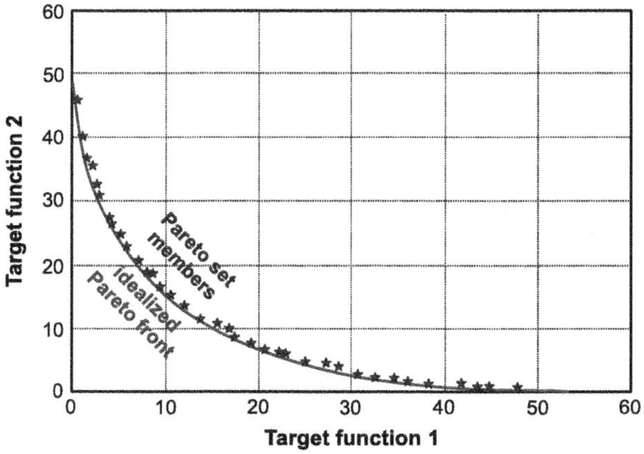

Fig. 2.14 Graphical representation of a typical two-dimensional pareto front. Both the actually determined target function combinations (*starred positions*) and an idealized, continuous front line are given for the non-dominated solution propositions.

(if there are arbitrarily adjustable configuration parameters such as temperatures or pressures in a technical system). If there is a large number of potential distinct settings the individual frontier points may be represented by an idealizing continuous boundary representation that may be handier for consecuting evaluation purposes. These pareto sets exhibit some properties that shall be discussed further in what follows.

If we begin with scrutinizing a single pareto front for a technical system with continuous variables, we frequently observe a qualitative target arbitration behaviour schematically shown in Fig. 2.14 for two targets. As an example we may take a residential heating system, with the inlet temperature into the hot water piping as the governing variable to trade heating costs, f_{costs}, vs. lagging of temperature adaptation in the heated rooms, f_{lag}. Both targets should be minimized. Raising the inlet temperature will yield a faster response of the system, i.e. f_{lag} will decrease. But at the same time energy losses rise and hereby produce higher costs, i.e. f_{costs} will increase, and vice versa. The dependence of both target functions on the temperature is, due to the underlying thermodynamic laws, smooth but conflicting.

As a certain setting must be chosen, at least for a given point in time, it is the question how the arbitration is individually performed. A classic method is the linear superposition of target function values, defined by a combined quality function Q_{eff} (see Fig. 2.15):

$$Q_{eff} = \frac{\alpha f_{costs} + \beta f_{lag}}{\alpha + \beta}.$$

Any such aggregation will define a linear superposition of the originally separate functions that may be expressed as a tilted new "optimization axis", or equivalently by arbitration iso-lines perpendicular to that axis. The angle of such iso-lines therefore may as well define the individual arbitration model. If, as an extreme example, almost only heating costs are considered, the arbitration axis would almost coincide

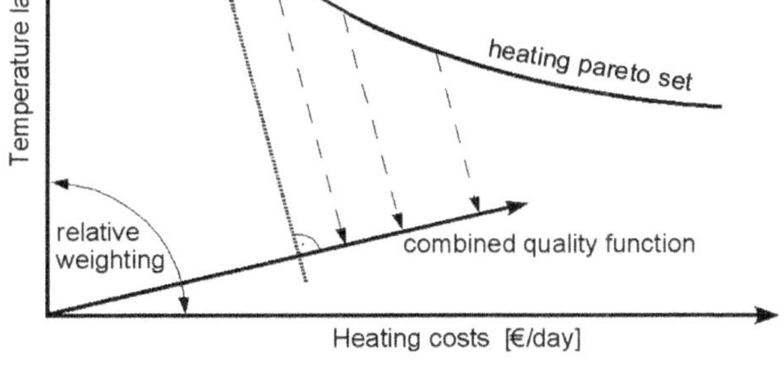

Fig. 2.15 Classical arbitration of two target functions by linear superposition. The general concept is transferable to more than two targets. Explanation see text.

with the abscissa of our plot, and iso-lines would be near to parallel to the ordinate. Every single pair of system responses in f_{costs} and f_{lag}, resulting from a respective setting of the piping inlet temperature, is projected onto the new optimization axis, leading eventually to the identification of an "optimal point". Changing the subjective focus and hence the relative weighting of the individual goals will result in a shifted optimum and, together with it, in a different temperature setting for the heating system.

If there are discrete configuration parameters in a pragmatically defined optimization problem, up to the point that strongly differing solution concepts compete with each other, the combined pareto front tends to become more complex and exhibits additional features. Let us investigate a rather simple but illustrative, idealized example: If we have to decide which means of transportation to use for traveling from a location A to a location B, we are interested in minimizing two target functions: the duration of the trip, represented by the inverse average velocity, and the costs of the trip. It is evident that taking a higher velocity transportation method will usually induce raised costs, as the investment, fuel, and maintenance costs of an airplane, for example, exceed that of a bicycle, or the repetitive consultation of a shoemaker's services. Mainly depending on the total distance to be traveled, quite different vehicles can be chosen for a certain trip. Only for very small distances the plane is no sensible option, and only for the really long distance trips bicycle riding and walking are ruled out. For each vehicle an individual pareto set may be expected: By applying a high technological input (like low friction bearings, aerodynamical spokes, streamlined frames etc.), an expensive bicycle can be made substantially faster compared to cheap ones from the superstore. But even for such high-tech bikes the ultimate limiting condition is the power input by the human, defining the maximum velocity even for highest-priced items. Similar arguments hold for the other vehicles as well, thus creating individual pareto sets of their own, expected to penetrate each other at specific, but case-dependent, front intersection points.

The resulting pareto set, shown in Fig. 2.16 as the purple-marked border line, exhibits some interesting properties. First, as a superposition of the individual pareto front lines, it is not necessarily differentiable, as the frontiers usually intersect with non-zero angles. Second, it does not even need to be steady, as adjacent partial pareto fronts need not touch each other. In our qualitative example, the pareto curve for the car starts already at higher velocities than a bicycle rider will ever reach. So, if at the position of that unsteadiness a small increase of trip costs is accepted there is a leap in transportation velocity by the switch to the other method of transportation. Such leaps can happen in any direction.

If we apply our method of linear superimposing target functions to this plot (as shown in Fig. 2.15) we will arrive at the conclusion to use either plane or bicycle. Tilting the superposition assessment axis from one extreme to the other will leave out any other means of transportation. Further this implies that rather extreme combinations of costs and velocity are regarded as equally acceptable at the switching point from one transportation variant to the other. Evidently, this rather unpractical result is an artifact of the linear assessment function. The problem is resolved if we

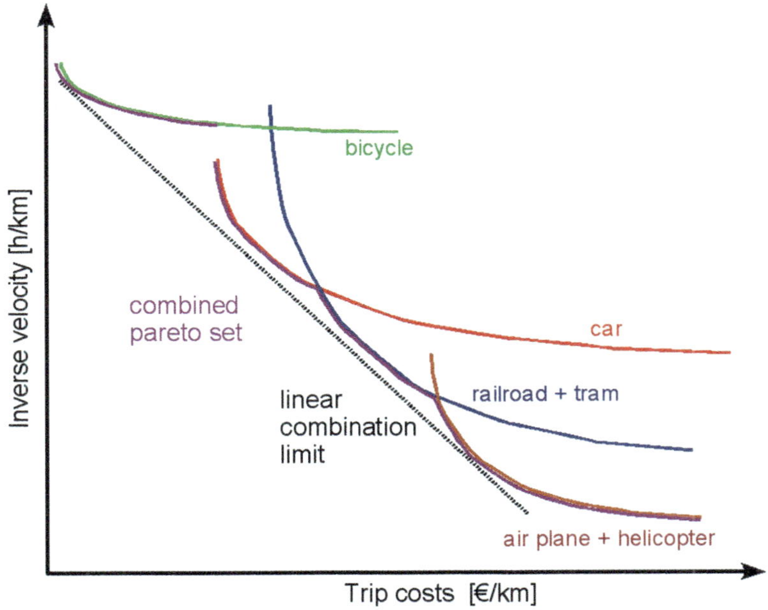

Fig. 2.16 Creation of a structured, qualitative pareto set of two target functions by superposition of individual sets of structurally different solution approaches. Explanation see text.

accept *non-linear* assessment functions i.e. we value the *simultaneous* fulfillment of target function qualities, as long as it does not lead to a major deterioration of the linearly superimposed solutions. Mathematically this may be represented by formulating the combined quality function Q_{eff} with an additional, non-linear term as

$$Q_{eff} = \frac{\alpha f_{costs} + \beta f_{lag}}{\alpha + \beta} + \gamma \cdot f_{costs} \cdot f_{lag} \quad (\gamma < 0 \text{ for minimization}),$$

With an increasing non-linearity factor γ we add stress to the simultaneity of advantageous individual target function settings. It is almost impossible, though, to express such subjective, nonlinear preference settings in an objectivized plot. Rather, it is sensible to report the complete (quantitative) pareto set calculated for an optimization problem to an individual or a group of human decision takers. On that basis every involved person can decide how far he or she would compromise on a back-laying, but simultaneously near-advantageous configuration of target values. In general, apart from regions of intermingled subsystem pareto fronts, most neighboring pareto set elements originate from neighboring configuration parameter settings. Small changes in target function values are usually caused by equivalently small changes in the system parameter settings, as is obvious in our residential heating example.

If a number of n target functions is considered the dimension of the "pareto front" is $(n - 1)$. Therefore, presenting pareto sets for two conflicting targets is not too elaborate: One can – in case of discrete system layouts – just either sketch the possible realizable target function combinations for the non-dominated solution suggestions, or present an abstracted "front line" of those solutions for densely populated sets (Fig. 2.14). However, addressing more than two targets is more difficult: Even if there still is a pseudo three-dimensional representation for three targets, its visualization and interpretation is not so simple. A fast display, freely rotatable in every direction by computer mouse interactions, can help to get a correct notion of the non-dominated front face items for the given problem. Even if an interaction of the decision maker with a computer is necessary for this, the display of the pareto front is basically static. Besides view angle changes no active work of the computer is required. Proceeding to four and more targets changes this situation significantly: There is no static possibility to display a "pareto front" with so many dimensions. How is a decision maker then able to adhere to the concept of pareto optimization; how can he or she get a notion on the pareto elements of a given problem, to be able to trade off the goals according to his or her individual preferences, taking into account the limits of feasibility? Here the decision maker support requires active work on the computer's side. As the pareto set cannot be displayed as a whole, individual interactions must help investigating it. A first approach might be to set all but two of the targets to fixed values temporarily and see what (partial) pareto front shows up for the two unrestricted values. The drawback of this approach is evident: For n criteria there are $\binom{n}{2} = n!/[2! \cdot (n-2)!]$ possibilities of combining them. So four criteria create six two-dimensional views, five criteria ten, and six criteria already 15. In addition this method only creates "cuts" perpendicular to the variables' axes, so results for combinations of criteria can not be considered.

An interesting alternative to those static display efforts are interactive spider plots or star plots as discussed in detail in the case studies 3.2 and 3.3. Here the identified pareto-optimal solutions of a given problem are displayed together with controls for each individual target function range. Operating these controls for one target immediately shows the consequences for the attainable ranges of the other ones, thus helping the human investigator to obtain "a feeling" on the arbitration space from what he or she can choose the favourite solution.

Finally, it must be realized that multi-criteriality always is a matter of subjective choice: If two or more conflicting goals cannot be resolved at the same time, with every objective criterion being already considered, the individual taste is the only guide to choose *the one* solution suggestion that is to be realized.

2.5.4 Objective and Subjective Quality

Up to this point only objective target functions have been discussed, being historically predominant in technical optimization contexts. But we should not close our eyes to the fact that there are many problems with intrinsically subjective goal functions in the technical sciences as well as in all other fields. Most of them involve human perception and senses, and subjective preferences play a crucial role.

Examples are the production of a coffee mixture with a special taste and smell, the various sounds produced by a vehicle, the development of perfumes, or digestibles with intended smells or tastes.

For illustration, a taste optimization is discussed briefly, based on [Her97]. How can a coffee mixture with optimal taste be created? As an appropriate methodology an evolution strategy has been identified, working on subjective assessments by comparative taste attributions of testers. There are no objective criteria which can be quantified, but coffee testers nevertheless have the task to produce a coffee with an intended taste. This presupposes a sense of taste which is trained in long times of practice. Additionally, the evolutionary procedure does not need a metric scale, instead it only relies on a *relative* comparative attribution of preferences. The only thing the test person has to do is to compare tastes pair-wise.

An evolutionary strategy for optimal taste determination may be set up by the following steps: Five randomly mixed coffee mixtures are given to the coffee testers. The precise compositions are not known by the testers. The task is to find out the coffee mixture coming closest to a reference mixture. The mixture of the coffee approaching the reference taste best is taken as a basis for five new mixture variations. This looped procedure converges after a few follow-ups – the testers cannot differentiate anymore between the quite similarly tasting test mixtures, although there are objectively differing mixing ratios, also with some distinct differences to the goal mixture. The relative assessments become ambiguous. Nevertheless the intended goal is reached: The desired taste has been approximated within the limits of subjective difference perception.

From this example we may abstract some characteristic conditions for optimization procedures with subjective goal functions valid in all areas of sciences and humanities:

- Optimization goals usually possess a considerable fuzziness.
- An optimization goal is crucially influenced by the group of assessing persons and their subjective notion of quality fulfillment.
- Subjectively shaped goal functions are usually multi-criteria functions (see below).
- Objectivity, which is often asked for and aimed at, can only be realized to a limited extent.

An important aspect of procedural development in treating subjectivity-biased melioration processes is the identification of more or less objectively determinable goal definitions that might replace the present subjective ones. If this succeeds even at least partially, some additional objectivity may be gained and the range of undecidedness be reduced somewhat. Some illustrating examples of presently subjectivity-dominated issues may be taken from the field of technical acoustics:

- Acoustic quality of larger rooms, such as concert halls, auditoria, or classrooms,
- adjustment of hearing aids for hear-impaired individuals,
- walking noise emission of modern laminate floors,
- acoustic quality of loudspeakers, musical instruments, bells, etc.,

- sound design as well as sound quality, used as acoustic visiting-card, for special technical products, like automobile doors.

Some subjectively perceived sound qualities are relatable to objective measurable quantities, although the overall perception of sound as an acoustic quality is very complex in nature. There are influential subjective factors as individual conceptions, hearing habits, etc. Consequently, sound perception cannot simply be described by objectively measurable sound pressure levels (or loudness), pitch (or timbre), and reverberation in a room. They correlate in some cases quite well, but in other cases there are striking discrepancies. As of today's knowledge objective assessments can replace subjective assessments only in exceptional cases. However, subjectively superior acoustic qualities often coincide with certain settings of objectively measurable properties.

Two facets of subjectivity play important roles in this respect. First, human beings do not perceive sound events in complex situations as a separable input. Instead, acoustic perceptions are embedded into non-acoustic environmental influences: The acoustic perceptions in a concert hall will also be dependent on optical perceptions, on architectural aesthetic effects, the expectations in acoustic performance, the comfort of the taken position (climatic conditions, disturbing influences, seat comfort), and on the psychosocial condition of the listener (missing acceptance of a musical piece, personal physical and mental condition). Second, the assessment is influenced by the group of test listeners and by its type. The assessment changes when performed by a group of laymen or experts, trained or untrained listeners, active or passive participants (listeners to music or musicians).

These insights from the field of acoustics may be transferred to other subjectivity-influenced domains, such as textual relevance attribution or climatic comfort issues. If we try, for example, to identify similar texts in a larger collection of writings, the *practical* attribution of similarity is strongly dependent on the relevance attribution of included passages as set by the assessor. Even if larger parts of two compared texts diverge, individual sections with subjectively decisive keywords found in both may lead to the conclusion that they are closely related. In the area of text comparison there is a chance of objectification by applying a frequency-of-incidence measure on individual words. But as ultima ratio only the valuating individual can judge the similarity.

The climatic comfort constitutes a more objective, although not sharply defined subjective target function, e.g. in the assessment of buildings. While the individual perception of coziness may differ substantially in a given room there are at least well-tested statistical comfort models, taking into account different dimensions such as environment temperature, radiation temperature imbalances of surrounding walls, environmental humidity, airflow around the human body, and some other influencing factors. Even though the resulting comfort model does not represent the perception of a certain single individual in a room, it very well describes the probability density distribution of the comfort attribution of a larger number of people. So, if a building is not designed for a certain small number of nameable persons, but to serve an initially not known number of anonymous inhabitants, such as office workers, these quasi-objective quality measures may well serve the needs of building designers in their effort of trading off different target qualities like costs, comfort, flexibility and such.

2.6 Model Optimization with Evolutionary Algorithms[4]

2.6.1 Introduction

Does one need more than one optimization method? Or, stated differently, is there an optimal optimization method? Following from the No Free Lunch theorem [WM97b]), in the general case – without a clearly specified task – there is not. For every single task, creating a specialized method would be advantageous. Unfortunately, this requires (i) a lot of effort, and (ii) extensive knowledge about the treated problem, and is thus not practiced. Alternatively, two strategies are usually followed when tackling a "new" optimization problem:

- Adapt an existing algorithm to the problem in its current form, and/or
- model/formulate the problem appropriately for an existing algorithm.

The first strategy modifies the algorithm design, whereas the second strategy modifies the problem design. These designs will be discussed in detail in the remainder of this section. "Traditional" mathematical optimization approaches mostly favor the second approach, which, however, may provoke unwanted side-effects: One has to make sure that the most important features of the original problem are taken over into the model. E.g., matching the problem to an existing algorithm may obscure its real global or good local optimizers so that they become unreachable for the optimization algorithm. Besides, many existing algorithms require the problem to fulfill properties it obviously or possibly does not, e.g. continuity and differentiability. Particularly, in cases where computing the quality value of a solution candidate requires running a complex simulation software, one seldomly knows in advance which properties the underlying (unknown) objective function possesses.

When nothing more than quality determining response values for any set of input variables are known for a problem, we speak of *black box* optimization. In the single-objective case, the common notion of an objective function and its global optimum/global optimizers – as given in Eq. (2.1) for unconstrained problems – is still useful. However, global optimizers, i.e. the set of input vectors **x** for which f(**x**) is optimal, cannot be determined analytically. An empirical trial and error method is the only way to find them.

$$f^{*G} = \min\{f(\mathbf{x})|\mathbf{x} \in X\} \tag{2.1}$$

The black box concept immediately leads to *direct search* methods – such a method only utilizes objective function responses and "does not 'in its heart'

---

[4] Contribution by T. Bartz-Beielstein, M. Preuss, K.-H. Schmitt and H.-P. Schwefel

develop an approximate gradient", as Wright [Wri95] puts it. As far back as in the 1960s, many direct search methods have been invented, e.g. the famous *Nelder-Mead simplex algorithm* [NM65]. At the same time, the first steps into the world of *evolutionary computation* (EC) were taken, presenting very simple versions of what is now subsumed under the unified denotation *evolutionary algorithms* (EA). These do not only use bio-inspired heuristics, they also employ randomness. Although the extensive use of random numbers and the fragmentary theory supporting EAs may be considered a drawback, these optimization methods have demonstrated their problem solving capability in numerous real-world applications.

Interestingly, in recent years, the mathematical optimization community has again shown increased interest in direct search methods, e.g. [KLT03]. This may have to do with (i) the fact that these techniques simply did not go extinct on the practitioners side, and (ii) improved theoretical analysis methods that now help tackling heuristic algorithms. In computer science, the growing field of *randomized algorithms* is exclusively dealing with algorithms employing random numbers – not only in optimization. Motwani and Raghavan [MR95] give an overview.

Although there have been precursors in proposing the utilization of evolutionary concepts for optimization tasks, as e.g. [Bre62, Fog98]), invention and development of the first evolutionary algorithms is nowadays attributed to a handful of pioneers who independently suggested three different approaches.

- Fogel, Owens, and Walsh introduced evolutionary programming (EP) [FOW65], at first focused at evolving finite automata, later on modified into a numerical optimization method.
- Genetic algorithms (GAs), as laid out by Holland [Hol73], mainly dealed with combinatorial problems and consequentially started with binary strings, inspired by the genetic code found in natural life.
- Evolution strategies (ESs) as brought up by Rechenberg [Rec73] and Schwefel [Sch75] began with solving experimental engineering problems by hand using discrete/integer parameters, but turning to real-valued representations when numerical problems had to be solved.

In the early 1990s, a fourth branch of evolutionary algorithms emerged, explicitly performing optimization of programs: Genetic programming (GP), suggested by Koza [Koz92]. These four techniques are collectively referred to as evolutionary algorithms, building the core of the evolutionary computation (EC) field.

This section targets at introducing the main EA concepts and specialized techniques for three important application areas: Multiobjective optimization, optimization under uncertainty, and multimodal optimization. These are relevant to the topic of this book as they are closely interrelated and often encountered conjoined in real-world applications.

What is an evolutionary algorithm? Today, there is little doubt about the components and the general structure of an EA. It is understood as a population based direct search algorithm with stochastic elements that in some sense mimics the organic evolution.

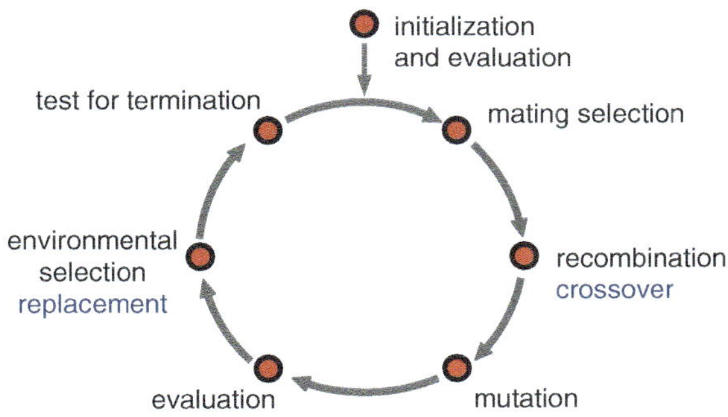

Fig. 2.17 The evolutionary cycle, basic working scheme of all EAs. Terms common for describing evolution strategies are used, alternative (GA) terms are added below.

Besides initialization and termination as necessary constituents of every algorithm, EAs consist of three important factors: A number of search operators, an imposed control flow (Fig. 2.17), and a representation that maps adequate variables to implementable solution candidates.

Although different EAs may put different emphasis on the search operators mutation and recombination, their general effects are not in question. Mutation means neighborhood based movement in search space that includes the exploration of the "outer space" currently not covered by a population, whereas recombination rearranges existing information and so focuses on the "inner space." Selection is meant to introduce a bias towards better fitness values; GAs do so by regulating the crossover via mating selection, ESs utilize the environmental selection.

A concrete EA may contain specific mutation, recombination, or selection operators, or call them only with a certain probability, but the control flow is usually left unchanged. Each of the consecutive cycles is termed a *generation*. Concerning the representation, it should be noted that most empiric studies are based on canonical forms as binary strings or real-valued vectors, whereas many real-world applications require specialized, problem dependent ones.

For an in-depth coverage on the defining components of an EA and their connection to natural evolution, see [ES02, ES03, BFM97].

2.6.2 Evolution Strategies

In the following, we introduce the most important canonical ES variants for single objective optimization, which serve as a basis for more specialized algorithms later on.

The (1+1)*-ES* The first ES, the so-called (1+1)-ES or *two membered evolution strategy*, uses one parent and one offspring only. Two rules have been applied to these candidate solutions:

1. Apply small, random changes to all variables simultaneously.
2. If the offspring solution is not worse (in terms of its function value) than the parent, take it as the new parent, otherwise retain the parent.

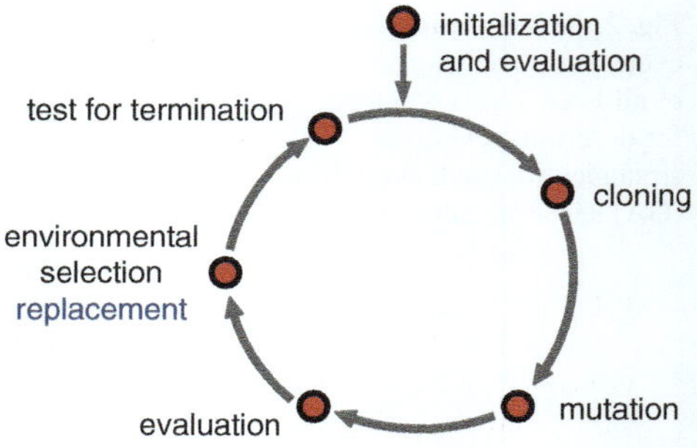

Fig. 2.18 The evolutionary cycle of a two-membered (1+1) evolution strategy.

Schwefel [Sch95a] describes this algorithm as "the minimal concept for an imitation of organic evolution." The $(1 + 1)$-ES (Fig. 2.18) is applied by many optimization practitioners to their optimization problem and included in this article for three reasons: (i) It is easy to implement, (ii) it requires only few exogenous parameters, and (iii) it defines a standard for comparisons.

The first $(1 + 1)$-ES used binomially distributed mutations for integer variables [Sch65]. These have been replaced by Gaussian mutations for continuous variables. Rechenberg [Rec71] already proposed a simple rule to control the mutation strength, the so-called 1/5 success rule. This simple ES requires the specification of four parameters (factors), namely the adaptation interval, the required success rate, the step size adjustment factor,[5] and the step size starting value.

Population Based ESs Population based ESs use μ parents and λ offspring. Rechenberg introduced the first multimembered ES, the so-called $(\mu + 1)$-ES. It uses μ parents and one offspring and is referred to as the *steady-state* ES. Schwefel introduced the $(\mu + \lambda)$-ES, in which $\lambda \geq 1$ candidate solutions are created in each generation, and the best μ out of all $\mu+\lambda$ individuals survive. He also introduced the (μ, λ)-ES, in which the parents are forgotten and only the best μ out of λ candidate solutions survive. These selection schemes will be discussed later in this section (p. 52).

A birth surplus is necessary for the (μ, λ)-ES, that is $\lambda > \mu$. Schwefel et al. [SRB95] and Beyer and Schwefel [BS02b] provide a comprehensive introduction to evolution strategies.

Note that whereas GAs rely upon a start population uniformly scattered in a closed search region, ESs – even if population based – may be started around any start vector like standard optimization algorithms, without lower and upper bounds for the variables.

Variation in ESs The use of populations enables an extension of the rather simple 1/5 success rule to control the mutation strength [Sch75]. Beyer and Schwefel

[5] This is a constant factor c with $1 \leq c \leq 0.85$, the lower bound being theoretically near-optimal for simple model problems like the sphere model.

[BS02b] propose some guidelines derived from the philosophy of Darwinian evolution to design these variation operators.

1. A *state* comprises a set of object and strategy parameter values $(x^{(t)}, s^{(t)})$. *Reachability* demands that any state can be reached within a finite number of iterations. This feature is necessary to prove (theoretically) global convergence.
2. Variation operators (mutation and recombination) should not introduce any bias, e.g. by considering only good candidate solutions. Variation operators are designed to *explore* the search space in contrast to selection operators that exploit the gathered information. Recombination works, according to Beyer [Bey95], mainly as gene repair operator, not only as a building block collection mechanism.
3. *Scalability* is the third criterion that should be fulfilled by variation operators: Small changes of the representation should cause small changes in the function values.

The standard ES recombination operators produce one offspring from a family of ρ parent individuals (usually $\rho = 2$). Consider a set of μ parental vectors of length N, representing either object or strategy parameters: $\{(x_{11}, \ldots, x_{1N}), (x_{21}, \ldots, x_{2N}), \ldots, (x_{\mu 1}, \ldots, x_{\mu N})\}$. Two recombination schemes are commonly used in ESs. Both use a set $\mathcal{R} = \{r_1, r_2, \ldots, r_\rho\}$, that represents the indices of the mating partners. It is constructed by randomly (uniformly) choosing ρ numbers (with replacement or not) from the set $\{1, 2, \ldots, \mu\}$. *Discrete recombination* selects the entries of the offspring randomly from $\mathcal{R}$, whereas *intermediary recombination* averages the ρ corresponding values of all mating pool members in each component of the newly generated vector.

Mutation is applied to the recombined intermediate solution. Mutation in multimembered ESs is a self-adaptive process that relies on the individual coupling of endogenous strategy parameters with object parameters. After being varied as described above, the strategy parameters (standard deviations, also called mean step sizes or mutation strengths) are applied to mutate the object parameters. To illustrate this procedure, algorithms with one common σ are considered first. To prevent negative standard deviations, mutation of this σ should be done multiplicatively. Beyer and Schwefel [BS02b] discuss an additional argument for a multiplicative mutation of the mutation strength on the sphere model. It can be shown, that in expectation σ should be changed by a factor that only depends on N. Therefore, the mutation operator can be implemented as $\sigma^{(t+1)} = \sigma^{(t)} \cdot \exp(\tau z)$, where z is a realization of an $\mathcal{N}(0, 1)$ distributed random variable. The parameter τ is the so-called *learning rate*. The object variables are mutated next: $x^{(t+1)} = x^{(t)} + w$, where w is a realization of an $\mathcal{N}(0, \sigma^{(t+1)})$ distributed random variable. The multiplicative mutation scheme for one σ can be extended to several strategy parameters $\sigma = (\sigma_1, \ldots, \sigma_N)$. Schwefel [Sch77] proposes the following extended log-normal rule: $\sigma^{(t+1)} = \left(\sigma_1^{(t)} \exp(\tau z_1), \ldots, \sigma_d^{(t)} \exp(\tau z_N)\right)$, where z_i are realizations of N standard normally distributed random variables, $1 \leq i \leq N$. This mutation scheme employs a single learning rate τ for all strategy parameters. An alternative procedure that utilizes a global and a local learning parameter τ_0 and τ, respectively, is

Table 2.1 Algorithm design of ES

Symbol	Parameter	Range
μ	Number of parent individuals	$\mathbb{N}$
$\nu = \lambda/\mu$	Offspring-parent ratio	$\mathbb{R}_+$
$\sigma_i^{(0)}$	Initial standard deviations	$\mathbb{R}_+$
n_σ	Number of standard deviations. N denotes the problem dimension	$\{1, N\}$
τ_0, τ	Multiplier for mutation parameters	$\mathbb{R}_+$
ρ	Mixing number	$\{1, \mu\}$
r_x	Recombination operator for object variables	{intermediary, discrete}
r_σ	Recombination operator for strategy variables	{intermediary, discrete}
κ	Maximum age	$\mathbb{R}_+$

suggested by Bäck and Schwefel [BS92]. Self-adaptive correlated mutations have already been introduced in 1974, see [Sch81, Sch87].

Selection in ESs Selection should direct the evolutionary search toward promising regions. In ESs, only candidate solutions with good function values are allowed to reproduce. The replacement (environmental selection) process is deterministic in contrast to the random processes used in GAs. This selection scheme is known as *truncation* or *breeding selection* in biology. The κ-selection scheme takes the age of candidate solutions into account: Only candidate solutions that are younger than κ generations may survive, regardless of their fitness. For $\kappa = 1$ this selection method is referred to as *comma-selection*: only offspring individuals can reproduce. The κ-selection is referred to as *plus-selection* for $\kappa = \infty$: Both the offspring and the parents belong to the mating pool. The plus-selection is an elitist selection scheme, because it guarantees the survival of the best individual found so far.

Table 2.1 summarizes important ES parameters [BB03]. These parameters build an *algorithm* design. In addition to algorithm designs optimization practitioners have to cope with *problem* designs which will be discussed next.

2.6.3 Ways to Cope with Uncertainty

In the following, we will distinguish three types of parameters that influence experimental results [SWN03a]. The first type of parameter to be mentioned is a *control* parameter. Control parameters can be set by an experimenter to "control" the experiment.

The second type of parameter, so-called *environmental* parameter depends on the environment at the time the experiment is performed. Some authors refer to environmental parameters as "noise" parameters. Note, that environmental parameters include measurement errors such as falsely calibrated measurement instruments, inexact scales, scale reading errors, etc. Data preprocessing techniques were developed to reduce this source of error, which occurs in nearly every field setting. In some situations, environmental parameters can be treated as having a given distribution that is characteristic for the given experimental setup.

The third type of parameter, the so-called *model* parameter, describes the uncertainty of the mathematical modeling. First, we have to take into account that computer simulations require a model which simplifies the underlying real-world scenario. Therefore, simulation results are only approximations of the corresponding real-world data. Next, if stochastic (and not deterministic) simulations are considered, the measurements may be exact (because there is no environmental noise), but some of the models' parameters are random parameters. In some cases, there is a known (subjective) distribution which describes this uncertainty.

As an example, we consider a sequence of traffic signals along a certain route or elevators' movements in high-rise buildings. *Optimization via simulation* subsumes all problems in which the performance of the system is determined by running a computer simulation. If the result of a simulation run is a random variable, we cannot optimize the actual value of the simulation output, or a singular performance of the system. One goal of optimization via simulation may be to optimize the expected performance. In addition, consider a field study which was performed to validate the results from the computer simulation. This field study includes environmental parameters.

Summarizing, there are two fundamental sources of uncertainty (or noise) that can be described by environmental and model parameters. Figure 2.19 illustrates these parameters in the context of algorithm and problem designs.

The efficiency of the evaluation and selection method is a crucial point, since averaging over repeated runs reduces the efficiency of the optimization process.

The Impact of Noise on EAs Noise makes it difficult to compare different solutions and select the better ones. Noise affects the selection process in evolutionary algorithms: In every iteration, the best μ out of λ candidate solutions have to be determined.

Wrong decisions can cause *stagnation* of the search process: Over-valuated candidates – solutions that are only seemingly better – build a barrier around the optimum and prevent convergence. Or, even worse, the search process can be *misguided:* The selection of seemingly good candidates moves the search away from the optimum. This phenomenon occurs if the noise level is high and the probability of a correct selection is very small.

One may attempt to reduce the effect of noise explicitly (*explicit* averaging). The simplest way to do so is to sample a solution's function value n times, and use the

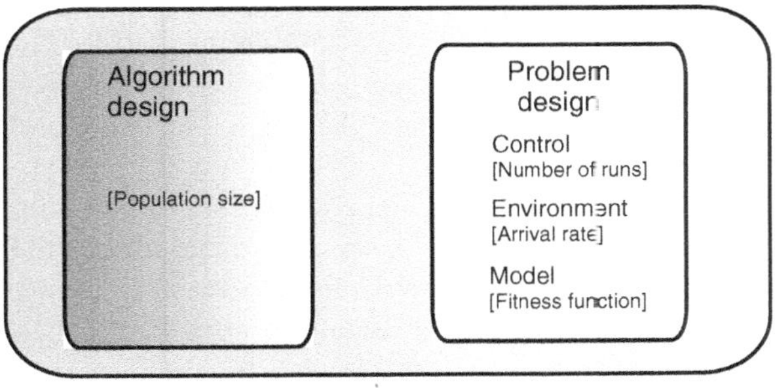

Fig. 2.19 Before an EA can be started, the optimization practitioner has to specify several parameters. Examples are shown in brackets. Environmental and model parameters can be affected by noise.

average as estimate for the true expected function value. This reduces the standard deviation of the noise by a factor of $\sqrt{n}$, while increasing the running time by a factor of n.

In contrast to explicit averaging, some authors proposed *implicit* averaging, i.e., increasing the population size to cope with uncertainty in evolutionary optimization. Theoretical results lead to contradictory recommendations: In [Bey93] the authors conclude that it is better to increase the population size whereas [FG88] shows that increasing the sample size is advantageous.

Further means used by evolutionary algorithms to cope with noise are averaging techniques based on statistical tests, local regression methods for function value estimation, or methods to vary the population size [Sta98, Bey00, SK00, Arn01, BSS01, BBM04, JB05b]. Because uncertainties complicate the selection process for direct search methods, some authors suggested modified selection operators.

A Taxonomy of Selection Methods As introduced above, noise affects selection. Following Bechhofer, Santner, and Goldsman [BSG95] and Bartz-Beielstein [BB06], we present a taxonomy of elementary selection methods. Depending on a priori knowledge, selection schemes can be classified according to the following criteria:

Threshold: subset selection – indifference zone.
Termination: single stage – multi stage (sequential).
Sample size: open procedures – closed procedures.
Variances: known – unknown, equal – unequal.

The goal of subset selection is the identification of a subset containing the best candidate. It is related to screening procedures. *Subset selection* is used when analyzing results, whereas the *indifference zone* (IZ) approach is used when designing experiments. The sample size is known in subset selection approaches, it is determined prior to the experiments in the indifference zone approaches. *Single stage* procedures can be distinguished from *multi stage* procedures. The terms "multi stage" and "sequential" will be used synonymously. The latter can use *elimination*: If inferior solutions are detected, they are eliminated immediately. Selection procedures are *closed*, if prior to experimentation an upper bound is placed on the number of observations to be taken from each candidate. Otherwise, they are *open*. Furthermore, it is important to know whether the variance is common or known. Bartz-Beielstein [BB06] discussed similarities and differences of these approaches. He also analyzed threshold-based procedures, which were successfully applied to noisy, dynamic functions, e.g., in elevator group control. Threshold rejection increases the chance of rejecting a worse candidate at the expense of accepting a good candidate. It might be adequate if there is a very small probability of generating a good candidate.

How can the experimenter cope with this multitude of selection methods? Surely, there is no general rule for the determination of the best selection method. Many theoretical results consider simplified sources of uncertainty, e.g. they regard environmental parameters as random with a distribution that is known. Performing experiments in a systematic manner might be useful. Modern approaches such as

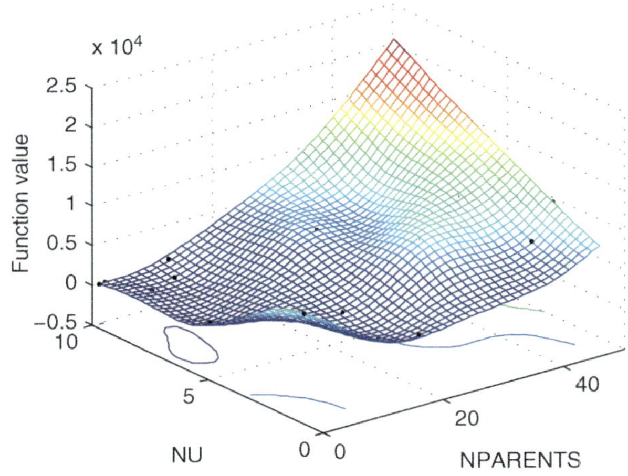

Fig. 2.20 SPO combines classical and modern statistical tools for the analysis of algorithms. Modifying population size (NPARENTS) and selective pressure (NU) can improve algorithm's performance significantly. Evolution strategies with small population sizes and moderate selective pressure perform best in this setting.

racing or sequential parameter optimization (SPO) can be recommended in this context [BSPV02, BBLP05]. A typical result from an SPO analysis is shown in Fig. 2.20.

Regarding the classification from Fig. 2.19, there are two starting points to cope with noise: (i) varying the algorithm design, e.g., choosing a modified selection operator or (ii) modifying the problem design, e.g., refining the fitness function. Evolutionary optimization itself can be considered as an evolutionary process. Based on results from previous optimization runs, the experimenter may gain insight into the behavior of the evolutionary algorithm and into the structure of the problem as well. He is able to modify (improve) algorithm and problem designs – black box situations turn into gray box situations. Combinations of classical and evolutionary methods (meta heuristics) may be useful in these situations.

2.6.4 Multiple Objectives

For many problems of high practical interest in science and engineering, several possibly contradicting objectives shall be pursued simultaneously. In daily life as well as in industry, we are confronted with many examples. E.g. in chemical process engineering, where the productivity of chemical reactors is in contrast to their loss during the start up and shut down phases. In the textile industry, a similar conflict arises for the production of fabrics. Figure 2.21 shows a simple discrete example. Total elongation (F_1) and extensibility (F_2) of the fabric shall be improved, by means of maximizing F_1 and minimizing F_2. All objectives are sufficiently defined and in this case pointwise quantifiable. Their values are determined by three adjustable control factors (decision variables): Number of knitting skewers (x_1), number of knitting rows (x_2) and number of weft threads (x_3) per inch. The challenge for a multi-objective optimization algorithm consists of finding decision variable value sets that fulfill all objectives as well as possible.

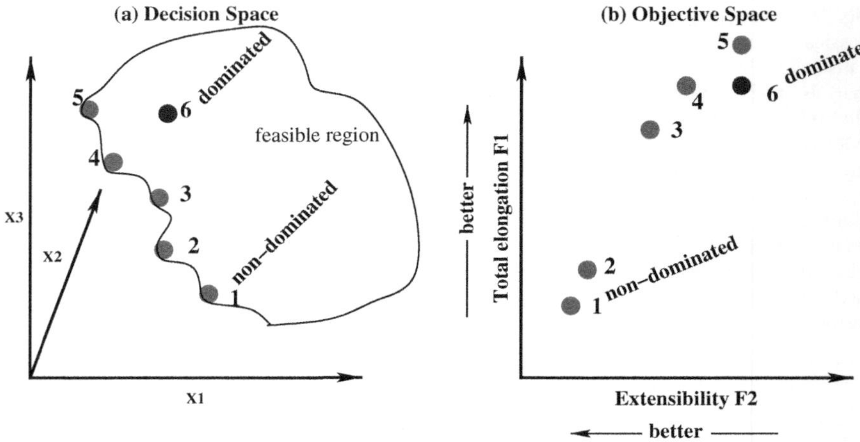

Fig. 2.21 The Pareto-dominance concept. (**a**) Decision space, (**b**) objective space

In this context, the Pareto [Par96] concept of optimality proved as suitable. During the beginning of an optimization run, it is often not hard to find solutions that simultaneously improve both objectives. However, if an objective can be improved further only by worsening an other objective, a solution is called *Pareto-optimal*. Due to different possible preferences concerning the single objectives, this leads to a set of Pareto-optimal solutions, each of them representing a valid optimal solution for the multi-objective problem (MOP). Figure 2.21 shows six solutions in the decision variable space (a) and the objective space (b) for the fabric improvement example. In this example, the decision variable space is discrete and constrained as indicated by the surrounding solid line. Consequently, there is only a finite number of possible objective value combinations. Direct comparison of solutions 5 and 6 shows that the former improves on F_1 without changing F_2. According to the Pareto dominance concept, solution 5 *dominates* solution 6. However, pairwise comparison of solutions 1–5 does not result in recognizing any such domination as improvement in one objective always comes along with worsening in the other. These solutions are therefore indifferent to each other, hence incomparable or *non-dominated*. If due to problem-specific constraints no further improvements can be obtained (solutions 1–5 are on the border of the feasible region) the set of all non-dominated solutions represents the *Pareto Set* in the decision space and the *Pareto Front* in the objective space. Since in each case only one solution can be realized, preference information of a decision maker (DM) must be used next to select the final solution of the MOP.

Why Use Evolutionary Algorithms? Problems with several conflicting criteria have been treated for many years, e.g. with a considerable variety of techniques developed in Operational Research. Concise overviews of existing approaches can be found in [AEN79, Mie98]. Usually one tries to reduce the MOP into a single-objective problem, so that it can be solved by means of methods from single-objective optimization. One possible approach consists of choosing a single criterion as main objective, and transform the other objectives to constraints with lower or upper bounds. Without specific knowledge of the problem, the choice of concrete upper and lower bounds suffers from arbitrariness. Alternatively, one may try aggregation-based approaches. These combine all criteria into a single, parametrized one. The aggregation can be accomplished by any combination of arithmetical operations (i.e. a weighted sum), according to some understanding of the problem. However, these techniques have several limitations. Some of them are e.g. susceptible to the shape (convex/concave) of the Pareto front, others to its continuity (connected/disconnected). In addition, most of the "conventional" approaches are only able to compute one single non-dominated solution per run. Searching for a representative set of non-dominated solutions requires a restart with different external parameter settings and different starting points for each run.

Evolutionary algorithms are robust search methods, whose success and failure is by far less susceptible to the shape or the continuity of the Pareto front. Their greatest advantage is that they are able to provide a point-wise approximation of the whole Pareto front in one go by employing cooperative search of a whole population.

Algorithm Design If one regards the development of the *evolutionary multi-objective (EMO) algorithms* within the last two decades, the increase of suggested approaches is impressing. The largest well-known collection of existing approaches was arranged by Coello Coello and contains over 1900 entries [Coe06]. A common classification of all EMO-algorithms comes from Masud [HM79]. Depending on the time at which the preference information from the DM is used, four classes can be differentiated: (i) Non-preference, (ii) a-priori, (iii) interactive, and (iv) a-posteriori. In the following, this classification is not discussed in detail as most EMO-algorithms can be assigned to the last category. The optimization process takes place before any preference information is incorporated. This entails a clear task definition: Find a representative set of non-dominated solutions as close (convergence) as possible to the *Pareto optimal set/front*. Additionally, the resulting approximation has to exhibit a good distribution of solutions in terms of both spread and uniformity – usually described by the term of *diversity*. The aim of this section is to give an overview of the main methods that have been developed in order to achieve these goals.

Fitness Assignment: When moving from single-objective to multi-objective optimization while applying EAs, the most important changes to be made concern the selection operator and especially the fitness assignment. In EAs, the fittest individuals have better chances to survive and reproduce. For single-objective optimization, only one scalar fitness value exists. However, in the multi-objective case we have to deal with a fitness vector. Since EAs need a scalar to work on, generally two design decisions must be made: On the one hand, this vector must be scaled to enable for EA selection, and on the other hand the two conflicting tasks of convergence and diversity shall be respected. But how to assign the fitness of an individual in order to express suitability towards both goals? We can roughly divide the existing answers into two categories:

Combined Fitness Assignment: Fitness is assigned such that the fitness value represents convergence and diversity at the same time.

Single Fitness Assignment: Fitness assignment respects only one goal. Usually, this is convergence, as in the single-objective case.

Aggregation-, performance-, and Pareto-based approaches belong to the first category. Aggregation-based approaches are the most traditional as well as simplest possibility. Recently, performance-based fitness assignment strategies are successfully used to evaluate the fitness of a new individual in relation to the entire population. For example, the S-metric selection (SMS)-EMOA utilizes the well-known S-metric (hypervolume) to calculate the fitness of an individual. This measure is commonly used to evaluate the performance of an EMOA. It respects proximity to the Pareto front as well as diversity of the solution set.

Pareto-based approaches use the Pareto dominance concept itself for fitness assignment. Differences between these approaches arise in the methods employed to exploit the partial order. According to Zitzler et al. [ZLB03], this kind of information can be divided into: (i) *Dominance rank:* The number of solutions in the population that dominate the solution under consideration, (ii) *dominance count:* The number of solutions in the population that are dominated by the solution under consideration, and (iii) *dominance depth:* The rank of the solution in the non-dominated sorted population. The latter approach is utilized by many successful algorithms, e.g. the *Non-dominated Sorting Genetic Algorithm (NSGA)-II* by Deb et al. [DAPM00b]. Dominance rank was first employed by Fonseca and Fleming in their *Pareto envelope-based algorithm (PESA)* [FF96]. Today, a multiplicity of methods are based on this principle, see for example [BT05]. Dominance depth and dominance rank are successfully combined in the *Strength Pareto Evolutionary Algorithm 2 (SPEA2)* approach by Zitzler et al. [ZLT01b].

However, most of these algorithms apply a *secondary fitness assignment strategy* that serves the goal of diversity. In most cases they try to incorporate density information into the selection process (mating/environmental), according to the rule: The smaller the density of individuals within a neighborhood, the larger the chance of an individual to reproduce. Figure 2.22 shows the three most frequently used methods: *Kernel-based*, *grid-based* and *nearest-neighborhood* measures. Fitness sharing, as e.g. used in NSGA, is a kernel-based strategy. The distance of an individual to all other individuals in the population is calculated and summed up. These values are then used to deflect the evolutionary search out of densely populated regions. Grid-based techniques as e.g. utilized by the *Pareto Archived Evolution strategy (PAES)* of Knowles and Corne [KC99], employ hypergrids to define neighborhoods within the objective space. The more individuals in a box, the heavier they are penalized (see Fig. 2.22). Nearest neighborhood techniques as used in SPEA2 and its variants calculate the distance between an individual and its nearest neighbor in order to estimate the neighborhood density.

Criterion-based approaches represent the second category of fitness assignment strategies. They all share the same basic idea: The fitness value of an individual is determined by only one of the criteria according to the goal of convergence. However, the choice of a single criterion for any individual shall be reconsidered

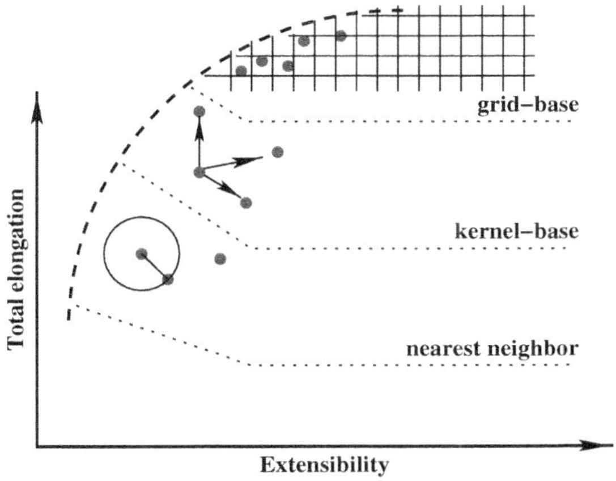

Fig. 2.22 Most common diversity preservation strategies in EMOA.

repeatedly (in each generation). As thereby parts of the population are selected according to different criteria, it is hoped that the goal of diversity can be achieved indirectly (see Schaffer [Sch84] and Laumanns et al. [LRS98]).

Representations and Variation Operators Design and analysis of representations and corresponding genetic operators is prevalent in the field of evolutionary computation. Often, an adept combination of all components determines the system's success or failure. This insight is ubiquitous in the case of single-objective optimization. However, in multi-objective optimization, the conceptual approaches are still mainly concerned with the selection operator. Research focusing on variation operators or representations remains rare. Some recent approaches are: Rudolph [Rud98] and Hanne [Han99] who investigate control mechanisms for the mutation strength in the multi-objective case. Grimme and Schmitt [GS06] focus on recombination operators that produce diverse offspring in each generation.

Elitism Elitism preserves previously attained good solutions from one generation to the next. The prime example of an elitist algorithm in the single-objective case is the "plus"-selection ES. In the multi-objective case two types of elitism are used: Maintaining elitism in the current population, as is already done in the single-objective case, or doing so in an archive (secondary population) that stores non-dominated solutions externally. Archive contents may or may not be integrated again into the optimization process (Zitzler et al. [ZLB03]). Of vital importance is the criterion used to control replacement of archive members, the most commonly used of which is the dominance criterion. It leads to an archive of non-dominated solutions, relative to all solutions generated during a run.

Future Perspectives As has been hinted to in the previous paragraphs, a lot of work remains to be done on EMOAs. We briefly discuss the currently most promising paths:

Investigating representations and variation operators: Büche et al. [BMK03] show that the interaction between selection and search operators is often not coordinated well, and that approximation of the Pareto front cannot be done with arbitrary precision. Further on, there is the dilemma of stagnation with good diversity of the solution set on the one hand, or arbitrarily exact approximation of a few points on the Pareto front. We conjecture that this trade-off between convergence and diversity can be attributed to the fact that variation operators cannot simply be taken over from the single-objective case and that changing only the selection operator is not sufficient to meet the requirements of multi-objective optimization.

Focusing on the region of interest (ROI): In the last years, most EMO researchers focus on algorithms that are able to find the whole Pareto front. However, in practice, the decision maker is only interested in a specific region of the Pareto-front. Focusing on a region derived from user preferences may help to increase convergence speed and/or quality and also simplify solution selection by the DM later on.

Parallelism: Considering the suitability of EAs working in a parallel manner, one should expect that the development of parallel approaches stands only at the beginning. Apart from first successful attempts to convert the state-of-the-art algorithms into a parallel version [OHMW02], an increasing number of parallel approaches has been published only recently [CC05, MMSK04].

Parameter tuning: Attaining good parameter settings for a given problem-algorithm combination currently is one of the hot topics in single-objective optimization [BB06]. It is necessary to adapt those techniques for the multi-objective case in order to avoid the commonly used manual parameter tuning and provide important insight into parameter interactions.

2.6.5 Multimodal Problems

Although, during the last decades, many empirical and most of the theoretical studies in EC have been devoted to simple test problems with only one extremal point, the great majority of practical applications requires optimization in far more complex fitness landscapes. Multimodality – the presence of more than one locally optimal point – requires a shift from a hill-climbing oriented towards a global perspective. At the top of the hill, the need arises to somehow "escape" the associated local optimum. This may be done in two different ways. Either, one tries to save as much positional and learned (step sizes/mutation strengths) information as possible and, preserving this information, attempts to jump over the neighboring valleys. Or, one completely gives up the current search space location and performs random initialization again. For mutation strenghts getting larger and larger, the former scenario more and more resembles the latter.

However, if the treated optimization problem is not available in a closed algebraic form, detecting the arrival at a local optimum may not be trivial, depending on the employed variable representation. Combinatorial and binary encoded optimization problems come with a natural minimal step definition which enables enumeration of the neighborhood. Nevertheless, the bounded but still infinite neighborhood cannot be completely explored efficiently and one has to rely on the strong causality assumption [Rec89]: similar causes entail similar effects) to identify local optima at least in probability.

Strongly related to the notion of local optima is the one of basins of attraction; these encompass the search space portion leading to an optimum if the steepest descent is followed. For this local search process, efficient approximation methods are known, e.g. quasi-Newton algorithms. However, identification of different basins is even more difficult than local optimum detection if no further information regarding size and/or location of the basins is available. The key property of multimodal optimization methods is thus how efficient they are in finding the different search space regions that contain the best local optima.

Canonical population based EAs perform global and local search at the same time, gradually narrowing their focus to the most promising regions, and more sooner than later to a single basin of attraction (e.g. Preuss, Schönemann and Emmerich [PSE05]). From the discussion above, it becomes clear that the ability to explore multiple promising regions – either concurrently or sequentially – is decisive for obtaining well performing EA variants. But for a given limit of available computational time, these always have to face the global vs. local search tradeoff like any other global optimization algorithm.

One possible way to speed up local optimization, so that more effort can be diverted to search space exploration, is to hybridize EAs with existing local search methods. These approaches are subsumed under the term *memetic algorithms* (MA) that was introduced by Moscato [Mos89]. A recent overview is given by Krasnogor and Smith [KS05], together with a suggested taxonomy.

Most other specialized EAs strive for enhanced global search capabilities by means of at least one of the following three techniques:

Restarts are utilized to enhance the chance of reaching the/a basin of attraction of the global optimum. As an example, an efficient restart CMA-ES for multimodal problems has been suggested by Auger and Hansen [AH05]. Multistart methods obtain potential solutions consecutively, and every new instantiation may be provided with search results of completed previous runs. They avoid the problem of jumping into a neighboring good region by giving up the current search space location completely.

Diversity maintenance aims for a uniform distribution of individuals over the whole search space. Comparing relative or absolute distances of solution candidates and applying clustering methods are common means to prevent overlapping search paths and promote good search space coverage. Diversity may be held up explicitly or implicitly. Following Eiben and Smith [ES03], explicit means that active measures are taken to model the distribution of search points in the desired way, whereas implicit stands for deliberately slowing down information exchange by restricting recombination or selection/replacement. Classical island models provide implicit diversity maintenance by building relatively independent subpopulations. Spatially structured EAs [Tom05] do so by restricting the effect of recombination and selection operators to the local neighborhood. *Shifting balance* GAs by Oppacher and Wineberg [OW99] exemplify explicit diversity maintenance as they prevent subpopulation overlap which is measured by absolute population distances.

Niching methods also strive for a suitable spread of search points, only on the level of basins of attraction. As Mahfoud [Mah95] points out, it is the aim of *niching* algorithms to detect separate basins and keep them in focus of the search. Unfortunately, *basin identification* within an EA is not easy and prone to error, so that endogenously retrieved basin information is highly unreliable and nonexistent when the optimization starts. Crowding by De Jong [De 75] and fitness sharing by Goldberg and Richardson [GR87] are regarded as the classical niching methods. The former employ relative, the latter absolute distances. These have been carried further e.g. [LBPC02, SSUZ03], and Shir [Shi05], but still the radii employed for detecting search points located together in a basin remain problematic. Only few approaches integrate fitness topology information into the basin identification process, e.g. the *universal evolutionary global optimizer* (UEGO) by Jelasity [Jel98], Ursem's multinational GA [Urs99], and the sample-based crowding method proposed by Ando et al. [ASK05].

It shall be noted that solving multimodal problems is related to tackling constrained or multiobjective ones. Removing constraints from a problem by transforming it by means of (metric) penalty functions (see e.g. [MS96, Coe02]) as commonly done in EC most often leads to multimodal problems even if the original problem was unimodal.

In multi-objective optimization, the focus has been mainly on the objective space for a long time. Today, it becomes increasingly clear that population movement in the decision (search) space heavily depends on the multimodal search properties of the applied optimization algorithms (Preuss, Naujoks and Rudolph [PNR06]).

2.6.6 Conclusions

May it be (or not) that one day there is no more need to invent new optimization tools because we have got the best tailored ones already for every possible real-world problem. May it be (or not) that then the dream of hardliners has come true that all of these best tailored methods can abstain from using pseudo random numbers for deciding upon the next iteration in the search for the solution. But, contemporary tools are still well advised not to rely on deterministic algorithms alone. That is why an idea from the early days of digital computers is still alive, i.e., the idea to mimic procedures found in nature that obviously have led to remarkably effective systems or subsystems. One may think that nature had enough time to achieve a good solution by means of pure chance, but time has always been scarce when there are competitors, and the way nature finds its way is much more sophisticated.

Anyway, it is a matter of fact that evolutionary algorithms have become widely used in practice since their invention in the 1960s and even found their way into articles in the field of theoretical computer science. Their domain of application are "black box" situations, where the analysis of the situation at hand does not help or is too costly or dangerous, i.e., in case of experimental design and even computer simulation of nonlinear dynamic systems and processes. However, situations may occur where the black box situations turn into gray or even white box situations. EAs can be combined with classical methods which leads to *m*eta heuristics, and the optimization practitioner can get the best from both worlds.

2.7 Structure Optimization by Decision Support Systems[6]

2.7.1 Introduction

Engineers, scientists and managers are trained to understand their physical, abstract or economic environments in which they act. In fact, *expertise* in a field of research or application results from having a good knowledge and understanding of the topic. In many applications of these fields, *models* can be built from this expertise, which convert "input" data to "output" data about a given problem. For example, modern Computer Aided Design (CAD) software can calculate various physical properties from a virtually drawn construction. In a similar way, many products such as engine parts, effects of a medical therapy, or returns from a stock can be *virtually simulated*. However, an expert can only use knowledge that can be automated to build the models composing such a virtual simulation. In the case of the CAD software, these models could be the basic physical laws of mechanics. In the end, to base a decision on output data from such a simulation, it is up to the expert to judge the virtual design according to various criteria. A financial expert, for example, can devise models to predict stock prices. The output is in this case an estimate of the probability distribution of future returns. The manager must then evaluate this forecast according to various risk measures, expected profits, integration into existing stock portfolios etc. into make a buying decision. This knowledge, however, which is often based on experience, can only seldom be included to expert systems.

2.7.2 Forward Planning in Decision Support

Software that facilitates a decision-making process by providing models based on knowledge that can be automated are called Decision Support Systems (DSS). Many existing expert systems and DSS can be used as *forward planning* tools. Classical forward planning consists of three steps that are repeated until the decision-maker is satisfied with the outcome. The first interaction with the system is to specify the design parameters of the product. Then the virtual simulation is run and the output data is generated. The third step is the evaluation of this data based by the expert. If the decision-maker is not yet satisfied, the input parameters are altered to achieve an improvement of the result, the simulation is repeated, and so forth (see also to Fig. 2.23).

In some cases, the user is offered the possibility to describe his quality measures using a function, which can be automatically optimized by the DSS. This aids the decision-making process because tuning and selecting the input parameters is automated this way. This way, the financial expert from the previous example could express his or her idea of balancing risks and rewards and let the system calculate the number of shares she has to buy so that her portfolio fulfills the specified criteria.

[6] Contribution by K.-H. Küfer and P. Süss

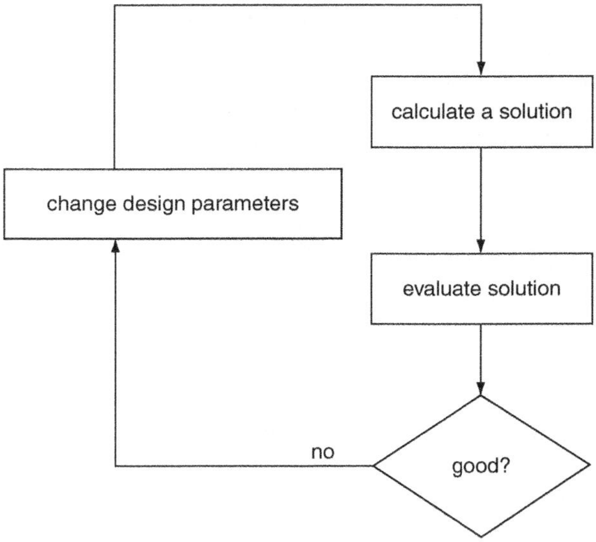

Fig. 2.23 Forward planning is a method of successive improvements and local improvement. It suffers from not conveying any tradeoff information between the different involved criteria and that the expert is forced to interact a lot with the DSS.

There are three main difficulties with this (partially automated) forward planning technique, which prevent the decision-maker to concentrate solely on his or her expertise. We now present these difficulties and suggest the *multicriteria decision-making* paradigm to overcome them.

First, there is the problem with *scalarizing* the preferences. To specify a mathematical function that is able to judge the quality of a product is practically impossible in most situations. There are often many criteria by which a design is judged that are not quantifiable. And even if all involved criteria are quantitative – like in the simplified financial example – it is difficult to specify a priori how these criteria should be balanced and traded off against each other.

The second problem with classical forward planning is that this approach can only deliver snapshots of the true variety of solutions which exist for all possible scalarizations of a problem. Let us consider the shape design of a wing of an airplane and assume that an engineer can describe the shape of the wing using a few geometrical parameters. This is the input to the simulation. Moreover, a CAD software can calculate the weight and the stability of the wing under some given circumstances when these design parameters are fixed. This is then the output of the model of the wing design. A big solid wing is probably very strong but is also heavy. Even if it was possible for the engineer to express his or her preferences about these two criteria in one formula, only one solution that fulfills this preference would be generated. However, there might be wing designs that exhibit similar characteristics to what the engineer specified, but are favorable in terms of ease of production or assembly to the body of the airplane – both quality measures which are difficult to express with a mathematical function. If the decision-maker is interested in performing such a "what-if" analysis on the current design, or if it is not possible to express the preferences precisely, it can become a tedious task to search for similar designs by altering the design parameters or the scalarization many times.

The third main difficulty with this approach consists of actually solving the scalarized optimization problems. Often, the outcome of the virtual simulations is used in yet another model that has a very simple structure so that a generic solver can perform the optimization of this *surrogate* [BW51]. However, such generic solvers are often not suited for the particular problem at hand. In many cases, the models used for the virtual simulation can be included in the optimization to create custom solvers for the design problem. This way, tremendous speedups in calculation time, during which the user has to wait, can be achieved. The final two sections contain several examples of this approach.

As a consequence, using forward planning, the expert is busy for potentially a long time producing a collection of alternatives in addition to evaluating them based on his or her expertise. If the decision-maker does not have to worry about the details of a scalarization of his idea of quality, and if a variety of solutions can be presented in a clever way, the expert can spend more time evaluating the designs and has a much stronger base to perform scenario analysis to test out similar alternatives. The aim of a DSS design should be to aid the user to come to a decision in one sitting. To achieve this, a DSS can be equipped with mechanisms of multicriteria decision-making.

2.7.3 Practical Multicriteria Decision-Making (MCDM)

With this new paradigm, the most important difference for the decision-maker is that he or she is not limited to one formula to judge the quality of a design. As the name suggests, many criteria to evaluate a design may be formulated at the same time. In theory, this is all a DSS needs to calculate and present so-called *efficient* solutions of the problem. A solution is called efficient if any particular criterion can only be improved by worsening at least one other criterion. If, for example, the effectiveness of a medical therapy can only improved by also increasing the likelihood of some unwanted side-effect, this therapy is efficient with respect to those criteria. The added benefit for the decision-maker is that the manual configuration of the design parameters is no longer necessary. Moreover, for every alternative that is presented to the user, the information about the *tradeoff* between the specified criteria is readily available, as was discussed before in 2.53 and 2.64.

Calculating or approximating all efficient solutions of a problem with many criteria is the objective of theoretical multicriteria optimization [Ste86]. For practical use, it is important that no relevant efficient solutions are withheld from the user. It is easily imagined that a large number of specified criteria lead to a large set of efficient solutions. For this reason, there exist methods that attempt to convey the possibly complicated shape of this set of efficient solutions to the decision-maker. Conveying the shape of this set is important because it contains the tradeoff information between the quality aspects of different alternatives. This is most often done implicitly, either by giving the decision-maker a myriad of solutions, or by interactively calculating representative solutions that are efficient. The difference between the latter alternative and forward planning is that in multicriteria decision-making, the information about the alternatives is also conveyed. Also, there are more efficient ways to explore efficient alternatives than by manually altering design parameters. Miettinen [Mie99] adopts the following classification of methods to attain practical solutions:

1. no-preference methods
   methods where no preference information is used, i.e. methods that work without specification of any preference parameters
2. a posteriori methods
   preference is used a posteriori, i.e. the system generates a set of efficient solutions of which the decision-maker selects one
3. a priori methods
   where the decision-maker has to specify his or her preferences through some parameters prior to the calculation and
4. interactive methods
   where preferences are not necessarily stated prior to calculations, but are revealed throughout.

According to Stewart [Ste92], the choice which type of method to use depends mostly on three considerations. The input data must be meaningful for the decision-maker, the mechanisms used to arrive at a solution must be transparent and it must be simple and efficient.

2.7.4 Ingredients for Practical MCDM

What is needed to equip a DSS with the multicriteria framework is a method of scalarizing a multicriteria problem, a numerical solver to optimize these scalarizations and the implementation of one of the four methods specified above.

The type of scalarization used depends on the type of method to assist the decision-maker with the selection of an efficient plan. The scalarization and its parameters must be fully specified if an a priori method is used, for example. For a posteriori and interactive methods, on the other hand, specifying the type of scalarization suffices in most cases and the system adapts the parameters according to the interaction with the user. One interactive method, to cite an example, is the so-called *reference point method* [Mie99]. Here, the user selects a set of quality scores (the reference point) which he or she would like to achieve and the system calculates an efficient solution that is as close as possible to the specified quality. One possibility to explore the set of alternatives around this solution is to vary the reference point and "get a feeling" for how the efficient solutions change in terms of their individual criteria. This approach lends itself to a very intuitive graphical user interface where the decision-maker can interactively vary the reference point. However, it requires that the resulting scalarized optimiziation problem can be solved in real-time.

Calculating the efficient solutions for some method happens in the backend of the DSS, hidden from the decision-maker. It is especially the interactive methods such as the reference point approach that demand a fast and reliable solver to determine efficient solutions according to some scalarization. But also for methods where a set of efficient solutions is calculated before the user interacts with the system, the solver should be fast to perform these optimizations as quickly as possible. In any case, the model equations are the determining factor of the speed of the calculations. Some simulations in CAD software to estimate physical quantities can take several hours. This leads to unacceptably long optimization times during which the system is not ready for interaction with the decision-maker. However, if a generic optimizer is used together with a surrogate for the original model, there is no possibility to reduce the simulation time. The reason is that the generic optimizer can only use the simulation as a "black box".

One approach to lower the calculation times is to couple the simulation with the solver. Often, the virtual simulations solve equations to determine all outputs with a high precision. In engineering applications, for example, the equations are often solved on a *discretization* of the underlying objects. Finer discretizations with a high resolution lead to more accurate simulations as the true underlying structures are usually better approximated. However, they are also more time-consuming to simulate. It has been recently shown for a certain type of cancer treatment planning that it is possible to construct such high-performance solvers by integrating the simulation with the optimization [Sch06b, Sue08]. The speed gains result from adapting the discretization to only simulate precisely where it is absolutely necessary, and to use very coarse representations where a fine resolution is not necessary to solve the optimization problem.

Finally, the method of choice to aid the decision-maker with the problem must be implemented and a user interface that is simple to use has to be installed. This implementation and its interface then determine what kind of problems are given to the solver. Designing such interfaces that are usable in practice is probably the most challenging of the ingredients for practical multicriteria decision-making.

Example 1: a DSS for the design of non-woven material The first example of a DSS implementing the multicriteria decision-making framework was designed by the Fraunhofer Institute for Industrial Mathematics in Kaiserslautern to support the configuration of a machine that produces *non-woven* material. In this particular production process, rotating spinnerets deposit a fluid raw material on a moving conveyer belt where the deposited material then hardens. Figure 2.26 displays this process.

The outcome is a fabric that is judged according to several quality measures. The homogeneity of the deposited material is critical for the stability of the fabric. Another quality criterion is the occurrence of any patterns in the density of the material. Stripes of a lower density, for example, also result in a fleece with unfavorable stability properties. Other criteria such as the stability of the production process are more difficult to quantify, but they are of concern to the production manager.

There are various parameters the manager can adjust to control the process to modify the fabric. Among the most influential are

1. the speed of the conveyer belt,
2. the speed of the rotating spinnerets,
3. the amount of liquid material released per unit of time, and
4. the angles of the nozzles at the end of the spinnerets.

A simulation that is based on models for the movement of fibers can be performed to predict the density of the fabric based on a configuration of these input parameters. This simulation is incorporated into the DSS to allow forward planning of the production process. Figure 2.24 displays the interaction with the forward planning part of the DSS.

The DSS is further equipped with a solver and a method to calculate a variety of efficient production parameters that are described in [FC08]. It is an implementation of an a posteriori method. Simulating the density of the fleece for a given input takes less than one second. However, an optimizer typically requires several iterations and evaluates the densities for a potentially large number of different input parameters. Therefore, even if the simulation is fast, real-time interaction with the system is not possible if used as a "black box". For this reason, the amount of material released from a spinneret is described in a heavily simplified model. This way, it is possible to predict the change in density by a modification of the amount released without running the simulation again. This allows to include this part of

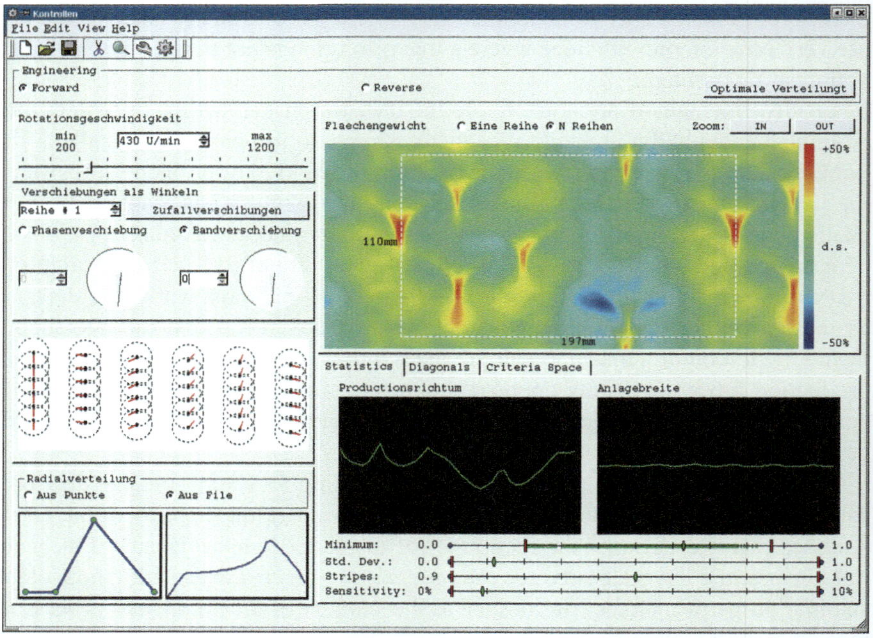

Fig. 2.24 Illustration of the production process of a non-woven material.

the fleece simulation directly into the optimization and results in running times that allow real-time smooth user interaction. This simplification is only used during the optimization. As soon as a solution is found, it is simulated using the finer simulation to avoid imprecision in the displayed fabrics.

In Fig. 2.25, the user interface to the multicriteria decision-making part of the DSS is shown. Here, the manager has an overview of the set of efficient parameter settings. The graph on the right in the screenshot displays the tradeoff between two criteria – the homogeneity and a specified measure for sensitivity (which include patterns) in this case. It is further possible to view the tradeoff between all possible pairs of the specified quality measures. The manager can use this information, and the illustration of the density on the top, as well as the input configurations displayed on the left to make a well informed decision. Using this DSS, it was possible to reduce the total mass flow by 28%. Moreover, the fabric of this improved input exhibits an almost constant density.

Example 2: A DSS for SiP design The second example for a DSS is to aid the design of System-in-Packages (SiP) designs in electronics. The DSS was also developed at the Fraunhofer Institute for Industrial Mathematics in Kaiserslautern. SiP is a novel integration approach where heterogenous electronic devices (actives, passives, microcontroller, etc.) of a circuit are included in a single system. It is found in many electronic applications such as cell phones, portable music players, etc. The electrical devices are placed on *modules*, which are either stacked and vertically connected as in Fig. 2.27 or folded on one substrate as in Fig. 2.28.

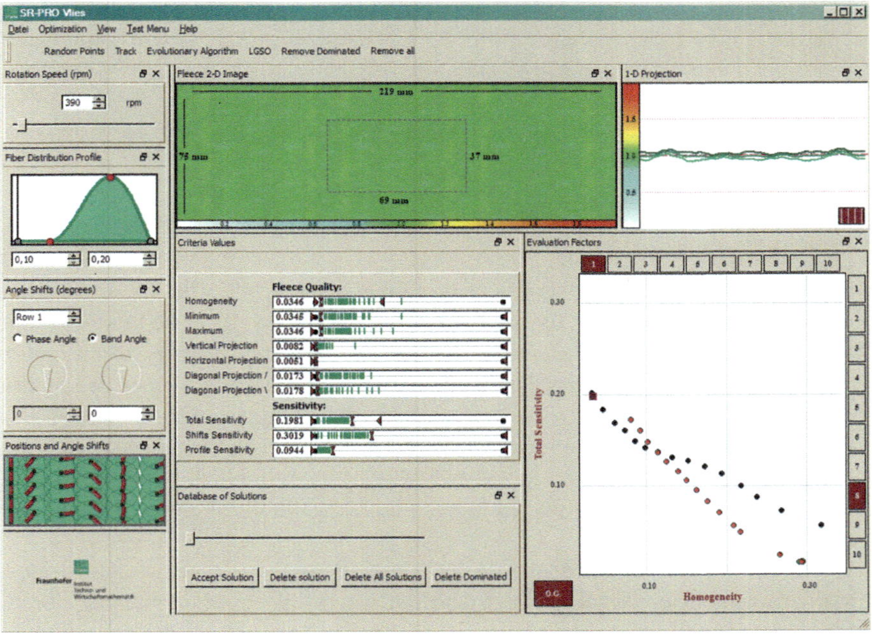

Fig. 2.25 The graphical user interface of the forward planning part of the DSS.

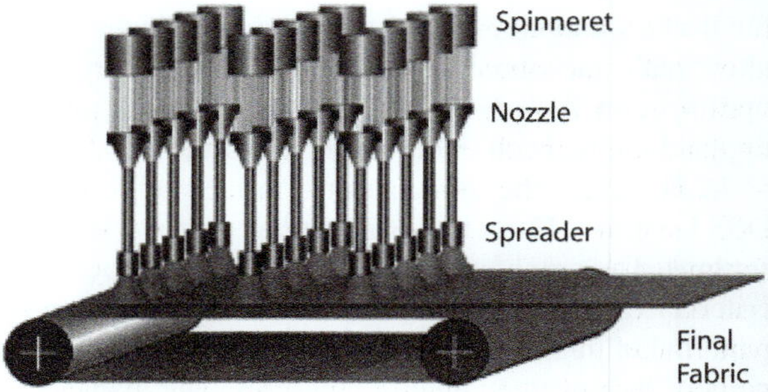

Fig. 2.26 The decision support interface for the multicriteria decision-making part of the DSS.

Fig. 2.27 An SiP where the modules are stacked on top of each other.

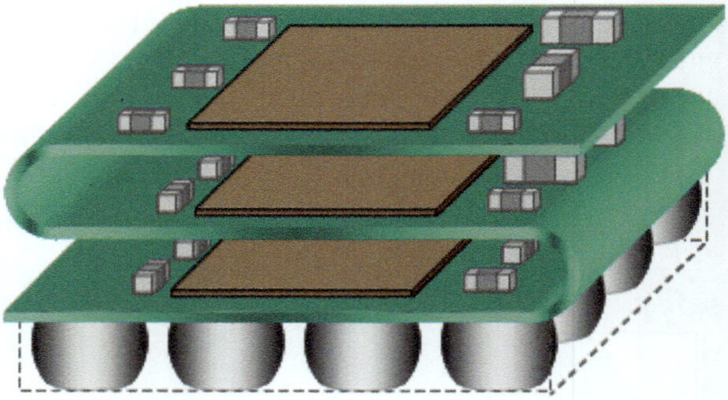

Fig. 2.28 An SiP where the modules are placed on one folded substrate.

To take advantage of a shorter time-to-market for SiPs, design automation becomes a key success factor for producers of such systems. Because they are found in many portable devices the aim of the design is to keep them as small and as light as possible. However, the goal conflicts with the ambition to keep production costs low and makes robust designs more difficult. In other words, the design of a SiP is truly a multicriteria optimization problem. Of critical concern regarding the robustness is, for example, the heat dissipation of the individual components. The closer the active devices are placed to each other, the more likely are hot spots in the SiP at any time.

Fortunately, many of the quality criteria lend themselves to mathematical modeling. The dimensions of the design and the weight are easily calculated. Some factors related to the production cost, on the other hand, can be well approximated.

The wiring to establish the connections between the circuits, for example, can be reasonably estimated by simple surrogate models which can be directly included into the optimization. The heat dissipation, however, is a very difficult physical property to simulate for a given design. As such, using a thermal simulation as a black box in the optimization, and performing surrogate optimization with a generic solver suffers from long calculation times. Once again, it was possible to avoid the detailed simulation during the optimization by adapting the simulation. The framework to aid the decision-maker is again an a posteriori method: a representative set of efficient solutions is calculated before the user starts the decision-making process.

One approach to optimize SiPs is to split the problem in two stages. In the first stage, the individual components are separated into groups that are placed on one module. At this point, the heat is simulated using a very simple model that runs very fast. It is possible to estimate approximately how much heat will develop based on this grouping of the components (Fig. 2.29). Other quality measures that are considered at this point are the number of vertical interconnects between the modules, the estimated total wirelength between the devices, the half perimeter of the modules, and the height of the SiP. By only estimating the heat dissipation, the required wirelength and the required dimensions of the modules (the devices are not yet arranged), this partitioning can be done very quickly. This is yet another example of adapting the virtual simulation to improve the performance of the optimization.

In the second stage, the individual components are arranged on the substrate. Now it is possible to determine the heat dissipation and the other estimated quantities more precisely. However, still only refined models of the first coarse approximations are used to solve the placement problems quickly. The optimization is done using a multicriteria metaheuristic presented in [RPH$^+$07].

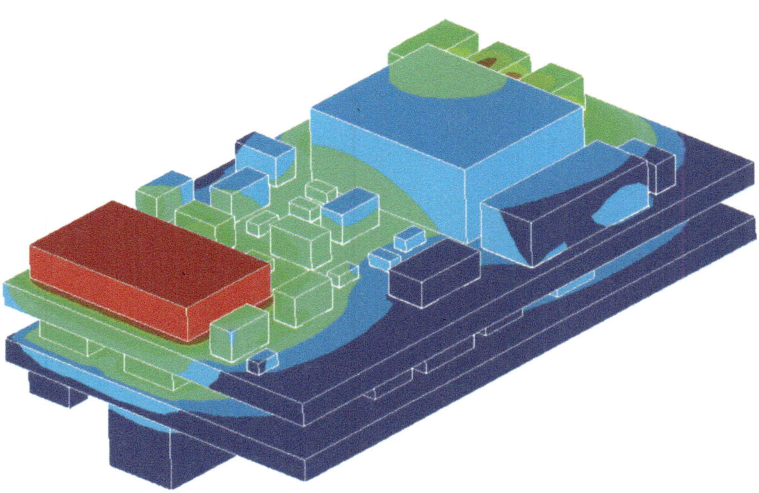

Fig. 2.29 Coarse estimation of heat dissipation of some components on a stacked module.

The solver calculates several hundred solutions to cover relevant ranges of objective functions by altering importance factors assigned to each individual quality score. Once the solver is done with the calculation of efficient solutions, the SiP-LayoutFinder is used to explore this set. As in the first DSS, the criteria can be compared by pairs and the tradeoff between the single criteria is visually displayed. Figure 2.30 shows a screenshot of this interface. Clearly, it would be naive to force the designer to review all the proposed layouts in the database. Rather, the LayoutFinder allows to structure the decision-making process by the following mechanisms. The designer can

- see the distribution of the calculated layouts with respect to a single criterion or a pair of criteria (the latter allows a trade-off analysis such as wirelength vs. half perimeter, shown at the top of Fig. 2.30),
- restrict the set of solutions to an interesting subset by setting limits on the preferred ranges for the criteria (show at the bottom in Fig. 2.30),
- compare technologies by restricting the database to the layouts of a certain technology (solutions using the folded flex technology are marked as red points in Fig. 2.30 and the green points correspond to stacked modules), and
- select a layout and explore it in a 2D- (to the right in Fig. 2.30) and 3D-visualization in a separate window.

If the user selects a single solution by clicking on the point, the image to the right displays the corresponding design. The decision-maker can now "navigate" through the calculated set of solutions and select a design based on the criteria except the heat dissipation. Once a good design is found, the detailed thermal simulation is run

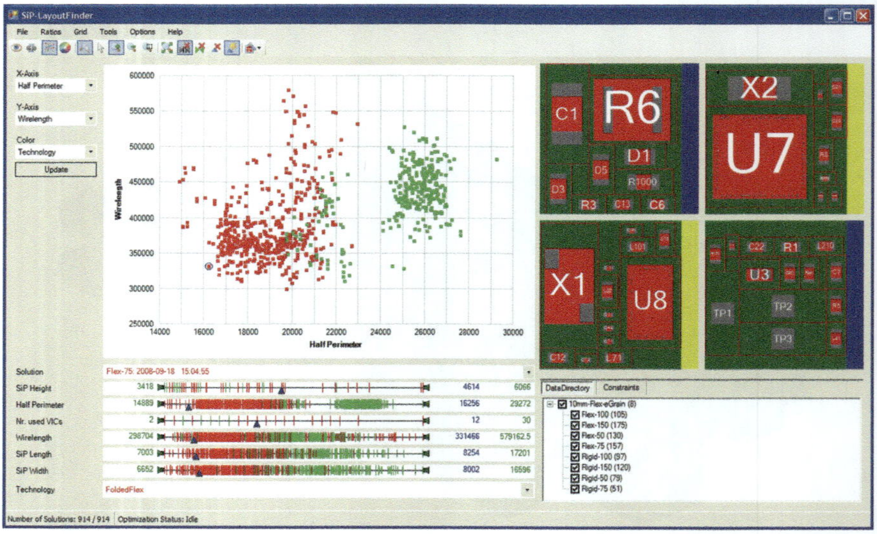

Fig. 2.30 The user interface to the decision support part of the DSS.

and the results are presented to make that last missing information available to the decision-maker.

2.7.5 Conclusion

Making a decision about any problem requires good information to base the decision on. The more information an expert has available about proposed designs or products and their alternatives, the more his or her expertise can be utilized to make a good decision. It is argued here that the classical forward planning strategy, which is supported in many existing decision-support systems, suffers from several pitfalls. The additional burden of having to generate virtual designs and alternatives is placed on the decision-maker. By adopting the multicriteria decision-making framework, this task can and should be automated by the system.

To equip a decision-support system with the tools necessary to realize multicriteria decision-making, it is necessary to calculate efficient solutions of an optimization problem resulting from the user-specified quality criteria. It is demonstrated here that the approach to first calculate a set of solutions that cover a wide range of different quality scores and then communicating this set to the decision-maker is a successful approach to support the decision-making process. In general, this approach requires a more complicated solver technology than what is mostly used in support systems today. However, the successful applications shown here demonstrate that it is well worth the effort to adapt existing simulation models to the optimization to realize fast and efficient multicriteria solvers which can produce valuable alternatives for the decision-maker to evaluate.

Chapter 3
Structures in the Disciplines: Case Studies

The study of structure generating processes is an interdisciplinary subject. Case studies from the various fields reveal similarities as well as differences in the methods of solution and also in the goals of analysis. The following selection serves to give an overview over the types of analysis typical for the various disciplines. It is by no means exhaustive. The individual articles are written by experts in the various fields with a strong emphasis on illustrating the points elaborated in the preceding chapter on the transdisciplinary foundations. Readers wishing to enhance their understanding and to proceed more thoroughly to the specific disciplinary details are put in the position to do so by consulting the cited literature.

3.1 Optimal Structures of Injection Molding Tools

Jörn Mehnen, Thomas Michelitsch, Klaus Weinert

Molding is an important and frequently applied production technique of mass production: Most plastic components from micro switches to bumpers of cars are manufactured via injection molding. Metallic structures such as aluminum gear boxes or stairs of escalators are produced via die casting. All these products are generated by injecting a hot material into a die. In zinc die casting the injected metal has a pouring temperature of about 435 °C and the pouring temperature of aluminum varies from 620 °C (thick walled) to 730 °C (thin walled) [Has06]. Aluminum and zinc are examples of the most prominent metallic casting materials. The material cools down, solidificates, and the workpiece is ejected from the die before a new process cycle can start. In this process the layout of mold temperature control designs is decisive for efficient die casting. The efficient cooling of a die is one of the main factors to reduce the cycle time and, therefore, to increase the cost effectiveness of a mass production tool. Improvements of up to 50% [Wie03] are possible, because today the layout of the bores carrying the cooling liquid in casting tools is typically done manually. The layout design is based on expert knowledge and rules of thumb. Experts have a good estimate about where to position cooling bores properly. Heuristic rules are quite common in practice. These rules are fast and work very efficiently. Unfortunately, only little of this knowledge is available in the form of

K. Lucas, P. Roosen (eds.), *Emergence, Analysis and Evolution of Structures*,
Understanding Complex Systems, DOI 10.1007/978-3-642-00870-2_3,

explicit mathematical rules [Zö97]. Solutions for complex problems may not be satisfying as existing heuristic rules deduced for simpler tasks may contradict each other. If problems with blowholes or insufficient surface qualities arise as a consequence of an inferior bore layout the die requires frequently very expensive and time-consuming remanufacturing. Due to the complex creation process die casting tools cost up to five hundred thousand Euro. Therefore, well designed cooling circuits help to significantly reduce manufacturing costs of the die, minimize cycle times, increase tool life and improve the workpiece quality (Fig. 3.1).

Systematic optimization requires quantitative measures for the quality of a proposed solution. Therefore, the expert knowledge together with physical properties have to be modeled, leading to the exact definition of quality criteria, the choice for efficient problem spaces, data structures, and efficient quality criteria evaluation methods. Due to the multiobjective character and the high complexity of the problem, powerful optimization algorithms have been used. Very useful techniques for solving difficult problems are evolutionary algorithms.

These algorithms are able to find surprisingly good solutions by clever analysis of the quality functions. Sometimes even experts are surprised about the new solutions found and get motivated to use new approaches. There are several advantages of evolutionary algorithms. The algorithms are rather robust against changes in the fitness functions.

Many classic optimization algorithms – this is also true for older evolutionary algorithms – are only able to solve single objective problems. Although often the single objective case is complex enough, many real world problems have a multi-objective character. Problems may have contradictory aims, i.e. one objective value cannot be improved without deteriorating another. The reduction of all criteria to only one value is possible only in special cases. In the case of uncertainty about the preferences of a user toward one or another criterion, a set of best compromise solutions should be presented, as already discussed in Sect. 2.5.3.

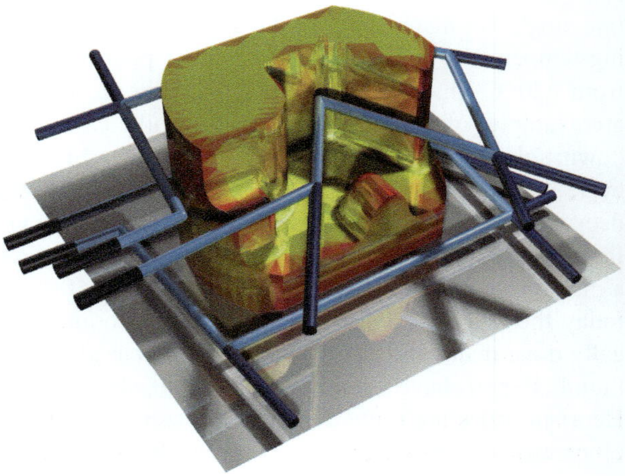

Fig. 3.1 Example of a cooling bore design in a mold with two independent cooling circuits. The die surface is shown *golden-red*, while the mold itself is only represented by the bores contained in it.

Multiobjective evolutionary algorithms can be used either with a posteriori or with a priori techniques. A posteriori approaches follow the idea of uncertainty about a perfect solution. Their goal is to generate best compromise solutions, so called Pareto optimal solution sets, from which an expert can choose according to non-formalized ad hoc knowledge. In case of well defined multiobjective goals sometimes also scalar evolutionary algorithms are useful.

Several multiobjective evolutionary algorithms are mentioned in literature. A very good evolutionary scalar optimizer is the Evolution Strategy (ES) [Sch95b]. It can be used for multiobjective problems technique via aggregation. State of the art for a posteriori multiobjective optimizers are the NSGA II [DAPM00a, DAPM00b] and the SPEA 2 [ZLT01a]. All three techniques have been tested on the mold temperature design problem.

Another important issue, when evolutionary algorithms are applied to real-world applications, is computational speed. Stochastic algorithms tend to need a lot of fitness function evaluations. Especially difficult problems with high dimensions, restrictions and a difficult multimodal character etc. lead to high population sizes or long runs of the stochastic search algorithm. Therefore, an efficient design of the evaluation functions is necessary. The definition of fast evaluable quality functions is particularly necessary in the case of mold temperature control design. Surrogate functions or a well motivated simplification of the problem can lead to a significant speed up without much loss in quality.

Finally a systematic search for good parameter adjustments of the multiobjective search methods is necessary to improve the search process itself and its results. The best algorithmic parameter settings for various problems often differ from the standard. In order to compare multiobjective optimization algorithms that generate Pareto fronts, it is common to introduce scalarization methods, so called metrics. The application of statistical design methods helps to systematically, but experimentally, meta-optimize the parameter settings, leading to robust, statistically sound and optimized parameter adjustments. Here, existing classical and new statistical approaches have been applied to multiobjective as well as to single criterion optimization of the mold temperature design problem.

3.1.1 Modeling Aspects

3.1.1.1 Geometric Aspects

Molding tools are composed of various parts such as sliders, bores, pillars, the die surface etc. The cooling or heating of a certain area of a die is realized by introducing channels into the tool that lead water or oil into the vicinity of the die surface. These channels form circuits. The number of bores and circuits depend on the complexity of the molding tool. Simple structures need few bores and a single circuit only. Complex tools may contain more than thirty bores and more than five circuits. The circuits can be operated independently from each other and can be used to heat up or cool down the tool. Heating up is important for reducing thermal stresses particularly for large tools in die casting. Dynamically changing strong thermal stresses can lead to increased tool wear or even to tool breakage. In the following, cooling and heating circuits are subsumed under the term cooling circuits because for the present discussion only the geometric structure of the bores is important.

The number of the bores should not be too high because manufacturing of deep hole drilling bores is cost and time intensive. Deep hole drilling machines for large molding tools show restrictions with respect to the orientation of the bores. Often only horizontal orientations of the bores are possible while the die can be rotated on a table freely around the z-axis. Although some modern deep hole drilling machines do not have these restrictions, a design tool for realistic bores should be able to respect them.

Another restriction in the bore directions is related to machining problems of intersecting bores. Bores that intersect each other in small angles cannot be manufactured safely by deep hole drilling. Therefore, it is preferable to design bores that intersect each other in large angles – ideally orthogonal.

The number of bores should also be kept small because any bore implies machining risks and costs and also weakens the molding tool. A certain wall thickness around each bore is necessary to reduce the probability of deformation or even breakage of the tool. In the design of large die casting tools rules of thumb are fast and well established but also very case depending and cannot be transferred linearly for small tools because the physics of the temperature follows nonlinear laws.

The simplest structure of a cooling circuit is a consecutive sequence of bores. This is also the most prominent structure. More complex cooling strategies such as the branching of bores or so called finger bores for local cooling are more complex but lead to a higher efficiency of the cooling. Shortcuts with other cooling circuits are not allowed. A grouping of inlets and outlets of several circuits is also desired from a practical point of view.

Dead ends of bores used in circuits with more than two segments are sealed with plugs. Dead ends appear because each bore of a circuit has to be machined by deep hole drilling from outside. Plugging is necessary to seal a circuit from leakage. Plugs can be placed in any part of a bore. After setting a plug it seals a hole permanently. In some cases a plug can also be used to seal a die surface that has been penetrated by a bore. In general, drilling through the die surface is not allowed and should be avoided whenever possible because the subsequent machining, e.g. grinding, is very cost intensive.

Typically, dies and molds are designed with CAD systems. There exist various techniques for describing the geometric elements of a tool in the CAD system. The most general way is the triangulation of the surfaces. CAD-surfaces can be exported from CAD systems, e.g. in standard STL or IGES format and can be imported into a simulation software. The consecutively used simulation software then utilizes triangulation for describing the complete molding tool. The bores are modeled as virtual cylinders. Actually, it is possible to use arbitrarily shaped triangulations of a die's surfaces. For simplification reasons the tools are generally assumed to be rectangular blocks. The cooling cycles are modeled as sequences of consecutive cylinders. In the mathematical abstraction these sequences are simple polylines. Each cooling circuit can be defined uniquely by the position of the vertices of the polyline. The start and end points of the polyline always lie on the border surface of the molding tool. The model contains the complete bores for the testing of collisions and efficient machining. Plugs are defined by their position at the polyline. The drilling direction for the machining of a bore is not unique. Therefore, always both possible drilling directions can be modeled and analyzed.

3.1.1.2 Quality Criteria and Evaluation

The cooling of an injection mold has to satisfy several aspects [Meh05]. The most important idea is to generate cooling circuits that keep the cycle times in the later usage in mass production as low as possible while the machining of the circuits should be as cost efficient as possible. Already these two basic aspects are contradictory. On the one hand the amount of bores enhances the cooling effect and hence reduces the cycle time. On the other hand the costs increase with the number of bores. The efficiency increases with the number of bores but the costs do also. Hence, the basic problem is multiobjective with conflicting criteria. The efficiency of the cooling of the circuits can be increased by changing the layout of the cooling bores. The corresponding design problem also has a multiobjective character because the cooling efficiency of each bore should be both uniform and strong. A uniform cooling distribution along the die surface is important e. g. to reduce internal stresses within the workpiece. An intense cooling is necessary for short cycle times. The relative distance and orientation of a bore to a point on the die surface is decisive for the cooling efficiency of the bore. Larger distances generate more homogenous temperature distributions on the die surface but impede the absolute cooling intensity and vice versa. The length of a bore also has a certain influence on the cooling effect. The longer the bore the better the cooling effect but the higher the costs for the manufacturing of the bore.

The complete expert knowledge for modeling the layout of cooling circuits can hardly be captured by an algorithm. Nevertheless, a lot of aspects of the current "fuzzy knowledge" can be formalized and made available for mathematical use. Many values can be deduced from physical or economical properties of the machining process. Some knowledge can be formalized as quality criteria, e.g. average temperature of the die, others as restrictions, e.g. possible bore directions of the drilling machine. Basic properties such as the bore directions may also be encoded directly in the data structures. For example if the bores would have been restricted to lie in a single plane the problem would have been encoded different from the arbitrary 3D case as in the model used here.

The definition of the fitness functions can also benefit from efficient encodings. Here, the meaning of efficient encoding refers to fast fitness function evaluation as well as to a reduced difficulty for the optimization strategy. A low dimension of the search space and a continuous fitness landscape with only few restrictions is preferable.

For a technical realization of the efficient description of the casting expert's knowledge, physical properties of the thermal flow in metal, economical assumptions about the manufacturing process as well as mathematical heuristics have been used. A realistic assessment about a sufficient level of detail is a helpful technique to reduce computation times. These approximations can be used to choose parameters of the geometric model of the tool such as the number and size of the triangles used to describe the die surface. Approximations have also been applied to obtain a very fast estimation of the cooling effect of the bores.

The quality of the cooling induced by a bore is determined by the local intensity of the cooling effect to the die surface. Additionally, a globally uniform heat extraction is important to get e.g. workpieces with a minimum of internal stress and, therefore, high qualities. Global and local cooling are conflicting goals. Additional conflicting goals such as the number and length of bores have to be taken into account. The layout of the bores has to fulfill technical and machining restrictions. The bores must not intersect neither with the die surface nor with any other bore. Of course, all bores have to be within the tool geometry. This restriction seems trivial in concept, but it is not at all so with respect to the simulation system boundary control implementation. Additional machining restrictions such as limited bore direction angles have been addressed in the optimization tool.

The cooling effect can be calculated quite accurately by Finite Element Methods (FEM). Unfortunately, this approach needs long calculation times. FEM has the advantage that it calculates the complete temperature distributions within the tool for all times. It is necessary, however, to use high numbers of nodes, because a low resolution can lead to relevant discretization errors when the bores change their positions only slightly. In a static case and a sufficient number of nodes FEM is a tool for precise calculations that can be very helpful for analyses of the final construction design. Actually, some companies use FEM to check the quality of the manually generated bores. This is especially necessary in die casting where blowholes – caused by insufficient cooling – have to be avoided.

In manual bore layouts often only the mean temperature in the time period between injection and ejection is used to estimate the actually dynamic temperature distribution in the tool [Zö97]. Using only the static case simplifies the analysis a lot and helps to speed up the calculations.

Due to the long evaluation times of the FEM, a fast surrogate model appears to be more attractive. The model described here approximates the actual heat flow mechanism by modeling a heat radiation case. The exponential decrease of the cooling properties with increasing distance between bore and die surface is reproduced by both models and the quality of the FEM solutions can be reached in many cases. Only when a bore is shaded by an object or two bores are lying very near to each other, the approximation becomes inaccurate. In most cases, however, the shading effect and the superposition problem can be disregarded because of the strongly decreasing local character of the cooling effect. Furthermore, the superposition of the bores introduces an overrating of the cooling effect. This leads in the optimization process to bores with larger distances to the die surface. The cooling distribution criterion introduces a separation of the bores and, hence, the overestimation error reduces exponentially again.

The basic idea of the radiation approach is to calculate the temperature distribution and intensity on the die surface radiated by a beaming cylinder. Each bore can be interpreted as a "neon lamp" illuminating the die surface. Mathematically, each bore is modeled as a straight line between each two vertices of the bore circuit, P_i and P_{i+1} respectively, $i = 1, \dots, n$, having n bores. The radiation $\tau_{i,j}$ effecting each point M_j of the die surface can be calculated by integrating the radiation intensity of each point along the line between P_i and P_{i+1}. The radiation intensity of a light point decreases with $1/r^2$ with increasing distance r. Accordingly, the intensity of the line of points between P_i and P_{i+1} effecting M_j is the integral [WMM+04].

$$\tau_{i,j} = |P_i - P_{i+1}| \int_0^1 \frac{1}{(P_i + t \cdot (P_{i+1} - P_i) - M_j)^2} \, dt. \tag{3.1}$$

The total cooling effect τ_j at point M_j generated by all the bores is the sum of all partial effects, i.e.

$$\tau_j = \sum_i \tau_{i,j}. \tag{3.2}$$

Each discrete point on the die surface M_j, $j = 1, .., m$ can be analyzed this way. Often a CAD surface is represented as a triangulation. In this case, the M_j can be taken as the centre point of each (small) triangle.

The global cooling effect f_l is calculated by the arithmetic average value of all τ_j. f_n is the normalized minimum effect value over all τ_j.

Function (3.2) can also be used to characterize the cooling distribution by calculating the statistical variances of the cooling values around each triangle τ_j. The global temperature distribution f_d is the mean of the variance values calculated over all triangles.

In the basic model a constant initial die surface temperature is assumed. In reality some areas are hotter than others, e.g. at the spray or where material accumulates. Demands for additional cooling is modeled by temperature values that are attached to the triangles. These cooling demands are integrated into the model (see Fig. 3.2).

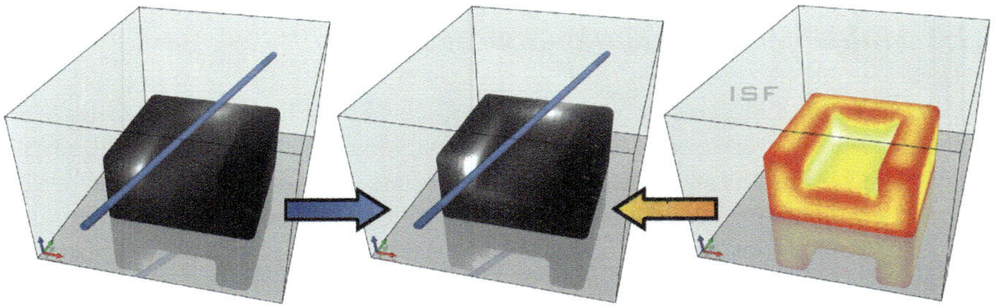

Fig. 3.2 Cooling effect calculated via the radiation approach (*left*). *Highlighted areas* indicate better cooling than *dark areas*. Specific cooling demands (*right*) can be defined manually. *Highlighted areas* indicate higher demands than *dark areas*. Superposition (*middle*) of cooling effect and demand yields an effective cooling quality that is used for the optimization.

The aggregated temperature effect f_t, which is used by the optimization process, is composed of three parts: the arithmetic average sum over all radiation contributions of all bores on all triangles (describing the absolute cooling effect) f_l, the temperature standard deviation (modeling the uniformity of the illumination) f_d, and the normalized minimum effect value f_n. All values are normalized and about similarly sensitive to changes in the bore design.

In multiobjective optimization it is possible to use a priori or a posteriori techniques. Both techniques have been used here. The a priori method is represented by an aggregation or scalarization technique. This method can be used adequately, if the relative weights of the contributing partial target functions can be defined a priori by the user. It is also known that linear aggregation should only be used for a convex Pareto front (cf. Sect. 2.5.3). Actually, in our problem domain there are only very few practical cases known where the Pareto front is concave. Nevertheless, we also applied an a posteriori analysis of the Pareto front of the mold temperature control problem that proves this fact again, as shown later.

The aggregated fitness f is a function of the standardized restrictions f_{pen}, the manufacturing costs f_c (i.e. the length of the bores), and the temperature effect f_t:

$$\begin{aligned} f &= f_t \cdot (1.0 + f_{pen}) \cdot (1.0 + d \cdot f_c) \\ f_{pen} &= a \cdot f_{bc} + b \cdot f_{sc} + c \cdot f_{ot} \end{aligned} \tag{3.3}$$

The allover penalty f_{pen} is a linear combination of the number of collisions between two bores f_{bc}, the factor f_{sc} describes the collision between the bores and the tool surface and f_{ot} is a penalty for the bores that reach out of the tool. a, b, c, and d are arbitrary real value weighting factors.

The radiation approach has the advantage of very short calculation times. Compared to the FE method, which takes about 10 minutes for a simple geometry, the calculation time for analysis of the cooling design using the radiation technique takes just about 1 ms on a standard PC. Short evaluation times and good approximation results allow to efficiently utilize stochastic algorithms and to solve the complex problem of mold temperature control design.

3.1.2 Multiobjective Evolutionary Algorithms

The field of multiobjective optimization (see Sect. 2.5.3) has been in the focus of increasing interest for some years. The classical single criterion optimization is a special subclass of the more general multiobjective optimization. In general, single criterion techniques cannot be applied to multiobjective cases without changes. Even the comparison of solutions is not as easy as in the scalar case, because sets of compromise solutions have to be compared with each other. Therefore, the optimization techniques have to be adapted to this class of problems. Optimization in this context means a gradual increase in quality for the solution sets, consisting of two or more partial target values. Furthermore, the calculated Pareto front, approximating the true Pareto set of the given problem, should be evenly spread at least in certain practically important regions. Additionally, of course, the approximation has to be calculated fast. In multiobjective optimization a lot of problems are very complex in an algorithmic sense, i.e. they cannot be solved in polynomial time. In combinatorial problems even the management of the solution sets can be NP-hard. Nonlinearity, a multimodal character of single or all functions, discontinuity and function ranges with no inclination, many nonlinear restrictions, high dimensions of the decision space and even dynamic features are characteristics of real world problems that make multiobjective optimization a very difficult field in mathematics [Meh05].

Most of the existing deterministic methods for multiobjective optimization are relatively old. Nearly all techniques find only one solution on the Pareto front per run. The theory in continuous multiobjective optimization is also not as well developed as in the single objective case, although there is very good literature available (see e.g. [Ehr00, Mie02]).

Multiobjective optimization techniques can be grouped into a priori, a posteriori and interactive techniques. Many approaches use the a priori approach, where all necessary knowledge about the fitness criteria and their relative importance is known in advance. In this case, typical approaches are scalarization of the multiobjective optimization problem via aggregation or lexicographical ordering. There are various aggregation methods that use e.g. weighting sums or sums of weighted factor combinations. Lexicographic max-ordering is an interesting alternative, because this technique allows to find points that are Pareto optimal [Ehr00]. This property is not fulfilled in linear weighting techniques. Today, a posteriori problems are generally solved by stochastic optimization as described later. Interactive methods are an interesting alternative to both extreme cases, where expert knowledge influences the search process directly [Meh05].

Deterministic techniques have favourable properties if the problem can be solved with these approaches. It is possible to give approximation estimations, and deterministic methods are also very fast. Of course it is not easy to find an efficient method for a specific problem. Furthermore, the NFL-theorem [WM97a] tells that there is no best optimization algorithm for any class of problems at all, although, reducing the general set of all problems to smaller classes, it is possible to find at least "appetizers" [DJW98].

Looking for algorithms that can find good solutions and also cover large areas of the Pareto front needing only one run leads to multiobjective evolutionary algorithms. In the single criterion case evolutionary algorithms have proved their effectiveness in many cases, as well in theoretical as in practical applications. They may not always find the theoretically best solution, but on the one hand in the multiobjective case it is difficult to prove the optimality of a solution anyway – especially if only simulations are available – and on the other hand in practice gaining a relevant improvement is often so difficult that already a certain melioration is highly desirable.

Multiobjective evolutionary algorithms (MOEA) are developing since the last decade [Gol89b]. A relevant increase in research and application came with the introduction of the Pareto dominance based MOEA. Algorithms such as the

NSGA-II [DAPM00b] or the SPEA 2 [ZLT01a] belong to the classics of the current state of the art. A description of current algorithms is available in [CvVL02] or [Meh05]. A brief discussion of MOEA is also given in the chapter "Evolutionary Algorithms" of this book.

Multiobjective evolutionary algorithms have the advantage that they do not have to be re-programmed when fitness functions are changed or new criteria or restrictions are introduced. They can be used either in scalar mode (classic single criterion EA) or in Pareto dominance mode (Pareto fronts and sets). Both approaches have been used for bore layout optimization.

In order to compare Pareto fronts with respect to their quality, various scalarization methods – so called metrics – have been introduced. For a well founded comparison see e.g. [vV99, CS03]. Metrics help to calculate a quality measure in the objective space of a Pareto front with respect to a given reference point, a known best Pareto front or a relative measure between two Pareto fronts, although the definition of the quality is not unique. Of course good approximations of the true Pareto front and well spread solutions are desired. Often the true Pareto front is not known. Only best solutions found so far may be at hand. Furthermore, practically obtained solution sets are discrete in nature. Metrics have to be able to cover all these multiobjective problems and to be as intuitive as possible. A technique that matches many of these issues is the attainment surfaces method [Kno02]. A variant of this method is the MMBBH-measure [MMBBH04]. This method has been used to assign a distribution measure and an approximation measure to one value which is needed for the comparison of algorithms. The solution sets for the comparison may come from different MOEAs, or the same MOEA with different parameter settings. This single value was necessary for the application in the univariate design of experiments as used here to meta-optimize the MOEA parameter settings.

3.1.3 Statistical Design and Analysis

The model of a physical or technical system is generally only an approximation of reality (cf. Sect. 2.4). It has to cover certain properties and should help interpreting aspects of reality. Uncertainty, over-adaptation and generalization abilities are conflicting properties in the modeling and design of experiments that are necessary for verification. Some models can also be used for extrapolation in time, space or contexts, but most of them are restricted to a certain area of interest and use data sets as a basis to describe input-output relations. Usually only a restricted number of data is available, some data is missing or sparse, or the data contains errors or noise. In general, the design of adequate models is a very difficult task. This is especially true for the optimization in real-world applications because often no models, i.e. no mathematical functions, exist at all. The selection of the data for the assessment of a model is also non trivial in general, because its adequate amount and distribution depends on the model and on the availability of the data.

A well established method for the analysis and optimization of problems, especially when the number of available data is low, is the statistical design of experiments. Design of experiments (DoE) methods need a minimum of data. The results can be used to derive functional dependencies between parameters and target values, and so support optimization purposes as well [WJ99]. Grid design plans have the disadvantage that the number of experiments grow strongly exponential with the number of dimensions of the parameter space. One-factor-at-a-time approaches are very popular but they do not allow to make statements about the interactions between parameters [WJ99]. Full factorial design plans use two or three parameters per dimension, fractional factorial design plans need even less experiments and allow to characterize factor interactions and optimization. The standard DoE models are linear or quadratic. Planar linear models, i.e. models without factor interactions, are used for screening. This means that in problems with high numbers of factors (parameters) the most relevant are filtered with respect to the influence on the responses (scalar outcome). In that phase Plackett-Burman design plans are used that can be calculated by systematic schemes [Mon01]. In the modeling phase a linear model with interactions is used to describe interdependencies between factors, i.e. factors can influence each other multiplicatively. After the screening phase, DoE modeling only utilizes the relevant factors because the design plans are a little more expensive. Here factorial design plans are used that also follow schemes listed in the literature [Mon01]. Central composite design plans are typical for optimization [WJ99]. In the optimization phase of DoE, a quadratic model is used to fit the measured data via regression since this type of supposed dependency can be solved efficiently. Algorithms such as the conjugate gradients method are frequently successful.

Design and Analysis of Computer Experiments (DACE) [SWN03b] is a technique for the generation of design plans and evaluation of experiments with deterministic behavior and quantitative character. The analysis of nominal factors, non-deterministic responses [BB05] and sequential introduction of new design points (SPO) [MMLBB05] are current state of the art extensions that improve this method.

DACE needs a space filling design. A typical plan is the latin hypercube design (LHD). Compared with DoE, LHD generates similarly few design points. The points are spread on a grid in a way that each column and line of the grid is occupied by one point only. Additionally, in the experiments the points are distributed in a maximum space filling fashion in the search space. The responses of the corresponding experiments following the LHD designs are interpolated by Kriging [Wik06a]. Kriging introduces a correlation between the measured points and yields a smooth interpolating response surface. The problem of noise in the responses can be solved by boot strapping [MMLBB05].

SPO (sequential parameter optimization) introduces an easy to use and efficient way to improve the design plans by iterative introduction of new design plans. New design points are set following a minimum trade off principle that minimizes the uncertainty of a prediction error (global search) and maximizes the local improvement to find better function values (local search) [SWJ98]. Starting from an optional DoE screening of the relevant factors, the first step in SPO is to perform usual DACE calculations. The Kriging model in the DACE approach plays the role of a statistical prediction tool. In the subsequent SPO iterations, only relatively few new points are introduced to improve the model fit step by step. Compared to the number of points used during the first SPO step, generally in the following steps only few (only 10% had been necessary in our experiments) new points are used.

The DACE/SPO technique is important when the number of fitness function evaluations has to be small. This is especially true when the evaluations are expensive. This technique can also be used for meta-optimization of EA parameters. Although the EA's behavior is relatively invariant against changes of the fitness functions, a tuning of the parameters can still help improving the quality of the results with a statistical significance.

3.1.4 Results

The evolutionary optimization of the model temperature control designs of casting dies starts with an initial definition of a set of possible solutions called a population. Each solution, called an individual is a vector of vertices of a polyline modelling the cooling circuit and the respective bore directions. The initial population contains values that are arbitrarily distributed in the volume of the tool. In the beginning it is possible that the bores may penetrate the surface of the die. These solutions have a relatively bad quality so these individuals are eliminated from the population during the evolutionary process rather soon. The initial parameters such as the number of bores, the geometric properties of the bores and the die, parameters that influence the optimization such as the population size, the selection pressure and many more values can be tuned with a graphical user interface controlling the underlying long-running evolutionary computation process. Selected bore designs can be visualized online during the run and can be modified interactively. The system supports a re-evaluation of modified solutions in real time. This allows the user to watch the optimization process and to tune the parameters of the simulation according to the individual impression interactively. In the analysis standard die geometries were used. Alternatively the system can import, visualize and analyze STL-CAD-geometries. Due to the modular structure of the tool, the evaluation process can be linked to an evolution strategy or to other MOEA-tools. The interfaces between the optimization strategy, the visualization and the evaluation module are realized as an internet socket connection. Therefore, several system components can be exchanged with respect to the respective aims of research.

The standard ES has been modified slightly. Numerical experiments showed that using correlated mutations yield a slight improvement in the convergence properties. A significant melioration of convergence probability towards good solutions was realized by introducing a limited minimum step size that is declining slowly. This variation was useful because the standard step size adaptation proved too fast for this application. An increase in the freedom of search, i.e using relatively large step sizes, helps to avoid getting stuck in local minima that are typical for the strongly restricted search space. Due to the fact that bores must not intersect each other, structures can get into constellations that cannot be resolved with small geometric step sizes. Large step sizes are an effective solution for this problem. Additionally the penalty functions for the intersection of fixed geometric restrictions such as pillars etc. that must not be intersected by bores can increase gradually in the simulation. This also improves the ability of the algorithm to explore the complete search space.

Uncertainties about the preferences of the user toward different quality properties can be modeled via a priori and a posteriori techniques. The ES approach aggregates the fitness functions with a product of individual target functions using normalized fitness values. This method allows to model fitness function interactions and helps to adapt the resulting scalar values according to the (subjectively motivated) preferences of the user. In this approach the single objective evolution strategy as described above has been used. This has the advantage that good initial parameter settings of the ES can be chosen according to good experience. The behavior of the ES is well known and the analysis of the optimization process can therefore be focused on the convergence behavior of the bore structures toward good solutions. The fast fitness evaluation of the radiation approach is a necessary precondition for the efficient optimization using population based techniques. FEM analyses in the ES loop have been tested but the long evaluation times allowed only very small population sizes and few generations.

The Kit for Evolutionary Algorithms (KEA) contains current state of the art MOEA algorithms such as the NSGA II and SPEA 2, as well as two multiobjective particle swarm algorithms for treatment of multiobjective problems. KEA is written in Java and communicates via an internet socket interface with the evaluation tool. It provides a graphical user interface (GUI) for choosing between different MOEA algorithms and various fitness functions. The GUI also supports the tuning of the MOEA parameters. The Pareto fronts of two criteria multiobjective functions can be displayed online during an optimization run.

A typical optimization of the mold temperature control designs using the ES working on the standard test problem of a spherical die takes about five minutes on a standard PC. The MOEA optimizers are similarly fast. Of course in the MOEA approach not every Pareto-optimal solution (i.e. bore structure) is displayed online during the run. Single solutions have to be selected from the Pareto front (cf. Fig. 3.3) by the user after the run and have to be displayed by the visualization tool of the evaluator separately. Figure 3.3 also shows that the Pareto front is constituted by clearly different domains defined by the number of bores. The general result is not unexpected: While a larger number of bores will generally deliver better thermal performance it also causes more costs. A smaller number of bores will be relatively cost efficient but is strongly outperformed with respect to thermal efficiency by the other bore variants.

The MOEA optimizers are not limited to two criteria. The mold temperature control design optimization typically uses four quality criteria (cooling distribution, cooling intensity, costs) together with an aggregated value of the restrictions. Using a higher number of criteria than four or five – actually more than twelve criteria are implemented – is not advisable due to practical reasons. The Pareto front becomes much too complex to be analyzed by the human user even though there is no principal upper bound for an interactive Pareto front display as shown in Sect. 3.3. In

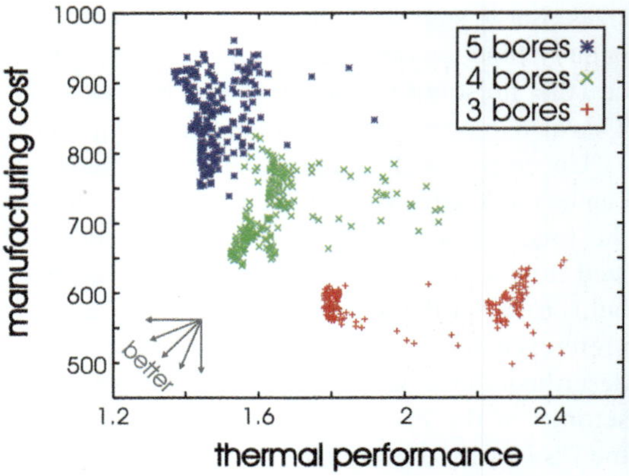

Fig. 3.3 200 solutions for runs with 3, 4, and 5 bores using the aggregating approach with an ES approximating a Pareto front. Inferior solutions relative to the pareto front with respect to the two target functions are also displayed as they show the population's remaining variation span in the target functions space at a certain time of the respective evolutionary optimization run.

the literature on realistic real-world applications a maximum of seven criteria is suggested [CvVL02].

Looking at the test case's geometric results of the bore designs for the half sphere die surface, it is striking that a relatively high number of local optima appear. The solutions can be grouped to clusters of designs with a similar structure (see Fig. 3.4). The structures between the clusters differ significantly. This is an indication that the fitness landscape is multimodal. The average quality of the results of each cluster also differ. Even for a human expert it is not easy to tell what structure has a better quality than another. Therefore, the objective a posteriori evaluation is a valuable help for the designer to compare designs and to learn more about efficient structures.

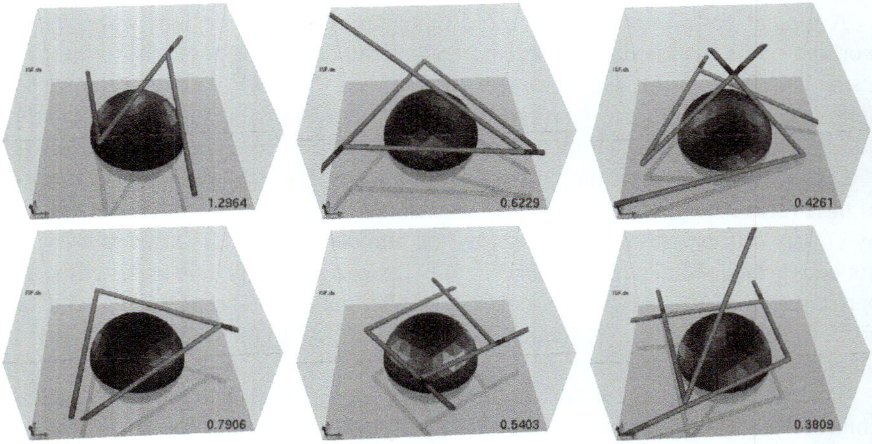

Fig. 3.4 Three design solutions (*first row*) using each three, four, and five bores. Alternative solutions using the same number of bores (*columns*) and having similar cooling qualities are shown in the corresponding second row. Alternative solutions with the same (*scalar*) fitness values emphasize the fact that optimal designs of the cooling bores are generally not unique. Geometrically different solutions with similar cooling values are indicators for a multimodal fitness landscape.

The application of the standard statistical design of experiments proved to be an adequate technique to improve the strategy settings of the ES as well as of the MOEA, leading to superior target function qualities, i.e. better approximations to the true weighted absolute optimum or the Pareto front. The MOEA results have been compared using the MMBBH-measure [MMBBH04]. The meta-optimization using DoE assumes a quadratic behavior of the response values. In many practical applications and also near the optimum this assumption is very useful and an optimum can be found efficiently. Often an optimum lies on the border of the min-max-limits of the factors. Then the area of interest should be shifted to a more promising region and the experiments have to be repeated. Applying DoE to the ES and MOEA increased the quality of the results significantly, i.e. the values of the measure qualities improved more than ten percent compared to results with standard settings.

A more efficient approach to the design of experiments is provided by the DACE/SPO-method [MMLBB05]. Although also here extrapolation is not allowed, a larger region of interest can be scanned and modeled with free form surfaces (see Fig. 3.5). The Kriging model shows very intuitively the interdependencies between the parameter settings and their influence on the quality. The strong stochastic character and the multimodal fitness function is a problem. DACE usually needs a deterministic behavior of the responses. The evolutionary algorithms generate a spectrum of results for the same parameter setting. The distribution of the experimental results can be handled efficiently with a boot strapping method. The number of experiments per parameter setting can be kept surprisingly small, although the variance of the results is quite high. The boot strapping method generates single

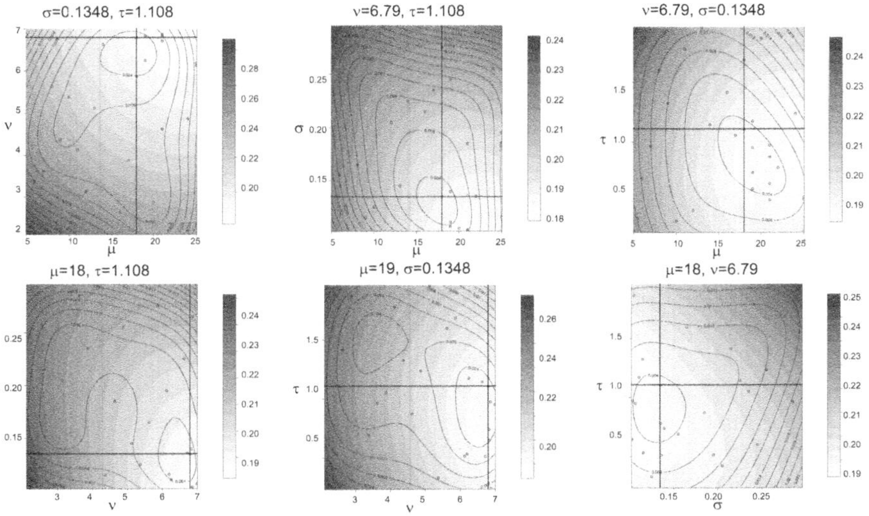

Fig. 3.5 DACE response fitness landscapes (*grayscale*) and errors (*contour lines*) for the parameter optimization of the ES. The best parameters are marked by a cross.

representative response values that were used in the DACE model as deterministic responses. The number of additional new design points that are used to improve the DACE model in the SPO step is also quite small. Therefore, the DACE/SPO technique is very efficient and supports an intuitive interpretation due to the flexible Kriging visualization. Roughly quadratic dependencies between the parameters are displayed for the spherical test design problem. In DoE only quadratic hypotheses can be confirmed or rejected. The application of the DACE/SPO technique to the mold temperature control design optimization algorithm improved the EA results significantly.

3.1.5 Summary

The mold temperature design optimization problem shows a high structural complexity, the fitness functions are multimodal, complexly restricted, high dimensional and multiobjective and the uncertainty about the users preferences make the search for good solutions extremely difficult. An adequate geometric modelling as well as the objective modelling of qualitative and quantitative quality criteria is an important precondition for realistic real-world optimization. This problem is even more challenging when the evaluation of the criteria is time consuming and many evaluations have to be performed.

The modelling of the fitness functions follows an approximation of the real temperature flow evaluation using a radiation technique and the modeling of practical heuristics. The multiobjective optimization is performed via evolutionary algorithms. Different techniques such as scalarization and Pareto dominance were applied alternatively to provide different solution methods to the user. Speed, visualization and interactive optimization are important issues for real-applications. The systematic improvement and verification of the results have been done using statistical approaches such as classical design of experiments and modern DACE/SPO methods.

3.2 Treatment Structures in Intensity Modulated Radiotherapy

Karl-Heinz Küfer

Radiotherapy is, besides surgery, the most important treatment option in clinical oncology. It is used with both curative and palliative intention, either solely or in combination with surgery and chemotherapy. The vast majority of all radiotherapy patients is treated with high energetic photon beams. In this technique, the radiation is produced by a linear accelerator and delivered to the patient by several beams coming from different directions (see Fig. 3.6).

In conventional conformal radiation therapy, only the outer shape of each beam can be smoothly adapted to the individual target volume. The intensity of the radiation throughout the beam's cross section is uniform or only modified by the use of

Fig. 3.6 The irradiation gantry moves around the couch on which the patient lies. The couch position may also be changed to alter the beam directions.

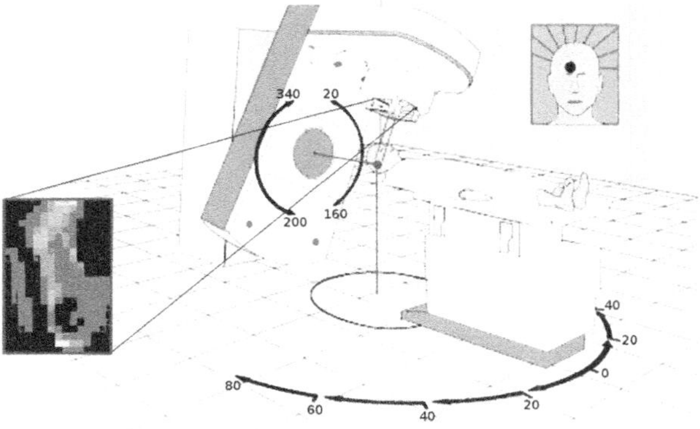

pre-fabricated wedge filters. This, however, limits the possibilities to fit the shape of the resulting dose distribution in the tissue to the shape of the tumor, especially in the case of irregularly shaped non-convex targets like para-spinal tumors, prostate carcinoma located close to the rectum, or head-neck tumors in the proximity of the parotid glands and the spinal chord.

A significant advance in treating such difficult cases was the development of intensity modulated radiation therapy (IMRT) where the intensities on the cross-section of a beam can be varied. Using multi-leaf collimators (MLCs) (see Fig. 3.7), the intensity is modulated by uncovering parts of the beam only for individually chosen opening times and covering the rest of the beam opening by the collimator leaves. This permits shaping highly irregular dose distributions in the target volume.

An IMRT plan is physically characterized by the beam arrangement given by the angle of the couch relative to the gantry and the rotation angle of the gantry itself, and by the intensities on each beam. The treatment aim is – as a pragmatic definition – to deliver sufficient radiation to the tumor while sparing as much of the healthy tissue as possible. A major challenge in IMRT planning is to cope with these fuzzy goals and demands.

Fig. 3.7 A Multileaf Collimator (MLC). The square opening in the center of the machine is partially covered by leafs, each of which can be individually moved. (picture from [SM01])

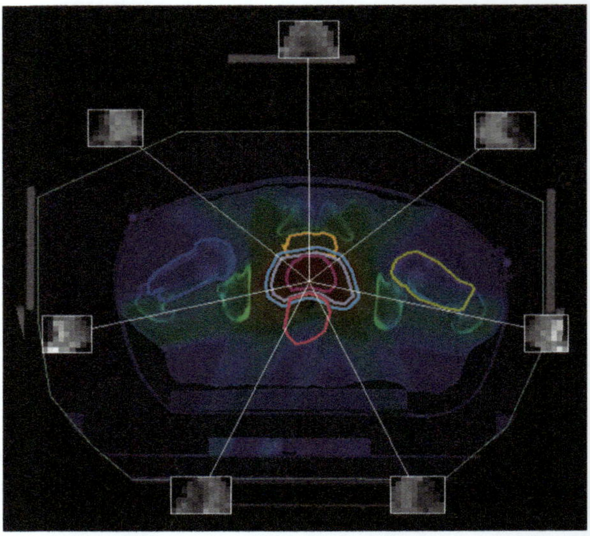

Fig. 3.8 Beam directions in a concentric irradiation geometry with an equiangular setup

Concerning the beam arrangement, in most cases a concentric irradiation geometry is chosen: the beams meet in one iso-center as depicted in Fig. 3.8. Additionally, the beams are generally chosen to lie in the same plane, as positioning the couch to realize a full 3D geometry increases treatment time substantially. Since the mathematical formulation of the planning problem including the irradiation geometry poses a much more difficult problem description,[1] the beam directions are usually fixed manually. In most cases, there are fewer than 10 beams and often they are arranged in equiangular distance around the patient.

The remaining planning parameters – the description of the intensity modulations of the individual beams – may now be determined by mathematical optimization procedures. However, the choice of an appropriate objective function to optimize presents one of the biggest hurdles in formulating a mathematical optimization problem for IMRT planning. Many quality measures for a treatment plan based on different approaches have been proposed by medical physicists and oncologists. Common to all measures is that they are based on the distribution of the realized dose in the patient's body.

Some measures typically used are, for example, the average dosages received by organs and tumor volumes. Often variations from ideal reference doses are measured. Yet another function is the variance around the mean dose in the target. Structurally, these commonly used quality measures are descriptive and comparable for the planner.

In most clinical cases, one or two objective functions are specified for the clinically relevant structures. This way, a multi-criteria formulation of the planning prob-

[1] This problem is not convex and hence can not be guaranteed to be solved exactly in reasonable time.

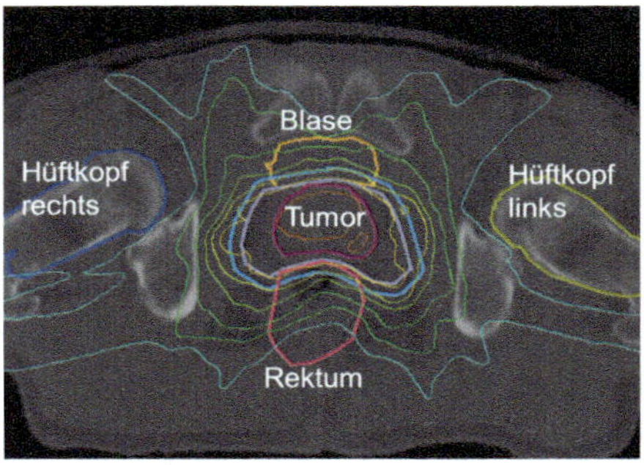

(a) The lines depict the dose distribution in the body

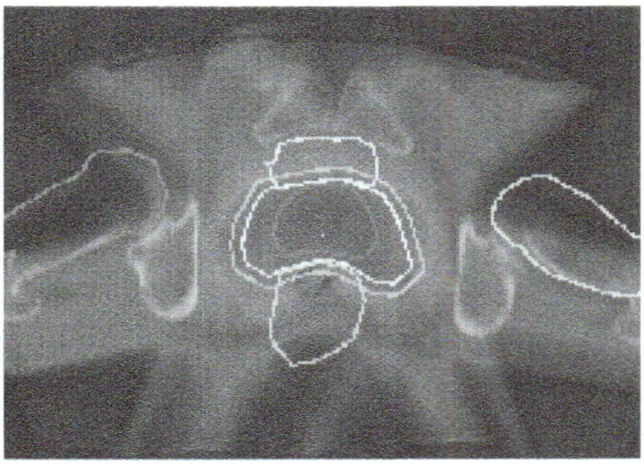

(b) The same distribution as in (a) with a more detailed resolution - the original picture shows different colors for each level of dose

Fig. 3.9 Visualization of the dose distribution in one of more than 100 CT-slices of a patient

lem arises naturally. The quest for a solution that has a single optimal quality score is thus extended to a search for solutions that are *Pareto optimal* or *efficient*. An efficient treatment plan has the property that none of the individual criteria can be improved while at least maintaining the levels of the others. In our context, given a Pareto optimal treatment plan, lowering the dose in one organ can only be achieved if the dose in the tumor is also lowered, or at least one more organ receives a higher dose, for example (Fig. 3.9). In general, there is a multitude of such solutions to any given multi-criteria optimization problem. We refer to all efficient solutions as the *Pareto set*.

The subjective nature of IMRT planning is in part due to the specific choice of objective functions by a treating oncologist. A different planner may prefer similar but different quality measures for a given case. Since all quality indicators in IMRT planning are designed to achieve the same fuzzy goal – allowing control over the spread of the tumor – they measure more or less similar effects. It may therefore safely be assumed that the quality measures are correlated with each other: good plans under one objective function are also good plans with respect to another objective function. In other words, a Pareto set obtained using one specific set of quality measures is most likely not altered too much if a different set of correlated quality indicators were used. This diminishes to some extent the severity of the fuzzy environment IMRT planning is situated in. As long as the planner uses quality indicators that correlate with commonly agreed on sensible measures, it is most likely that the treatment plans he obtains are of high clinical quality.

The software MIRA (Multi-criteria Interactive Radiotherapy planning Assistant) developed by the Fraunhofer institute for industrial mathematics (ITWM) in Kaiserslautern is designed to calculate Pareto sets and, more importantly, provides an interactive environment to select an efficient plan that is according to the notions of the oncologist. The latter is the implementation of a support system for a highly complex decision problem. Detecting planning possibilities and physical limitations in a database of efficient solutions while considering multiple clinical preferences at the same time is a very challenging task.

The optimization algorithms implemented in MIRA calculate a database of Pareto optimal treatment plans that cover a clinically meaningful range. As soon as the decision-maker has selected some quality measures, MIRA begins with the calculation of solutions. A high number of criteria leads to a relatively large Pareto set. Since this set is continuous, the possibility of displaying the infinitely many solutions is not an option. To still select a treatment from the Pareto set, it suffices to approximate the set appropriately. For this a so-called planning horizon is determined to exclude clinically irrelevant solutions. The Pareto set will contain a lot of plans where the trade-off between overdosage in the organs and a specific dose in the tumor will be too high to consider. MIRA deploys *extreme compromises* as corner points for the planning horizon in the Pareto set. They are the compromise of all possible combinations of the objective functions. The idea and its mathematical consequences are described in more detail in [KMS$^+$05].

After this planning region has been marked, individual solutions are placed between the extreme compromises to approximate the Pareto set in more detail. Calculating an extreme compromise or another solution from the Pareto set requires solving a high dimensional optimization problem. Specialized optimization routines developed at the ITWM enable efficient calculations and creation of the databases in reasonable time. The strategy called the *adaptive clustering method* solves a sequence of approximate problems, which iteratively converge to the original problem. The solution of each previous problem is used to adaptively refine the approximation. The method is explained more fully in [SKM$^+$04].

After the calculation of a database of treatment plans, the interaction with the decision-maker begins. MIRA has an interface called the *Navigator* (see Fig. 3.10) which enables the user to explore the Pareto set using visual controls. All user interactions with the Navigator are internally translated as optimization problems on the Pareto set and solved in real-time (see Fig. 3.11). This allows a smooth interaction with MIRA and explains the names "assistant" and "Navigator".

The star on the left hand side is composed of axes for the different quality indicators. The objective functions associated with the organs at risk are combined into a radar plot, whereas the functions associated with tumor volumes are shown as separate axes. The interval on the axes corresponds to the range of values contained in the database for the individual objective functions. The white polygon marks the indicator function values for the currently selected plan. The shaded area represents the planning horizon.

The interaction with MIRA is characterized by two fundamental mechanisms which are patented for the use in IMRT planning by the ITWM:

- *restriction*, to alter the planning horizon, and
- *selection*, to alter the current solution.

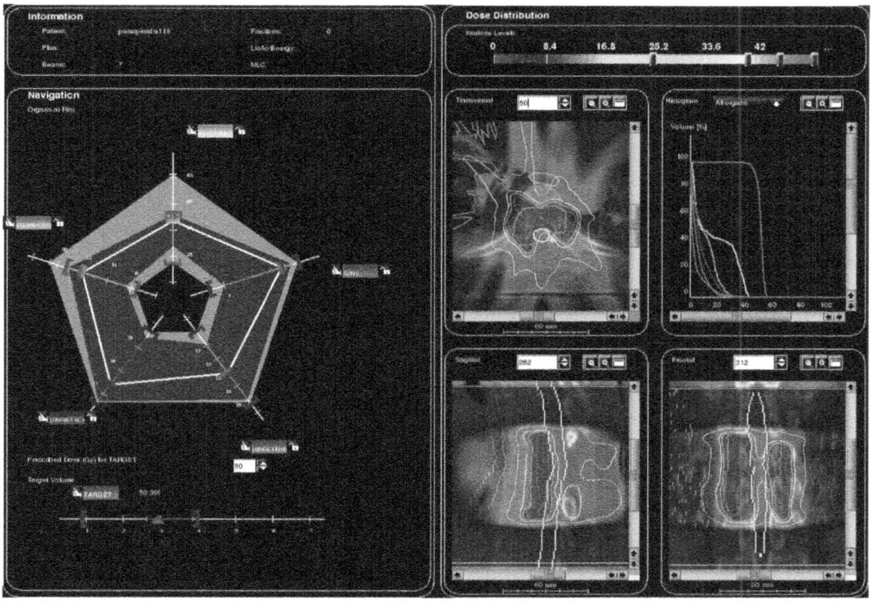

Fig. 3.10 The Navigator. The star on the *left hand side* shows the planning horizon and the current solution. On the *right* the dose visualization and other statistics for the current solution are shown.

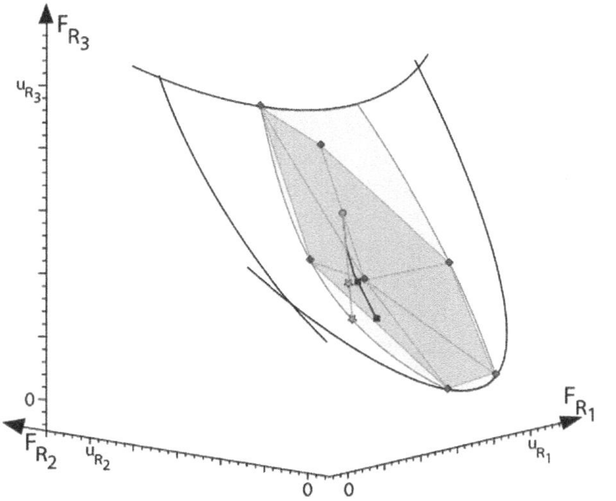

Fig. 3.11 A conceptional view on the functionality of the Navigator: while the user changes the current solution, the "distance" of the current solution on the surface to the *point* represented by a *star* is minimized. The *line* connecting the *stars* shows the path the user has dragged in the navigation by moving a slider. This is where the users would like the solution to be. The shaded region depicts the planning region on which the path casts a "shadow". The coordinates of this shadow are displayed in the Navigator.

Restriction allows the user to set the bounds of the planning region in both directions: to narrow it and to expand it. Unwanted parts of the Pareto set can be excluded from planning this way. The planning horizon displays a valuable piece of information for the decision-maker: the inner boundary close to the center of the star represents the highest ambition in an individual objective the user may have, while the outer boundary depicts the absolute highest "cost" in a criterion that has to be paid to improve another objective.

The line representing the currently selected plan has handles at each intersection with an axis and triangles for the tumor related axes. Both can be grabbed with the mouse and moved to carry out the selection mechanism. Internally, optimization routines ensure that the changes in the other objectives are at a minimum. This enables the greatest possible control over the navigation through the Pareto set: MIRA tries to change the current solution only in the direction the decision-maker wants it to.

The right hand side of the screen displays the corresponding plans concurrently. It is the visualization of the dose distribution parallel to the interaction with the Navigator that conveys the possibilities and limitations in a clinical case.

Several optimization problems are solved each second to allow a smooth interaction with the user. The sensitivity of the quality measures that are not actively changed to the objective that is currently altered is shown by the magnitude of the changes in the current solution while the handle is pulled. If small changes in one quality measure lead to rather large changes in some other criterion, the objective functions are particularly sensitive to each other in the region around the current solution. The decision-maker gathers a lot of insight and feeling for the clinical case using this valuable information and is thus able to accurately determine the trade-offs involved with each treatment plan.

We now demonstrate a small planning example for a head-neck case. These are typically difficult to plan, since the primary tumor can be located anywhere in the naso- and oropharyngeal area, and regularly the lymphatic nodal stations have to be irradiated because they are at risk of containing microscopic tumor spread. This results in big, irregular shaped target volumes with several organs at risk nearby.

The salivary glands are such organs at risk that are quite radio-sensitive. The tolerance dose of the biggest salivary gland, the parotid gland, is approximately a mean dose of 26 Gy. The goal should be to spare at least one of the parotid glands. Otherwise the patient might suffer from xerostomia (a completely dry mouth) which can significantly reduce the quality of life. Other normal structures that have to be considered are (depending on the specific case) e.g. the brain stem, the spinal cord, the esophagus and the lungs.

The objective functions to two tumor volumes are the homogeneity of the dose around a specified mean. The axes of the star depict objective function values of the spinal chord, unclassified tissue, parotid glands, the left eye, the right eye and the brain stem. Now the interactive part of the planning process begins. The required compromises between the selected objectives can be evaluated by pulling the handles as described above. A first intuitive feel for the possibilities in the case is developed this way.

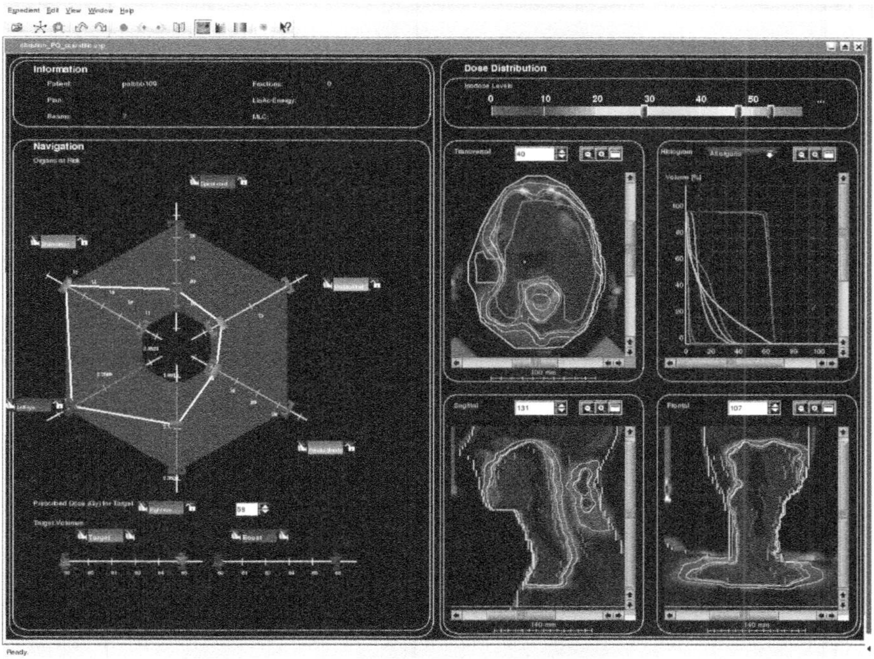

(a) Navigator at the beginning of the planning process

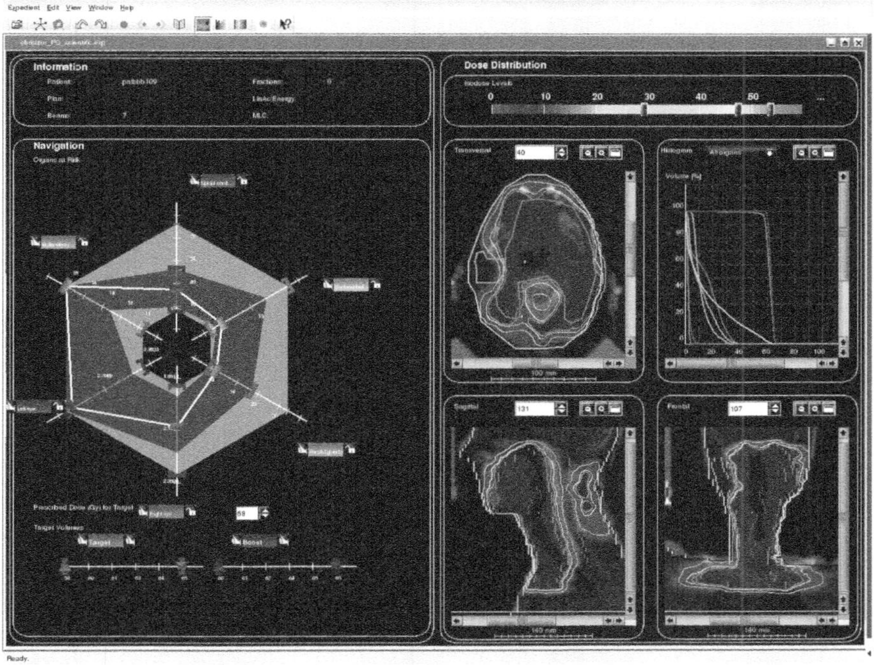

(b) The Navigator after some restrictions have been made – note the significant difference in the remaining planning domain.

Fig. 3.12 The Navigator for the head and neck case

While the user alters the solution, the Pareto set and its dose distributions are explored in real time. MIRA also provides a locking mechanism for an organ. The user can fix a value on an axis, which will henceforth not be breached unless the lock is removed. All treatment plans that score worse than the fixed values are excluded from the planning region. This effect can be seen by the reduced planning horizon in Fig. 3.12(b).

Complex planning scenarios can be interactively explored in this manner and the best clinical treatment can be found within shortest time.

The design of the Navigator and the functionality of MIRA are targeted to spare the user from mathematical formulations of optimization problems, and rather provide clinically descriptive information that is easy to interpret by experts. These aspects play a decisive role in acceptance of MIRA among treatment planners.

MIRA and the navigator present a tremendous opportunity for improvement of the planning workflow in hospitals all over the world. The graphical user interface for navigating a complex solution set and the flexibility in modeling provided by MIRA also facilitate interaction between physicians and medical physicists during the planning stage.

3.3 An Interactive Multicriterial Decision Tool

Peter Roosen and Karl-Heinz Küfer

Decision making is a mental structure generation process. What are the strategies to arrive at good decisions? Decisions are in most practical cases depending on many objectives. How do "uneducated, ordinary people" tackle multi-criterial decision making? What strategies do they pursue to "obtain what they want"? Professionals regularly and explicitly dealing with this kind of problems know several methods of approach, each of those being thoroughly investigated and assessed with respect of their predications and limitations. In spite of the frequency of such problems also in everyday life this knowledge is not widely spread. So, it is interesting to investigate whether non-professionals can make use of a support tool oriented towards this decision domain. The general idea is that the last resort of objectivity is the Pareto set of a given problem. Its proper display will yield the best possible decision support.

While most practical problems present more than two target qualities to be optimized there is no static presentation tool that would provide an intelligible, ideally printable, selection display for the decider to work on. Therefore a computer-based tool was created to support interactive, visual pareto set testing. As the principal display mode is based on a star plot, the system was called "Pareto Star". Its applicability is not limited to pure pareto sets, though, as in practical applications there is not necessarily a sharp distinction of important decision criteria vs. irrelevant ones. Instead, a kind of continuous transition of criteria more or less crucial for the decision is frequently observed.

The designed decision support system is not a tool to *identify* a pareto set. – It presupposes the existence of a database of distinct realizable solutions with their individual target function values, from which pareto-optimal individuals can be found by enumeration based on partial settings of criteria. There are cases, though, that exhibit continuous settings or operation modes and hence a dense pareto set of an unlimited number of members exists. In such cases no distinct, objectively chosen points in target function space can be provided. However, small changes in the relative trade-off target settings usually do not make large differences in the eyes of the decider. So it suffices to identify sensible representants of the originally dense Pareto set and present them as individual target function (and accordingly configuration parameter) combinations in the selection tool (cf. Sect. 3.2).

As a practical test bed the selection of a handheld GPS receiver was chosen. There are several practical selection criteria, and no all-purpose device serving the whole range of private navigation tasks exists. So the potential buyer is confronted with the selection problem, being an ubiquitous obstacle when scanning respective electronic discussion fora for answers and opinions. The tool devised as decision support is implemented as a web service, extendable to almost arbitrary multi-criterial decision problems. It operates on three basic tables, implemented by means of a database system: (i) the list of potentially interesting criteria, (ii) the list of available devices, and (iii) the cross-combination table of their individual characteristics.

For the given decision domain there are about 20 parameters potentially relevant to different usages. Those may be as different as navigating a sensible automotive way in the streets, with upcoming traffic jam information being considered in a dynamic re-routing along the way, and trying to find small treasury caches somewhere in nature as recreative outdoor activity [Gro05]. As the respective items tend to be relatively costly (ranges in autumn 2005 between about 200 and 1000 EUR) many users try to cover different application domains by choosing just one device, sensibly supporting potentially conflicting usage scenarios. This in turn individualizes the target function weighting to an always new set.

3.3.1 Decision Support Tool Setup

The decision support system was set up as a (German) web service, to reach widely distributed people as best as possible. It was placed on a non-commercial, topic related website of one of the authors with sufficient traffic to be found by chance. Additionally it was announced in various fora concerned with recreational GPS use, to attract the largest number of people possible. As there is no chance of forwarding a detailed usage manual to the casual visitor, the entry page was set up as such (Fig. 3.13).

In a very brief explanation the user is informed how to choose his favorite target values and how to maneuver in the star plot. Finally a hint is given that the resulting

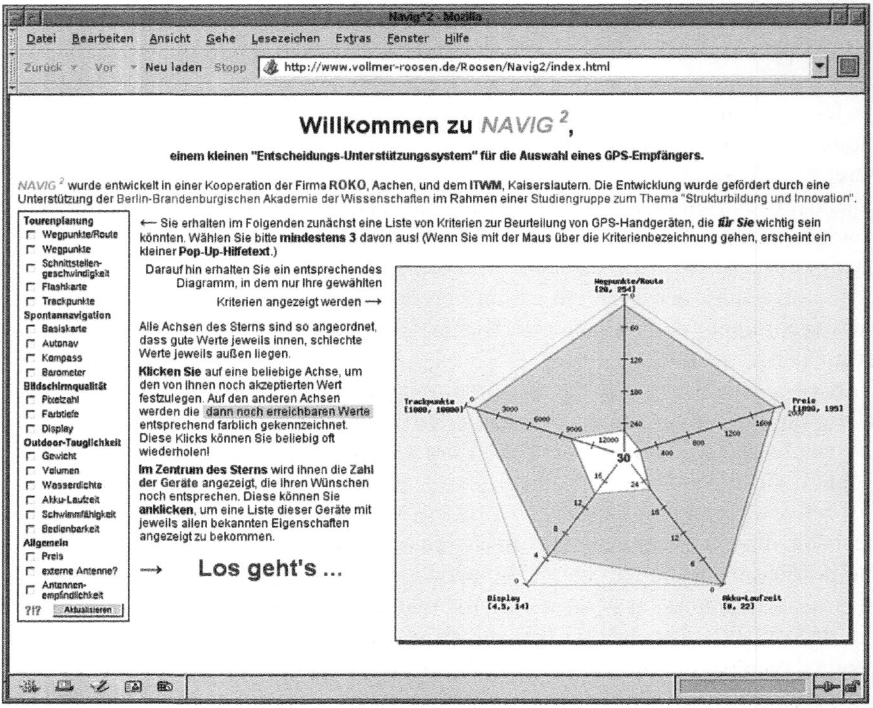

Fig. 3.13 Introductory page for the "Pareto Star" GPS handheld receiver selection aid. A small explanation of the usage is given.

list of complying receivers is available via the central number in the star, indicating how much of the basically available devices are still left.

Starting the application by clicking the appropriate link (Los geht's) leads to the personal criteria definition page (Fig. 3.14). Here the user is requested to select at least three criteria that he regards as important. For every criterion a short information box, shortly explaining what is evaluated and which possible values can appear for it, pops up if the user moves his mouse pointer on its name. After pressing the "Aktualisieren" (update) button the first star plot is presented (Fig. 3.15).

The axes of the plot are directed so that the better values always lie in the center of the star. (Non-orderable properties, like housing colors or geometries, cannot be handled within the scope of the tool presently.) The different colors of the plot areas indicate different things. On any axis there are transitions from light blue to yellow to light brown to white. The transition from light blue to yellow indicates an individually set limit of acceptance (see below) that the user tolerates for that criterion. Without any prior interaction that limit is automatically set to the lowest value the respective axis can display for the given set of choices. The consecuting yellow part on any axis indicates the range of values that is not covered by the choices' list, without being limited by a user's action though.

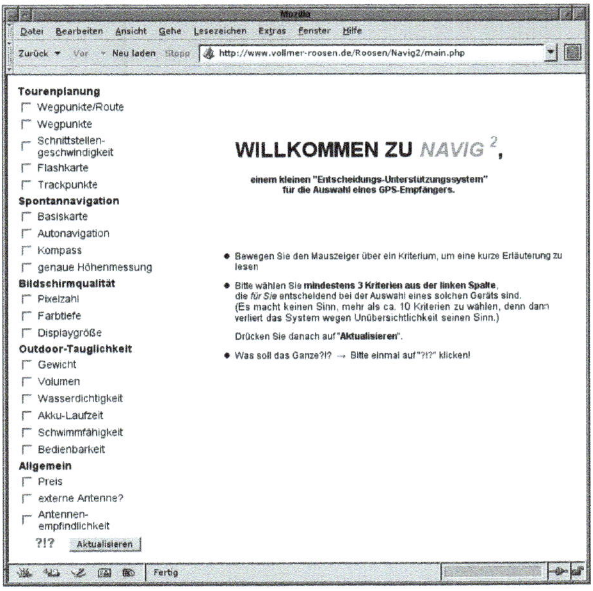

Fig. 3.14 When entering the GPS receiver selection aid the user is asked for his personally most important properties of the devices. A small step-by-step guide is presented to help him starting. By clicking on the blinking "?!?" he obtains information on the background of the service.

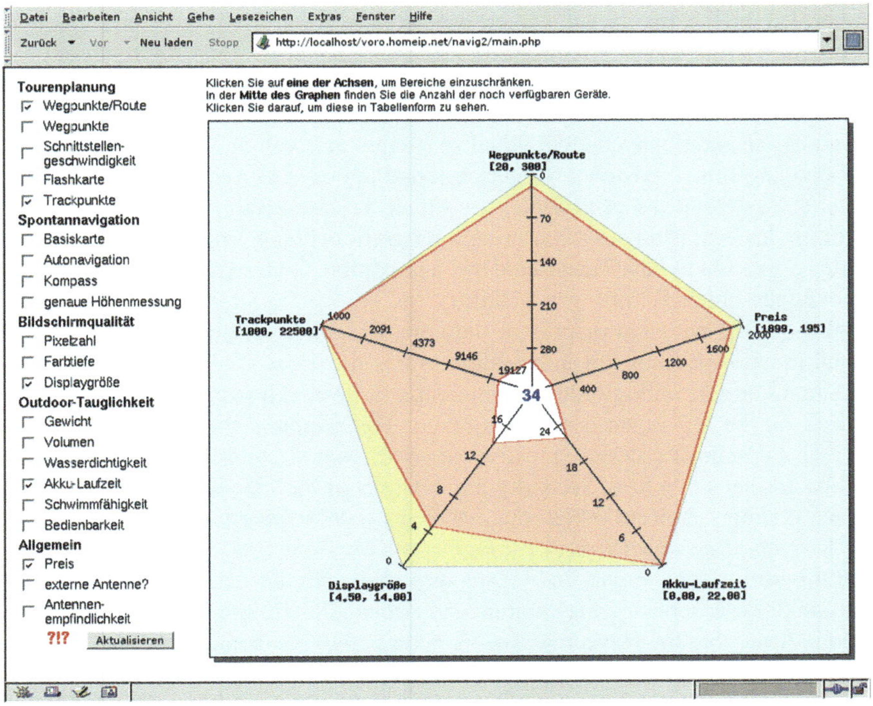

Fig. 3.15 After entering at least three personally relevant criteria. In this example the five criteria "waypoints per route", "track points", "display size", "accumulator runtime" and "price" were chosen.

The light brown part on the axis is the range that the real properties of the discussed GPS devices cover. As the display represents *all* devices not excluded by an individual acceptance limit, only the range for the whole set, not the individual properties of the devices, are displayed. The premise here is that the decider should concentrate on his basic individual demands, not on the concrete criteria of the devices.

The white range on any axis finally indicates a range that is not or no more accessible, either due to generally limited performance in the whole set of choices, or to a limitation that was defined for other criteria by the decider.

In the center of the star plot there is a thick blue number. It displays the number of choices left when the Pareto Star system honors the limitations set by the decider. For the GPS handheld devices there is an unlimited number of 34. The definition of a personal acceptance limit for any criterion selected beforehand will reduce the count of remaining devices respectively. The limits of the other target functions will change as well, both for the lower and the upper bound of the light brown range, as is evident, for example, when investigating the principally counteracting physical properties of (largest possible) display size and (smallest possible) device weight. Specifying a low value for the maximally accepted weight will reduce the maximum available screen sizes of the remaining items.

The point and click procedure on the accepted values, representing immediately the preferences of the decision taker, is re-iterated for arbitrary axes until the number of remaining items is reduced to a tolerable quantity, quite often around 5–8. During this process the decision seeker may as well drop criteria from his selection or choose additional ones from the continuously displayed left criteria list column. Choosing an additional one will initialize another axis without a set boundary value, but obeying limitations on its criterion range imposed by set limitations on other criteria. After that the central number should be clicked to obtain a simple table of those devices, listing *all* their criteria properties (Fig. 3.16). In this table pareto-optimal individuals have been found by enumeration, but also dominated individuals meeting the imposed limits are included.

Besides the potential display of the criteria description popups there are additional functions and bits of information in this list. First of all it may be sorted in ascending or descending order for any criterion by clicking on the respective small arrows on the left of the criterion's name. This supports the notion that usually there is a dominant criterion (mostly: the price) even if a multitude of relevant ones has been selected before. Secondly the columns of the list are colored differently. Green columns indicate pareto-optimal items *with respect to the criteria set and the individual accepted ranges settings of the decision seeker*. It must be stressed that this pareto set is highly individual, as a change in the criteria list or alternative thresholds of acceptance may significantly change it. Columns colored white and blue indicate that the respective item is not a pareto set member but still complies with the imposed limits. In each non-pareto column the dominating devices are shown at the bottom of the table (row "dominierende Geräte"), so the decider can immediately compare them with each other. The reason to include the non-pareto optimal devices in the list is the lack of a practical sharp distinction between relevant and irrelevant criteria in the selection process. Even if a criterion has not been

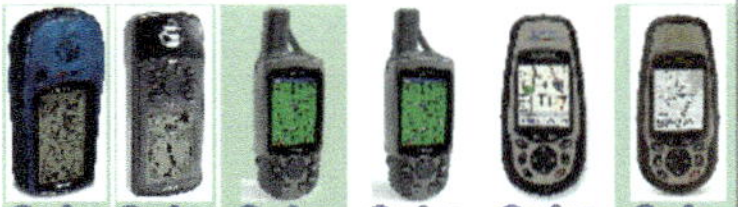

Bild	Großes Bild	Großes Bild	Großes Bild	Großes Bild	Großes Bild	Großes Bild
Name	Etrex Legend	GPS 72	GPSmap 60C	GPSmap 60CS	Meridian Color	Meridian Gold
Hersteller	Garmin	Garmin	Garmin	Garmin	Magellan	Magellan
Wegpunkte/Route	125	50	250	250	50	50
Wegpunkte	1000	500	1000	1000	500	500
Schnittstellen-geschwindigkeit	115200	115200	1e+06	1e+06	115200	115200
Flashkarte	0	0	0	0	4	4
Trackpunkte	10000	2047	10000	10000	2000	2000
Gewicht	150	215	200	200	240	240
Volumen	171	390	312	312	378	378
Wasserdichtigkeit	7	7	7	7	7	7
Akku-Laufzeit	18	16	20	20	14	14
Schwimmfähigkeit	0	1	0	0	1	1
Bedienbarkeit	1	2	2	2	2	2
Preis	289	259	559	639	578	438
externe Antenne?	0	0	1	1	1	0
Antennen-empfindlichkeit	0	1	1	1	1	1
Dominierende Geräte				GPSmap 60C	Meridian Gold, Meridian Platinum	

Fig. 3.16 Display of the properties' list of the remaining items, after reducing their number via the interaction with the star diagram. A combined image is shown because the computer screen is not large enough to contain the table as a whole.

included into the interactive Pareto Star selection that does not indicate that it's value is completely irrelevant. If, for a given device, such a factor becomes very favorable while the chosen "important" ones strongly resemble each other, it may as well serve as a secondary, hierarchically less stringent decision argument. This leads to a discussion on parity and hierarchy of optimization goals and will be revived in a later section.

3.3.2 Feed-Back from Practice

The decision support system was designed in order to monitor "uneducated" users' reactions on the availability of an individually determined pareto set filtering system since this kind of decision domain is ubiquitous in normal life. To attract a sufficient number of users the tool was placed on a well-frequented web site dealing with the application of smaller GPS devices. By means of small introductory notes it was announced on several related web-based fora.

On the web site itself a visually quite dominant teaser (red, bold and blinking "?!?" click field, placed directly next to the main display refresh button) was placed to motivate users to express direct responses. If a user clicks it he or she becomes a short note on the background of the tool and a text entry field to submit his or her impressions on the tool. In addition to those (rather few) direct responses the usage of the system was monitored by analyzing the server log files over an interval of about 3.5 months.

About 30 direct responses were received. Half of them just helped correct some faulty decision data on the ever-so-changing specifications of the individual receivers (due to frequent firmware revisions that the manufacturers provide for their customers). Those responses indicated that the users observed the provided data with thorough scrutiny and were eager to clean up the data base in the sense of a community effort. The other half stated that they thought the tool really helped deciding on a device to be bought. Some of those responders remarked, though, that they already bought their devices prior to finding the decision support tool. The retrospection of their purchase decision was mixed: Some stated they might have decided otherwise if they had known about the Pareto Star beforehand, some reported an affirmation of their stomach based former decision. No response discussed the tool and the method by itself, though. Obviously the content aspect was too dominant to focus on the methodology.

In order to obtain more of those methodological data the server log files have been analyzed. Due to the setup of the service as LAMP system (containing the components Linux, Apache, MySQL, PHP) and the input data submission mostly via the http GET method most of the user settings with respect to choice and number of criteria could be reconstructed (Table 3.1). The total number of 2600 visitors should provide an adequate sample size to consider the data as trustworthy.

The anticipated essence of the tool – the toying with the star diagram – was used rather sparsely with the averaged 2.7 interactions per visit. Here a definitely larger number was anticipated. There is an explanation, however, in the supposition that the users tend to be somewhat result-oriented (in the sense that they want to decide on a concrete device) and that the playing around with the tool did not satisfy their immediate curiosity just *which* devices were still in the race and which had already been left behind.

Another point may be the non-familiarity of the users with the density distribution of the individual criteria. If a scale indicates that a large range of values is

Table 3.1 Statistical access values for the Pareto Star decision support system for GPS handheld receivers

Number of visitors (total)	2600
⟶ number per day	24
Star interactions per visit	2.7
Result table requests per visit	1.8
Mean number of criteria chosen for the star diagram	5.5

available it does not show that perhaps 95% of the devices are clumped in a (possibly non-favorable) small part of the span. This soon leads to rather sparse result lists. The average number of less than two result table requests per visit once again shows that the users do not tend to play with the tool, but instead are very target-oriented. After receiving a result table they rather seldomly reiterate the star plot selection process with changed criteria settings.

This in turn may be interpreted as result of economizing the selection process in the sense of a frugal decision method: The users seem to obtain a convincing final selection list by just focusing on the first result table. The fact that usually familiar brands and types are listed will enhance that notion, once again supported by the Bounded Rationality view of typical human selection methods, here "take the best known" (cf. Sect. 3.12).

These interpretations, derived from the average system's usage, do not hold for some very eager users who indeed played around quite intensely with the tool, reaching well above ten or fifteen interactions with the star diagram and a respective number of result lists.

As a second step towards feed back from practice the tool was presented to several hardware and service providers in very different application domains, such as used automobiles reselling, real estate brokerage, recreational accomodation services, or last minute flight ticketing. In almost every discussion the tool was honored as a very interesting method of information retrieval and assessment. Nevertheless the tool was not used for a respective innovative information preparation because a conflict of interests was regularly perceived. This may well be explained by discussing a potential application in the automotive reselling business.

For an internet-based vehicle reseller the usual business model is to provide a presentation platform for an arbitrary number of private vendors that publish their offers on it. For larger sites well above a million vehicles are offered at a given time. So even if the list size of criteria usually deemed relevant is estimated to five or six there are lots of offers not nearly approaching pareto optimality in the sense of the potential purchasers. If the trade platform offers a method to easily identify the most interesting ones in the sense of constructing a pareto front, most offers would never (or at least only very rarely) be shown. Vendors intending to use the platform would turn away from using it. This in turn conflicts the business model of the web site provider who would be the one to offer the additional pareto set filtering service – hence the aversion to put such a tool to work. It may be guessed that most potential vendors have a certain kind of qualitative notion about the competitiveness of their offers, so the availability of a selection tool would shy them away to other platforms *not* offering it.

This situation holds for most points of electronic sales, therefore it is not to be expected that this methodology of multi-criterial decision support will find its penetration into a broader usage. More promising fields of application are areas where all participating players do have a sincere interest in an objective target criteria assessment. One of them is presumably the consumer consulting service that does not earn its money by selling items but with offering the best possible objectivity in consumer information requests.

3.3.3 Refinement of the Selection Tool

The experiences with the usage behaviour led to the conclusion to redesign the tool. It should both familiarize people with a multi-criterial selection process support and better meet their expectations with respect to their desire to obtain almost immediate practical suggestions. The new "Pareto Star" system is consequently not a star plot diagram any more, but rather a semi-hierarchical table of bar charts with similar, but enhanced colorization ranges relative to the existing implementation.

While the final result table display will not be changed very much, the multi-criterial selection process is significantly changed towards a faster "real results" display, partially implying a partially hierarchical expectation concept. Starting with an empty list of selection-relevant criteria the user will be asked to choose a first criterion (Fig. 3.17).

After selection of one or several personally relevant criteria the result bars (Fig. 3.17, left side) display a differentiated colorization (see below). A set of adequate receivers is presented on the right side of the interaction window, being

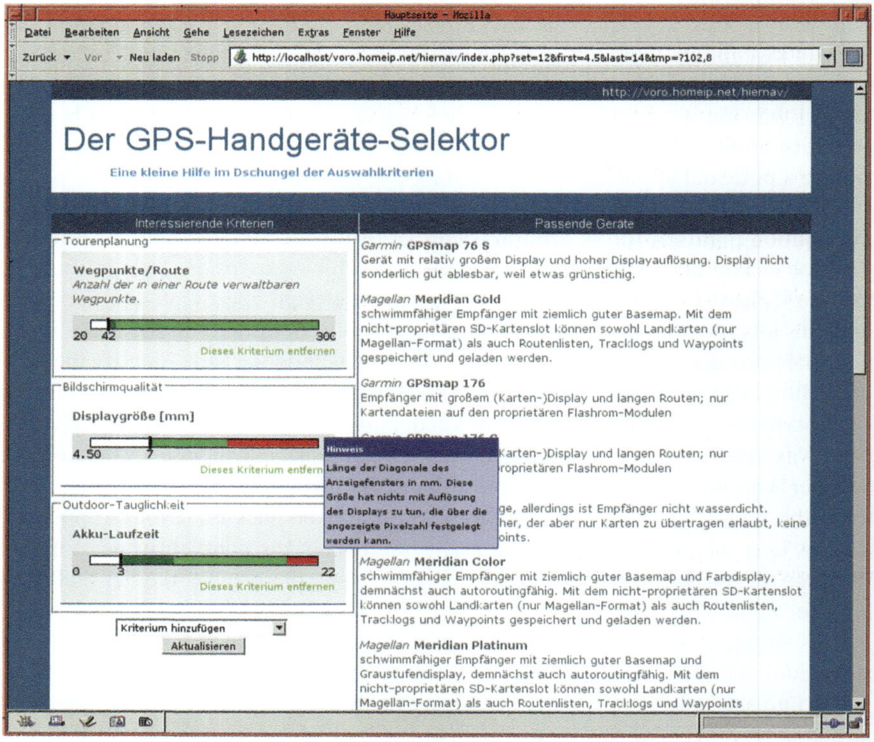

Fig. 3.17 New multi-criterial decision support system with instantaneous semi-hierarchical result display. (The visual appearance of the interface itself represents an intermediate development phase).

background-colored in red to display a Pareto set membership as in the older version of the Pareto star system final result table.

The order of parameter bars on the left side is treated hierarchically, with the importance of the criteria decreasing from top to bottom. Accordingly the order of actual receiver displays is sorted in a strictly hierarchical way. Receivers exhibiting identical settings for the most important goal are sorted by the second important one and so on. In case of an additionally chosen criterion the list on the right is resorted accordingly. It is as well possible to switch places and hence the order of importance for the selected target criteria, with a respective effect on the order of the actual receiver display.

The graphical bars describing the single criteria contain additional information relative to the Pareto Star. Favourable settings are always on the right side of the bars, so the user will tend to click into the bar to limit the respective value with increasing rigidity from left to right. His personal limit is indicated by a thick black vertical bar in the horizontal criterion distribution, with the chosen value being displayed on top of it. On the left of that marker there is a white background. On the right usually a light green range is situated, indicating an interval that does not contain items with respective values. A light blue range follows, hinting towards actually available items. Contrary to the Pareto Star an additional background colored range follows: The red section indicates the interval of target value settings that will lead to a complete emptiness of the receiver list. This range is influenced by the user's limit settings in other criteria, due to the fact that those restrict the available choice in others.

In the former Pareto star system the user did not receive information about the density and distribution of the criteria values independent of his personal choices. Therefore an additional piece of information was introduced in the new hierarchical system: Small black dots in the center of the horizontal bar indicate the actually available values in the underlying database.

In the end the user will choose to display the resulting items table that is presented in a similar fashion as in the presently available Pareto Star system.

Finally, the new system contains visual procedures to work on discrete and unsortable list criteria. Usually there are values that cannot be ordered by their values, like the color of a car body, or the selection of edge shapes for a chair or a desk. Here the user will be able to define his or her personal preferences, with the displayed order of potential values reordered upon each new change of choice, separating the desired ones from the undesired. As soon as this selection system is in a stable condition it will replace the present Pareto Star, and its availability will be announced once again in the respective fora to attract new users.

3.4 RNA Structure Optimization on Large Neutral Networks

Peter Schuster

Structure formation in biology is basic to life and at the same time the prototype for molecular complexity. Folding biopolymers into stable structures is visualized as an optimization process that commonly follows the criterion of minimizing free energy. Function in nature is designed and optimized by evolution, which can be understood – in form of the Darwinian mechanism – as an optimization of a cost function. In natural ecosystems, in particular when they are free of human intervention, the cost function for evolution is called fitness and measures the mean number of fertile progeny in the next generation. The Darwinian mechanism operates in populations with a dispersion of fitness values: Each variant with a fitness larger than average will be present more frequently in future generations whereas the frequencies of all variants with fitness values below average will decrease. Fitness counts offspring only and does not deal with the complexity of the objects, which are optimized. Hence, the rules for Darwinian evolution are the same whether we are dealing with molecules, with organisms, with groups of individuals or with any other variable object.

Optimization of molecular properties by means of evolutionary methods based on variation and selection (SELEX)[2] is an established technique in biotechnology. Out of all classes of biomolecules ribonucleic acids (RNA) (Fig. 3.18) are best suited for the evolutionary design of products with predefined properties, because they can be easily amplified either directly by replication or via reverse transcription (RNA→DNA), DNA amplification, and transcription (DNA→RNA). The latter three step process is often applied because DNA amplification is easily achieved by means of the polymerase chain reaction (PCR). In addition, RNA molecules have a richer repertoire of functions than their DNA analogues. Various classes of RNA molecules with specific properties have been produced by evolutionary techniques. We mention here three examples: (i) Molecules that bind specifically and with high affinity to targets called *aptamers* [Klu06], (ii) molecules that catalyze biochemical and chemical reactions called *ribozymes* [WS99], and (iii) molecules that have more than one long lived structure called *riboswitches* [WB03, VRMG04]. Aptamers, in particular those with high specificities and affinities, are very useful tools for highly specific detection of the presence of their target molecules and are used for analytical and diagnostic purposes (For an example of an aptamer discriminating between the two closely related molecules caffeine and theophylline see e.g. [JGPP94a]). Ribozymes were discovered first in nature and later on produced in the laboratory through modification of RNA molecules by variation and selection. Alternatively they were also evolved from molecules with random sequences. The repertoire of chemical reactions, which are known to be catalyzed by ribozymes is incredibly rich. Riboswitches, finally, are RNA molecules with two or more long-lived (meta)stable[3] conformations. RNA switches may be self-regulated or the conformational change may be triggered by a small molecule binding to the RNA. In nature the small molecules are often metabolites, which regulate the pathway

---

[2] SELEX stands for selection by exponential amplification [ES90b, TG90b].

[3] Metastable states are local minima of the conformational energy surface. They have a characteristic life time that is determined by the height of the barrier connecting the state with a conformation of lower free energy. RNA molecules having a long-lived metastable state in addition to the minimum free energy conformation are commonly called self-regulated RNA switches.

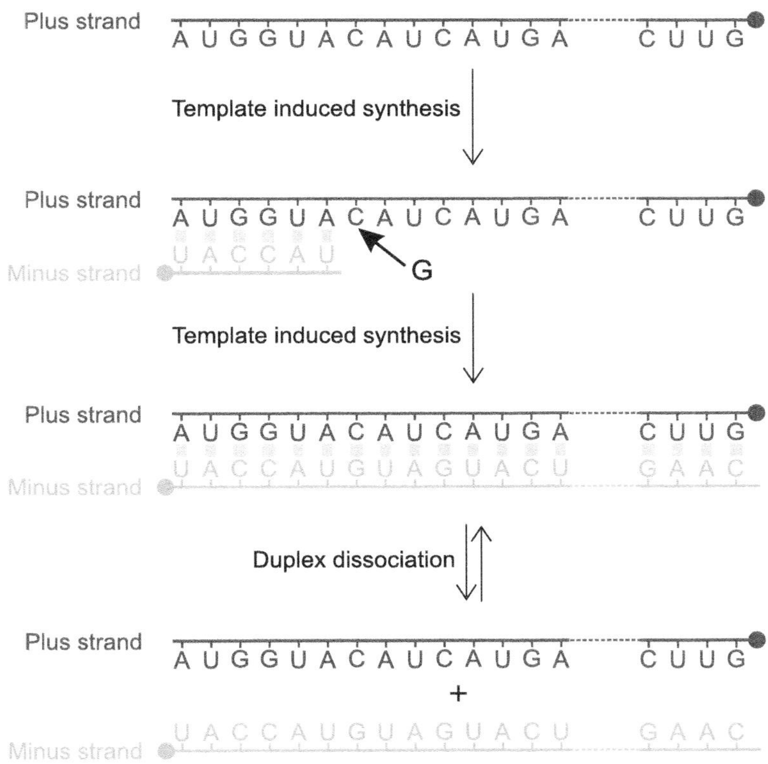

Fig. 3.18 Basic mechanisms of RNA replication. RNA replication commonly follows a complementary mechanism: A double-helical (plus-minus) duplex is synthesized from a (single) plus strand by making use of the complementarity of Watson-Crick base pairs. The critical step in replication is the separation of the duplex into two single strands because long double helical stretches are bound strongly. Then dissociation of double strands requires raising of temperature. In RNA evolution experiments separation into single strand is fulfilled by the replicase that prevents the formation of long double helical stretches by separating them into the two singe strands that form their own structures.

leading to them through inactivating translation via a conformational change in the messenger-RNA.

3.4.1 The Darwinian Mechanism

Optimization in the sense of Charles Darwin as laid down in his seminal book on the origin of species [Dar59] is a universal phenomenon taking place in populations. It can be cast into the interplay of three processes:

i. multiplication, in particular reproduction of organisms or replication of molecules,

ii. variation in the form of mutation or recombination, and
iii. selection as a consequence of limited resources providing upper bounds to population sizes.

Since the conditions sustaining these three processes can be fulfilled likewise by polynucleotide molecules in cell-free assays, cells or multicellular organisms, evolution in the sense of Darwin occurs in populations of very simple as well as well as very complex objects. Multiplication, variation, and selection take place in specific environments to which the populations adapt through a Darwinian process. Studying evolution is commonly complicated by two problems: (i) The times required for selection and adaptation in populations of higher organisms that have generation times from weeks to years are too long for direct observations, and (ii) changing environments make the situation very complex. This is particularly true for evolution in nature were sometimes large numbers of species evolve together or coevolve in the same ecosystem.

Encoded information on the reproduction of individuals is stored in the genotype[4] being a DNA or RNA molecule. The genotype is unfolded to yield the phenotype which is imagined best as the fully developed individual in its appearance and with all its functions and properties. Success and efficiency of evolutionary optimization are based on the different roles played by genotype and phenotype. Variation operates on the genome (Fig. 3.19) whereas selection operates on pheno-

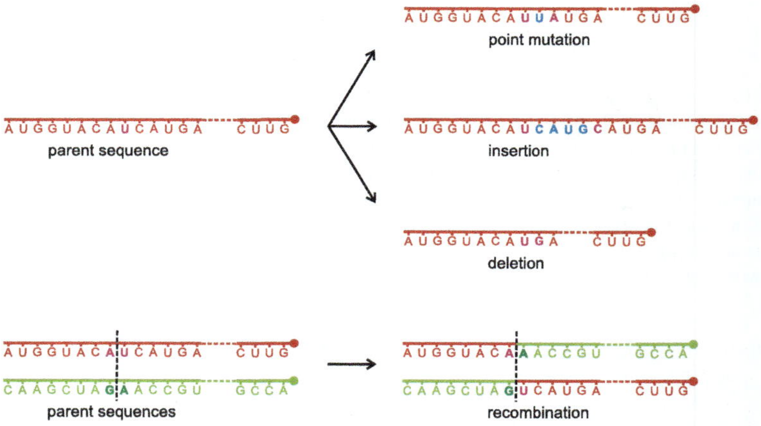

Fig. 3.19 Basic mechanisms of sequence variation. The upper part of the figure sketches three classes of mutations: (i) A point mutations where a single digit is changed and the sequence length remains constant, (ii) an insertion where part of the sequence is replicated twice, and (iii) a deletion where part of the sequence is not replicated. In the lower part we show a case of symmetric recombination between two sequences of equal lengths leading to two new sequences both with the same number of nucleotides. Asymmetric recombination (not shown) leads to sequences of different chain lengths.

---

[4] In molecular genetics the genotype is called the *genome*.

types exclusively. What counts in the selection process is the number of (viable and fertile) offspring in the next generation. This quantity is the fitness, f. In nature, and particularly for higher organisms, fitness is a highly complex function of the phenotype and the environment, which also includes the other organisms living in the same ecosystem. All species evolve together, co-evolution superimposes its dynamics upon selection and populations may never reach their optimal states. Evolution experiments were conceived for constant environments that reduce this complexity. In test-tube evolution of RNA molecules further simplification is possible: Genotype and phenotype are different properties of the same molecular species – sequence and structure, respectively – and fitness can be reduced to an (over-all) rate parameter under suitable conditions.

The dichotomy of genotype and phenotype is fundamental for the success of Darwinian optimization. It introduces a random element into the evolutionary process since variation of the genotype and evaluation of the phenotype are uncorrelated.[5] In other words a mutation does not occur more frequently because its phenotype has a selective advantage. Similarly the random element is indispensable for the Metropolis search method since it enables the algorithm to scan the entire solution space.

3.4.2 Evolution in the Laboratory

In order to render the evolutionary process easier intelligible optimization has been studied under simplified and controlled conditions. Three examples are mentioned here: (i) Bacterial evolution [ECL96, PSM$^+$99, EL03], (ii) an example of arms race in virus evolution [WWB05] and (iii) evolution of RNA molecules in serial transfer and automated evolution machine experiments [Spi71, MPS67, BG97, SE97]. The principle known as serial transfer is the same in both types of experiments: A Growth medium[6] is prepared and infected, either with bacteria or with suitable RNA molecules. Then, reproduction starts instantaneously and after a defined time interval (Δt) during which the population has grown to a certain size a small fraction of the sample is transferred into a fresh growth medium. The procedure is repeated and continued until evolutionary phenomena can be detected. In the experiment with bacteria a transfer was made every day and the series runs now over more than twenty years already. In RNA test tube evolution the number of transfers is typically around one hundred.

Bacterial evolution is dealing with whole cells whose genotype is a DNA molecule with a few million digits called nucleotides taken from a four letter

---

[5] Mutations are often considered as random. This is not entirely correct since an individual mutation step is a chemical reaction as correct replication is. Small rate parameters are common for mutations and then the occurrence of a particular mutation falls into the domain of stochastic events.

[6] The medium is either a nutrient for bacteria, agar-agar for example, or a solution containing everything that is needed for replication.

alphabet, $\mathcal{A} = \{\mathsf{A}, \mathsf{C}, \mathsf{G}, \mathsf{T}\}$, and attached to a one-dimensional periodic 2'-deoxyribose-phosphate backbone. The phenotype is the bacterial cell with its highly complex structures and functions. Two results obtained in the experiments carried out with *Escherichia coli* bacteria in the laboratory of Richard Lenski are important for the comparison with the evolution of molecules: (i) Changes in the phenotype occur stepwise and not gradual [ECL96]. The phenotypic changes concern, for example, the size of the bacteria, which increases in steps under the laboratory conditions. The larger variants reproduce faster and grow out the smaller ones. (ii) Changes in the DNA sequence caused by point mutations being single digit exchanges progress continuously in the sense that the rate of change is a constant [PSM$^+$99]. A strong indication for adaptation to the conditions of the serial transfer was derived recently: Parallel experiments were carried out in twelve populations or clones that were recorded over 20000 generations [CRL03]. Comparison with the ancestors showed characteristic changes in gene expression data[7] but roughly the same increase in fitness. Eight out of the twelve parallel experiments revealed mutations in one key regulatory gene. On the genomic level, however, no two point mutations were identical.

Out of many experiments that have been performed on viruses one recent study performed in the laboratory of James Bull is remarkable [WWB05], because it provides a laboratory case study on co-evolution of attack and defense mechanisms in the sense of an arms race. A population of DNA bacteriophages ϕX174 has been grown for 180 days in a suspension of *Escherichia coli* bacteria corresponding to 13000 phage generations. This phage has a small genome of 5386 nucleotides, which is only one thousandth part of the genome of the bacterial host. Nevertheless, the phage parasite is able to cope with the evolving defense mechanism of the bacterial host. This experiment is an excellent example of co-evolutions under laboratory conditions. The authors identified mutations in the phage genome that increase its virulence. The arms race will continue until one of the two partners reaches the end of its capacity to adapt.

Evolution of RNA molecules in the test tube was studied first by Sol Spiegelman and his group [MPS67]. The nucleotide sequences of RNA molecules are the genotypes. In particular, they are strings built from the four letter alphabet, $\mathcal{A} = \{\mathsf{A}, \mathsf{C}, \mathsf{G}, \mathsf{U}\}$,[8] attached to a backbone that is almost the same as in the case of DNA. The lengths of RNA molecules with functions in nature genotypes vary from between twenty to about thirty thousand. Small interfering RNAs with regulatory functions are commonly twenty to thirty nucleotides long, artificially selected small aptamers have about the same lengths, and RNA molecules with catalytic functions derived through selection experiments have chain lengths from one hundred

[7] In cells not all genes are active or *expressed* that means transcribed and translated into protein. Mutations in regulatory genes may lead to different activities of the corresponding regulatory proteins, which in turn modify the availability of structural or metabolic genes.

[8] Considering the chemical formulas of DNA and RNA, thymine (**T**) is a derivative of uracil (**U**); **U** compared to **T** differs only by the lack of a methyl group in position 5. The backbone of RNA carries an additional hydroxy group in position 2' of the ribose moiety.

to several hundreds [WS99, Jäs01, Klu06]. RNA genomes in nature range from several hundred in viroids [FHM+06] up to thirty thousand in the largest known RNA viruses [GEZS06].

The phenotype in evolution experiments with RNA molecules is the spatial structure of the molecule, which is the carrier of all its functions and fitness relevant properties. In Spiegelman's experiments selection is optimizing replication rates, which are indeed the molecular equivalents to fitness values. In this way these experiments correspond to natural selection where fecundity and viability of offspring, subsumed together in the fitness values, are the only criteria for survival. Darwin himself used the results of animal breeding and plant crossings as an important support of his theory of evolution. In this case survival is not only a result of unguided fitness because human intervention introduces new criteria into selection. The molecular counterpart of animal breeding is directed evolution. It is carried out with the goal to optimize other properties than replication rates. Such properties are, for example, binding affinities, catalytic efficiencies, specificities, thermodynamic stability or robustness against mutation. Optimization following such criteria requires a selection procedure that eliminates molecules, which do not fulfil the criteria set by the experimenter.

3.4.3 The RNA Model

In simple evolution experiments with RNA molecules genotype and phenotype are two different properties of one and the same molecule: The genotype is the sequence, the phenotype is the RNA structure with minimum free energy (mfe) and accordingly, unfolding of the phenotype is encapsulated in structure formation under thermodynamic control. Figure 3.20 presents a sketch of RNA structure formation, which occurs in two steps: (i) The single stranded molecules fold into a *secondary structure* through forming double helical stacking regions (shown in color) through base pair formation (Watson-Crick, $\mathbf{G} \equiv \mathbf{C}$ and $\mathbf{A} = \mathbf{U}$ as well as $\mathbf{G} - \mathbf{U}$ pairs),[9] and (ii) the secondary structures fold into a full 3D structure on addition of two-valent cations [TLWK01]. RNA secondary structures have physical meaning as coarse grained versions of full structures.[10] At the same time they are sufficiently simple combinatorial objects and allow for an equivalent representation as a string of chain lengths n over an alphabet with three symbols $\mathcal{C} = \{(\ ,\ )\ ,\ \cdot\}$. The parentheses denote base pairs and fulfill the common rules of mathematics, and the dots symbolize unpaired bases. Accordingly, only a subset of all strings represents allowed structures, since (i) the number of left-hand parentheses has to

---

[9] The number of lines between the two nucleotides is a measure for the strength of the interaction: **GC**>**AU**>**GU**.

[10] Coarse graining of RNA structures can also be interpreted in terms of free energy contributions since the formation of stacked base pairs is by far the largest stabilizing contribution to the total free energy of RNA structure formation.

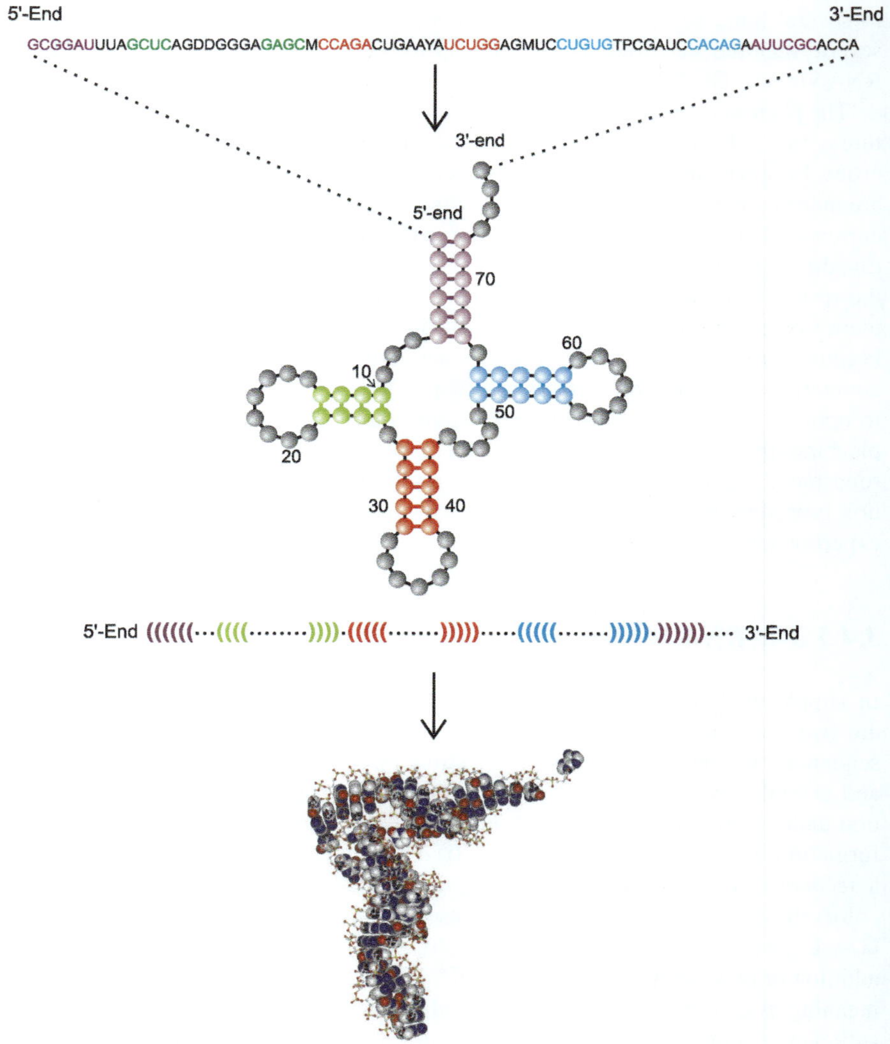

Fig. 3.20 The structure of phenylalanyl-transfer RNA from yeast. The sequence containing several modified nucleotides (**D**,**M**,**Y**,**T**,**P**) is shown on *top*. Under suitable conditions the sequence folds into a cloverleaf structure (drawing in the *middle*) called secondary structure that consists of rigid double helical parts and flexible loops and free ends. The secondary structure is equivalent to a string representing base pairs as parentheses and single nucleotides as dots. Changes in conditions, addition of $\mathbf{Mg}^{2\oplus}$ cations, completes the rigid 3D-structure (For more details on RNA secondary structures see [Sch06c]).

match exactly the number of right-hand parentheses, and (ii) no parenthesis may be closed before it was opened (e.g., "·) · · · (·" is an invalid structure). The symbolic representation of secondary structures is particularly well suited for investigations on sequence-structure relations (See Fig. 3.20). Algorithms for fast computation of

RNA secondary structures are available [ZS81, Zuk89, HFS+94] (For an on the internet server performing RNA folding see, e.g., [Hof03]).

The relation between genotypes and phenotypes is visualized as a mapping from a metric space of genotypes onto a (metric) space of phenotypes (Fig. 3.21). The space of genotypes of chain length n over an nucleotide alphabet $\mathcal{A}$ is denoted by $\mathcal{Q}_n^{(\mathcal{A})}$ and defined as follows: Every sequence of chain length n is a point in genotype or *sequence space*, and the distance between two sequences is given by

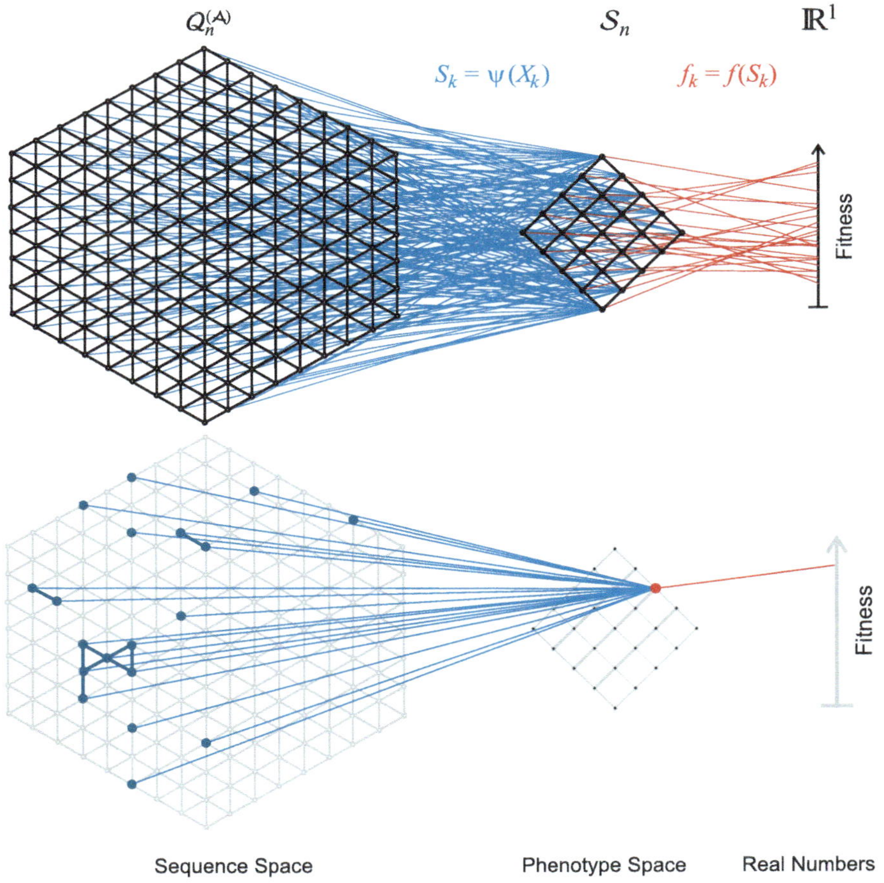

Fig. 3.21 Sketch of the maps from genotype space $\mathcal{Q}_n^{(\mathcal{A})}$ onto a space of phenotypes $\mathcal{S}_n$ and further into real numbers. Mapping sequences into structures of minimal free energy, Ψ as defined in Eq. (3.4), is many to one since for the natural nucleotide alphabets the number of sequences, $|\mathcal{Q}_n^{(\mathcal{A})}| = \kappa^n = 4^n$, exceeds the number of structures, $|\mathcal{S}_n| < 3^n$. Structures are evaluated by means of the function $f_k = f(S_k)$ (3.5). Here neutrality arises when $f(S_j) = f(S_k)$ holds for certain structures. The lower part of the figure highlights the part of the mapping concerning a single structure S_k. The preimage of the structure in sequence space is converted into its neutral network – shown in dark blue in the figure – by connecting all sequences of Hamming distance one.

the Hamming distance[11] $d_{\mathrm{H}}(X_i, X_j)$. In the RNA model the phenotypes are the secondary structures, which are represented by points in phenotype or *shape space* $\mathcal{S}_n$. Since secondary structures are equivalent to strings over the three-letter alphabet $\mathcal{C}$ it is straightforward to define a metric on shape space being, for example, the Hamming distance between the string representations of structures, $d_{\mathrm{H}}(S_i, S_j)$.

The mapping from RNA sequence space onto RNA shape space can be written as

$$\begin{aligned} \Psi &: \{\mathcal{Q}_n^{(\mathcal{A})}, d_{\mathrm{H}}(X_i, X_j)\} \Rightarrow \{\mathcal{S}_n, d_{\mathrm{H}}(S_i, S_j)\} \qquad \text{or} \\ S_k &= \Psi(X_j)\,, \; j = 1, 2, \ldots, n_k\,. \end{aligned} \tag{3.4}$$

By choosing the mfe criterion for RNA structures the assignment of structures to sequences is unique.[12] The number of sequences over the natural alphabet, $|\mathcal{Q}_n^{(\mathcal{A})}| = 4^n$, always exceeds the number of structures, $|\mathcal{S}_n| < 3^n$, where "smaller" indicates the restriction on acceptable strings mentioned above. Accordingly, the mapping Ψ is many to one and not invertible. Sequences folding into identical structures are neutral with respect to the map Ψ and give rise to *neutral networks*. In the RNA model fitness is derived from structure as expressed by a mapping from shape space into the real numbers

$$f : \{\mathcal{S}_n, d_{\mathrm{H}}(S_i, S_j)\} \Rightarrow \mathbb{R}^1 \quad \text{or} \quad f_k = f(S_k). \tag{3.5}$$

Although the relation between structure and fitness, encapsulated in $f(S_k)$, is not (yet) accessible by calculation, fitness values can be measured and obtained from plausible models. Different structures may have the same fitness value giving rise to a kind of neutrality in evolution that is superimposed upon the neutrality of sequence-structure map Ψ discussed before.

3.4.4 Modelling Evolution

Simulation of RNA structure evolution provides detailed insights into the optimization process that cannot be obtained by the currently available experimental techniques. An individual simulation run monitors the search of a population of RNA molecules for a predefined target structure by replication, mutation, and selection. The computer experiments simulate a flow reactor (Fig. 3.22), in which the material consumed by the replication process is steadily replenished by an inflow of a stock solution. The population size is regulated by the outflow of reaction mixture that compensates the volume increase resulting from the inflow. The time development

[11] The Hamming distance between two end-to-end aligned sequences of equal length is the number of positions, in which the two sequences have different digits.

[12] The minimum free energies of RNA molecules will be almost always different provided the energy model has sufficiently high resolution. Exceptions are symmetric conformations with palindromic sequences, which are rare and which shall not be considered here.

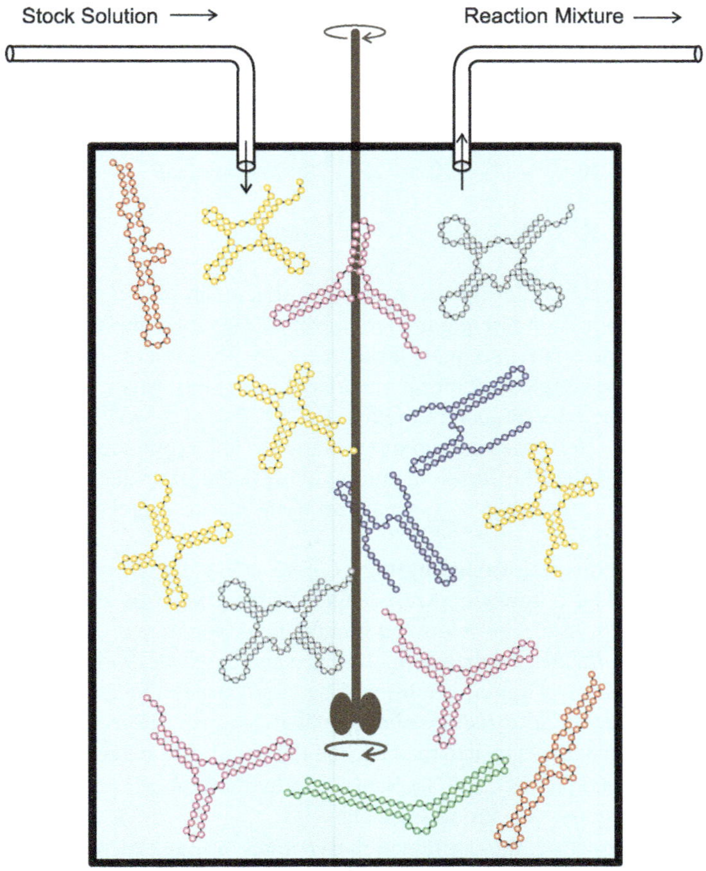

Fig. 3.22 The flow reactor as a device for RNA structure optimization. The reactor maintains off-equilibrium conditions by means of a flow, which supplies material for replication through inflow of stock solution. Molecules produced in excess are removed from the reaction mixture through an unspecific outflow. A population of N RNA molecules is subjected to replication and mutation. Minimum free energy secondary structures are computed and evaluated by means of the function $f(S_k)$ for all mutant sequences. The clover-leaf shaped yeast tRNA^{Phe} (*grey* shape in the reactor) was chosen as target structure. Inputs of an evolution experiment *in silico* are the parameters (i) population size N, (ii) chain length n of the RNA molecules, (iii) the mutation rate p, and (iv) the initial population.

of the system is modelled by coupled chemical reactions. Conventional chemical reaction kinetics is based on ordinary differential equations (ODEs). Accordingly, conventional kinetics is unable to describe properly processes at low concentrations, since it neglects fluctuations that are important at small particle numbers. This is particularly important for evolutionary optimization, because every mutation starts inevitably from a single copy. The model is therefore based on a stochastic process in many variables that is simulated by means of a straightforward algorithm developed in the nineteen seventies [Gil76, Gil77b, Gil77a]. In particular, the individual reactions are:

$$\begin{aligned} \mathbf{X}_i &\xrightarrow{f_i \cdot Q_{ii}} 2\,\mathbf{X}_i && \text{correct replication,} \\ \mathbf{X}_i &\xrightarrow{f_i \cdot Q_{ij}} \mathbf{X}_i + \mathbf{X}_j && \text{mutation, and} \\ \mathbf{X}_i &\xrightarrow{r} 0 && \text{dilution.} \end{aligned} \tag{3.6}$$

The indices "i" and "j" run over all molecular species in the reactor, f_i represents the individual replication rate parameters of the RNA molecule X_i, and the factors Q_{ii} and Q_{ij} are the relative frequencies of correct replication events and mutations, respectively. In the simplest mutation model assuming uniform error rates p per site and replication the two frequency factors are: $Q_{ii} = (1-p)^n$ and $Q_{ij} = (1-p)^{n-d_{\mathrm{H}}(X_i,X_j)} \cdot p^{d_{\mathrm{H}}(X_i,X_j)}$ for single point mutations. The replication rate parameters are modelled such that the fitness increases with decreasing distance from target, for example $f_i = 1/\big(\alpha + d_{\mathrm{H}}(S_i, S_T)/n\big)$ with S_T being the target structure and α an empirical parameter.

A simulation consists of building the average of a sufficiently large number of computed individual optimization runs called *trajectories*.[13] A single trajectory is shown in Fig. 3.23. In this simulation a homogenous population consisting of $N = 3000$ molecules with the same random sequence and the corresponding structure (S_0 in Fig. 3.24) is chosen as initial condition. The target structure (S_{44} in figure 3.24) is the well-known secondary structure of phenylalanyl-transfer RNA (tRNA^{phe}) shown in Fig. 3.20. The mean distance to the target of the population decreases in steps until the target is reached [FSS89, FS98a, Sch03]. Individual (short) adaptive phases are interrupted by long quasi-stationary epochs.

In order to reconstruct the optimization dynamics, a time ordered series of structures was determined that leads from an initial structure S_0 to the target structure S_T. This series, called the *relay series* [FS98b] is a uniquely defined and uninterrupted sequence of shapes. It is retrieved, through backtracking, that is in opposite direction from the final structure to the initial shape. The procedure starts by highlighting the final structure and traces it back during its uninterrupted presence in the flow reactor until the time of its first appearance. At this point we search for the parent shape from which it descended by mutation. Now we record time and structure, highlight the parent shape, and repeat the procedure. Recording further backwards yields a series of shapes and times of first appearance which ultimately ends in the initial population.[14] Usage of the relay series and its theoretical background

---

[13] In computer simulations different trajectories are obtained by repeating identical runs with different seeds of the random number generator. These seeds determine the sequence of stochastic events.

[14] It is important to stress two properties of relay series: (i) The same shape may appear two or more times in a given relay series. Then, it was extinct between two consecutive appearances. (ii) A relay series is not a genealogy which is the full recording of parent-offspring relations a time-ordered series of genotypes.

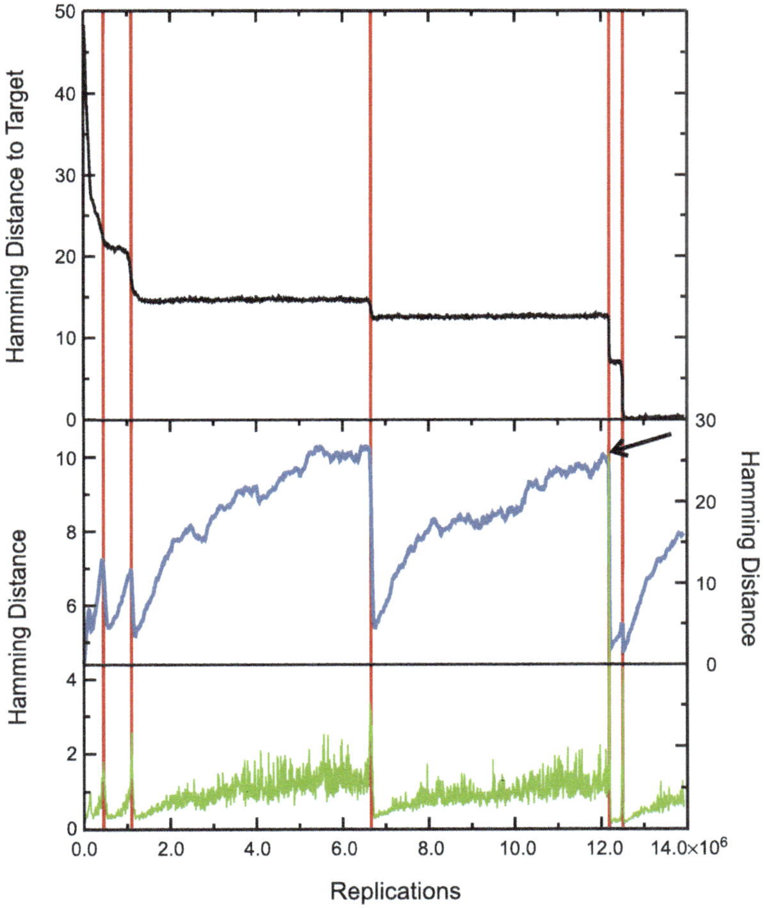

Fig. 3.23 Variability in genotype space during punctuated evolution. Shown is a single trajectory of a simulation of RNA optimization towards a tRNA^{phe} target with population size $n = 3000$ and mutation rate $p = 0.001$ per site and replication. The figure shows as functions of time: (i) the distance to target averaged over the whole population, $\overline{d_H(S_i, S_T)}(t)$ (*black*), (ii) the mean Hamming distance within the population, $\overline{d_P}(t)$ (*blue, right* ordinate), and (iii) the mean Hamming distance between the populations at time t and $t + \Delta t$, $\overline{d_C}(t, \Delta t)$ (*green*) with a time increment of $\Delta t = 8\,000$. The end of plateaus (*vertical red lines*) are characterized by a collapse in the width of the population and a peak in the migration velocity corresponding to a jump in sequence space. The *arrow* indicates a remarkably sharp peak of $d_C(t, 8000)$ around Hamming distance 10 at the end of the second long plateau ($t \approx 12.2 \times 10^6$ replications). In other words, every adaptive phase is accompanied by a drastic reduction in genetic diversity, $d_P(t)$. The diversity increases during quasi-stationary epochs. On the plateaus the center of the cloud migrates only at a speed of Hamming distance 0.125 per 1000 replications.

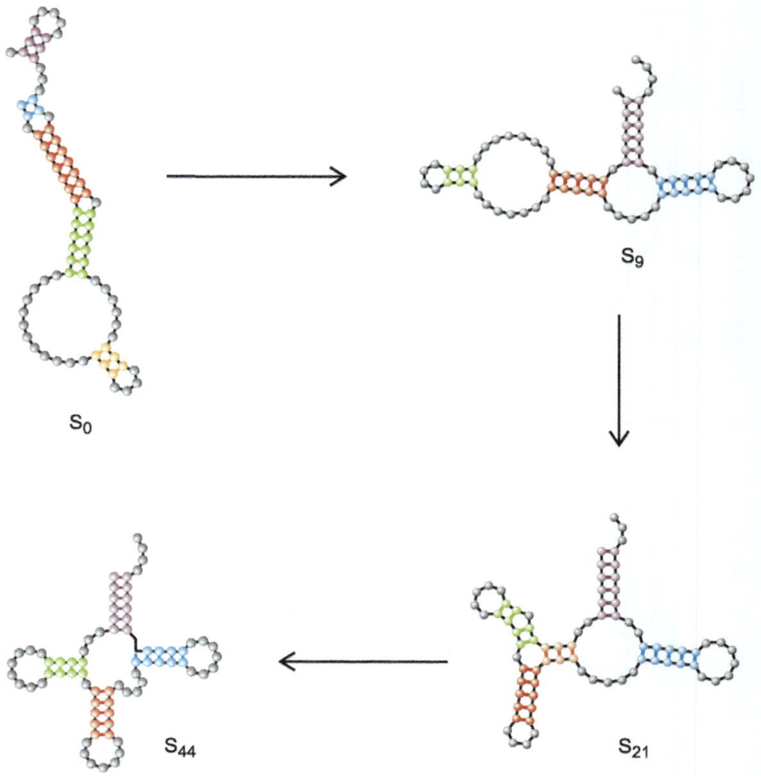

Fig. 3.24 Intermediate structures of structure optimization in the flow reactor. We show the initial structure S_0, structure S_9, which is characterized by a particularly long plateau in the trajectory, structure S_{21} that is one step before the formation of the cloverleaf, and the target structure S_{44}.

provides further insight into the evolutionary optimization process, allows to define nearness of phenotypes in evolution, and yields the basis for classification of transitions into minor and major events [FS98b, FS98a, SSWF01]. Figure 3.24 contains four selected structures of a relay series: Besides the start and the target structure, S_0 and S_{44}, we show S_9 being present for very long time in the relay series because there is no easy escape to a nearby structure with shorter distance to target, and S_{21}, which is one step before the formation of the cloverleaf through a major transition. Inspection of the relay series together with the sequence record on the quasi-stationary plateaus provides hints for the distinction of two scenarios:

(i) The structure is constant and we observe neutral evolution caused by neutrality of the map Ψ from sequences into structures. In particular, the numbers of neutral mutations accumulated on plateaus are proportional to the number of replications in the population, and the evolution of the population can be understood as a diffusion process on the corresponding neutral network [HSF96] (See also next section and Fig. 3.23).

(ii) The process during the stationary epoch involves several structures with identical replication rates and the relay series reflects a kind of random walk in the space of these neutral structures. This case corresponds to neutrality in the map f from shapes into fitness values.
Both classes of neutrality give rise to neutral evolution in the sense of Kimura's theory [Kim83].

The diffusion of the population on a neutral network or in a subspace of neutral structures is illustrated by the plot in the middle of Fig. 3.23 that shows the width of the population as a function of time [Sch03, GML$^+$06]. The distribution of the population in sequence space broadens during a quasi-stationary epoch and then sharpens almost instantaneously after a sequence was created by mutation that allows for the start of a new adaptive phase in the optimization process. The scenario at the end of plateaus corresponds to a *bottle neck* of evolution. The lower part of the figure shows a plot of the migration rate or drift of the population center and confirms this interpretation: The drift is almost always very slow unless the population center "jumps" from one point in sequence space to another point in sequence space where the sequence is located that initiates the new adaptive phase. A closer look at the figure reveals the coincidence of the three events: (i) beginning of a new adaptive phase, (ii) collapse-like narrowing of the population spread, and (iii) jump-like migration of the population center.

3.4.5 Neutral Networks

The graph derived from the subset of sequences that are neutral with respect to the mapping Ψ through connection of all pairs of sequences with Hamming distance one is the neutral network. The neutral subset also called the pre-image of S in sequence space is defined by

$$\mathbf{G}[S] = \psi^{-1}(S) \doteq \{X \,|\, \psi(X) = S\}. \tag{3.7}$$

Global properties of neutral networks may be derived by means of random graph theory [Bol85]. The most important property of neutral networks is connectedness (Fig. 3.25), which is determined by the degree of neutrality $\bar{\lambda}$. In particular, $\bar{\lambda}$ is defined and obtained by averaging the fraction of neutral Hamming distance one neighbors, $\lambda_X = n_{\mathrm{ntr}}^{(1)} / \big(n \cdot (\kappa - 1)\big)$ with $n_{\mathrm{ntr}}^{(1)}$ being the number of neutral one-error neighbors and κ size of the nucleotide alphabet, over the whole network, $\mathbf{G}[S]$:

$$\bar{\lambda}[S] = \frac{1}{|\mathbf{G}(S)|} \sum_{X \in \mathbf{G}[S]} \lambda_X . \tag{3.8}$$

Connectedness of neutral networks is, among other properties, determined by the degree of neutrality [RSS97]:

$$\text{With probability one a network is} \begin{cases} \text{connected} & \text{if } \bar{\lambda} > \lambda_{\mathrm{cr}} \\ \text{not connected} & \text{if } \bar{\lambda} < \lambda_{\mathrm{cr}} \end{cases}, \tag{3.9}$$

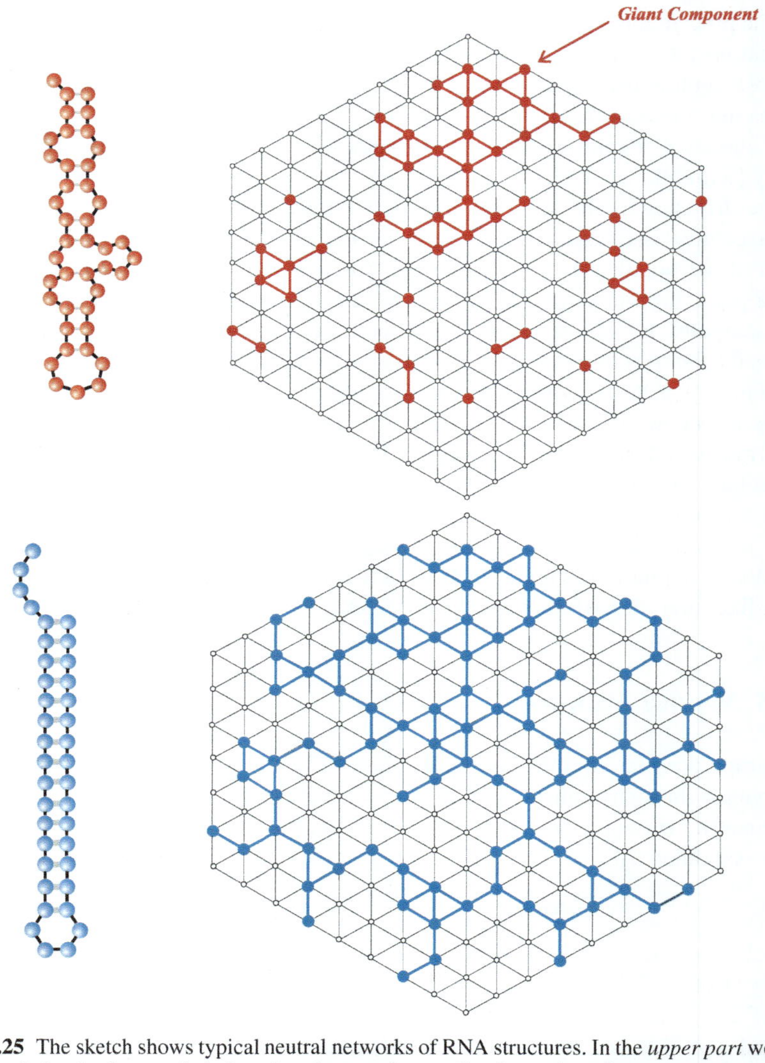

Fig. 3.25 The sketch shows typical neutral networks of RNA structures. In the *upper part* we show a network that is characteristic for a low degree of neutrality. It consists of many components and, if random graph theory applies to the particular network, one component is much larger than the others (giant component). The *lower part* shows a network for a degree of neutrality above the connectivity threshold, which is fully connected.

where $\lambda_{cr} = 1 - \kappa^{-\frac{1}{\kappa-1}}$. Computations yield $\lambda_{cr} = 0.370$ for the critical value in the natural four letter alphabet. Random graph theory predicts a single largest component for non connected networks, i.e. networks below threshold, that is commonly called the "giant component". Real neutral networks derived from RNA secondary structures may deviate from the prediction of random graph theory in the sense that they have two or four equally sized largest components. This deviation is readily explained by non-uniform distribution of the sequences belonging to $\mathbf{G}[S_k]$ in sequence space, which is caused by specific structural properties of S_k [GGS$^+$96b, GGS$^+$96a].

The existence of neutral networks is highly relevant for optimization on rugged landscapes.[15] As indicated in Fig. 3.26 adaptive walks minimizing the cost function would soon end in a local minimum. They are unlikely to reach the globally lowest point of the landscape. Depending on population size small and narrow local peaks

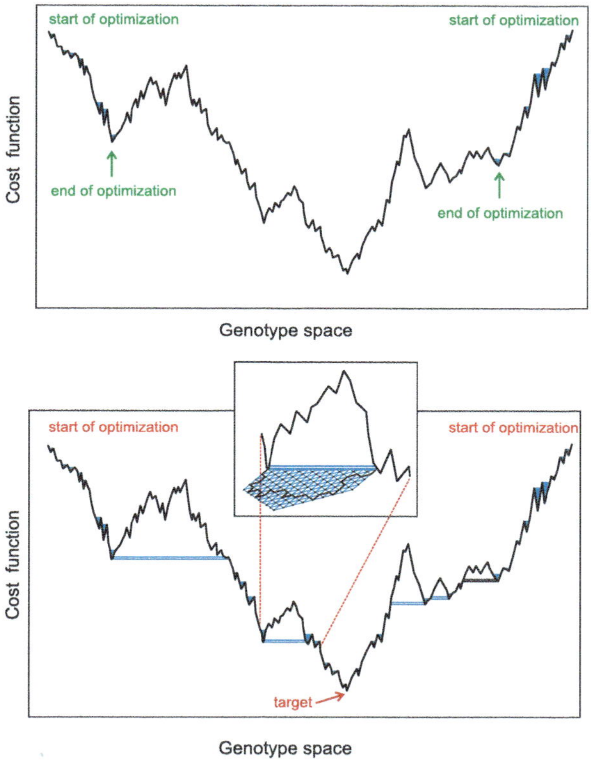

Fig. 3.26 The influence of neutral networks on optimization. Adaptive walks minimizing a cost functions on rugged landscapes soon end at a minor locally lowest point. The population size determines the widths of clefts that can be bridged by accumulation of point mutations. Commonly, this width is rather small (*upper picture*). The *lower part* shows the influence of neutral networks that allow for "tunnelling" beyond mountains through escape into another dimension where the network extends. The magnifying insert illustrates the escape into an orthogonal dimension and shows how a point that allows for further descent can be reached by a random walk on the neutral network.

[15] Rugged landscapes are characterized by a large variation of the values of the cost function in the neighborhood of almost all points. Landscapes derived from biopolymer structures are typically rugged.

can be overcome by accumulation of mutants. Higher and broader peaks, however, are unsurmountable obstacles for the optimization process. Neutral networks change this situation drastically since the sequence corresponding to the lowest point in one dimension is commonly surrounded by neutral sequences corresponding to points of equal values of the cost function. As illustrated in the figure the population escapes from the trap by performing a random walk in an orthogonal dimension.

Connectedness of neutral networks allows for migration of the population over large portions of sequence space whereas neutral drift on disconnected random networks is confined to the individual components. Simulations of optimization runs with nonnatural nucleotide alphabets that do not sustain extensive neutral networks have shown that evolutionary optimization is indeed much more difficult on sequence spaces that do not support connected neutral networks (For example on the sequence space over the {**G**, **C**} alphabet [Sch03, Sch06c]).

In Fig. 3.27 we present a sketch of evolutionary optimization on sequences spaces with extended neutral networks. The population performs an adaptive downhill walk that is characterized by a relatively fast decrease of the cost function until it reaches a point in sequence space from where no improvement is in reach by mutation. If this point is part of a neutral network the population proceeds by spreading on the neutral set (See the broadening of the population in Fig. 3.23). Neutral drift is slow compared to adaptive selection and takes place on a different timescale. In case the

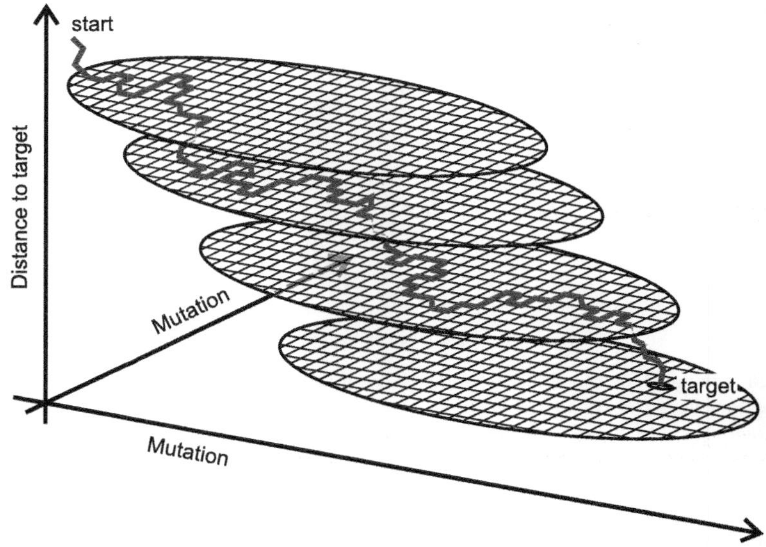

Fig. 3.27 An optimization trajectory in RNA secondary structure space. A target structure is approached in a stepwise procedure: Short periods of fast approach to the target are interrupted by long quasi-stationary phases during which the trajectory performs extended random walks on neutral networks. When a position is reached where a downhill path leads closer to the target, the trajectory leaves the neutral networks and approaches the next one. Such a position corresponds to a "tunnel" in Fig. 3.26.

population reaches a sequence that has sequences with lower const function in its neighborhood a new adaptive phase may begin, which leads to a network with a lower cost a function. The stepwise approach to the target is continued until the population eventually reaches the target or it stays confined to a neutral network from which no further improvement is possible.

3.4.6 Multiple Optimization Criteria

Structure is only one property of the phenotype of an RNA molecule. Other fitness relevant properties are thermodynamic stability, stability against conformational change, stability against mutation or existence of a second long lived conformation.[16] The goal of an actual optimization process may include several criteria for the molecule to be designed. Optimization according to multiple criteria is straightforward when we can assign importance to them in a hierarchical order (Fig. 3.28).

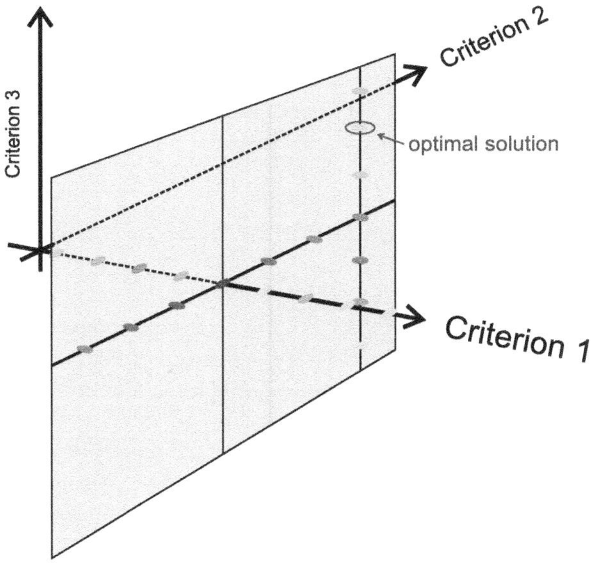

Fig. 3.28 Hierarchical optimization using neutral networks. The sketch shows optimization according to three criteria. Criterion 1 defines a large neutral network that sets the stage for further optimization. Here we assume criterion 1 to be most important or to have the highest waiting factor in the cost function followed by the criteria 2 and 3. The procedure indicated leads to the global optimum if and only if the three criteria are independent. Otherwise the procedure has to be iterated and may lead to a Pareto front. An example from RNA structural biology is discussed in the text where the three criteria are secondary structure, thermodynamic stability, and uniqueness of minimal free energy conformation.

---

[16] Thermodynamic stability means low free energy of structure formation. Stability against conformational change is tantamount to a large free energy gap between the minimum free energy structure and the first suboptimal conformation. A structure is stable against mutation when the degree of neutrality, $\bar{\lambda}$,is large.

In the design of catalytic RNA molecules with predefined functions it would make sense, for example, to assign highest priority to structure in the first step. Next, thermodynamic stability can be added as the second criterium for a search on the neutral net of the target structure and, eventually, one could introduce the requirement of stability against conformational change as the last property to be optimized. The result would be the most stable RNA sequence that forms a predefined structure with the largest possible gap between ground state and first suboptimal conformation.

Dropping or relaxing the hierarchical order of criteria leads to the well known problem in simultaneous optimization according to multiple criteria: There need not be a unique solution and one has to deal with a Pareto front or Pareto surface. For the properties of RNA molecules this problem has been addressed recently [SF03]. In reality, independence of optimization criteria will be the exception rather than the rule. Asking for the most stable molecule without the structural constraint will almost certainly yield a sequence that does not fold into the desired structure. In other words, molecules that form the right mfe structure will presumably have a higher free energy than the most stable molecule. Analogous considerations show that many other pairs of properties do not allow for independent optimization. There is also a counterexample in the realm of RNA: Efficient or fast kinetic folding and thermodynamic stability seem to form an almost independent pair of properties [Sch06c].

3.4.7 Applications of Optimization on Neutral Networks

The occurrence of neutrality with respect to optimization is by no means restricted to a world of RNA molecules. Other biopolymers – proteins, for example, show very similar degeneracy with respect to the formation of coarse grained structures. Application of the concept of optimization on neutral sets of solutions is straightforward.

A simple example for neutrality based optimization following hierarchically ordered criteria is taken from a fictive hiring strategy. We assume a very large number of applicants from which one person should be chosen. The first criterion obviously is qualification and usually a relatively large subset of candidates fulfils this requirement equally well – we may consider these people as neutral with respect to this criterion. The typical second criterion is the performance in an interview. A third criterion could concern the personal skills of the applicant, which are useful for the work the person is thought to do. If still more candidates are left, a very last criterion might be avoidance of coincidence of names in order to be able to distinguish people more easily.

Other examples that start from a large number of coarse grained, equally suited initial solutions, which are split according to secondary or higher order criteria, are readily constructed. The major question concerns the usefulness of the neutral network approach. Indeed, we could optimize according to hierarchically ordered criteria without knowing about solution spaces and neutral networks. In case independent information is available on the nature and the structure of the corresponding neutral networks in sequence space, it is straightforward to derive optimal population sizes, hints on nature and optimal frequencies of variations. Point mutations are suitable on relatively dense neutral networks whereas insertions and deletions may be required in case of sparse networks with many components. Similarly, we would be able to answer the question whether recombination is useful or not.

3.5 Employment of the Darwinian Principle for the in vitro Evolution of Molecular Structures

Jens P. Fürste and Volker A. Erdmann

RNA technologies are a fast developing field that shows until recently not foreseen potentials in molecular medicine and biotechnology [BE03, EBB06, EPB08]. This chapter will deal specifically with methods of molecular evolution in which nucleic acids (SELEX method) and proteins (ribosome display) are generated with high affinities or other new functions previously not observed in nature.

With the discovery of ribozymes by Cech [KGZ+82] and Altman [GTGM+83], who showed for the first time that RNA molecules are capable of enzymatic reactions, it became clear that the earliest organisms on earth were most likely those of an RNA world and that the central dogma of molecular biology had to be rewritten. RNA in form of a mRNA was not merely the carrier of the genetic information from the DNA to the ribosome, the protein synthesising factory of the cell, but it could virtually have also all other functions attributed to DNA and proteins.

This chapter will deal with a special aspect of RNA technologies: the in vitro evolution of RNA molecules by the SELEX method (Systematic Evolution of Ligands by Exponential Enrichment) as discovered by Tuerk and Gold [TG90a] and Ellington and Szostak [ES90a] in 1990. The method described by both laboratories allows the development of nucleic acids (aptamers) with high affinities to virtually any target in a similar way as it is known for antibodies. The basic principle of the SELEX method can also be applied to specific mRNAs coding for a protein, in order to have this protein bind to a specific target, or to alter its biological function (Hannes and Plückthun [HP97]). There are no doubts that the employment of molecular evolution for the generation of newly generated nucleic acids and proteins will have great impacts on the future developments in biotechnology, molecular biology and molecular medicine.

In principle the SELEX method permits a Darwinian evolution in a test tube in a matter of hours. In this evolution it is the experimenter who finally determines for what type of a target the selection is to be made. The selection can be for small molecules like the amino acid alanine or the explosive TNT, a protein like a tumor growth factor, a whole virus, a whole bacterial cell, or even a human cell infected by a virus. As indicated in Fig. 3.29 a nucleic acid sequence, and therefore a structure, is mutated to many different variants, usually to numbers which may reach theoretically 10^{15}. From these variants the most suitable one is selected for its affinity to a specific target or for a biological function. After isolation these

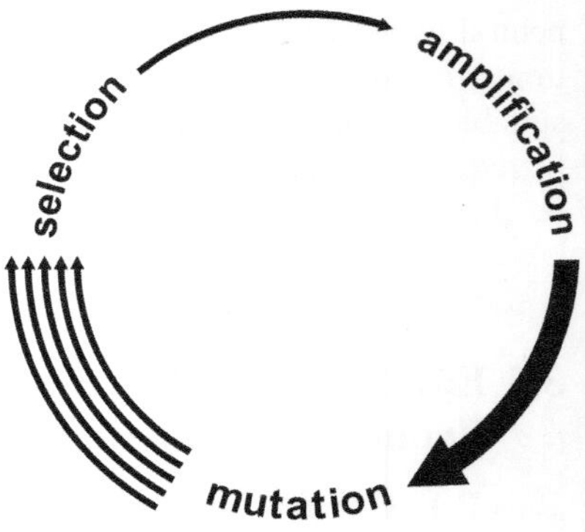

Fig. 3.29 Cyclic optimization of properties as proposed by Charles Darwin (1852). The cycle comprises three functions. First, a population (of molecules) is amplified through replication. Second, new variants with mutated sequences are generated through errors during the copying process. Third, fitting variants are selected as a consequence of limiting conditions.

molecules are amplified, mutated and again isolated according to the criteria set up by the experimenter. The cycle is repeated (usually between 10 and 15 times) until a molecule with the desired characteristics has been obtained. From this principle it is quite clear that the in vitro evolution described here in this chapter is an experimental way of producing molecules with new characteristics which did not exist previously.

3.5.1 RNA Aptamers

The power of SELEX derives from the inherent nature of RNA molecules to combine both a genotype (nucleotide sequence) and a phenotype (ligand binding or catalytic activity) into one molecule. Taking advantage of this unique property, oligonucleotide ligands with high affinities for a striking variety of molecular targets have been identified through directed evolution. The specific oligonucleotides emerging from this method have been called "aptamers" (Ellington and Szostak [ES90a]). The in vitro selection of RNA aptamers in the general SELEX scheme is shown in Fig. 3.30.

The SELEX procedure starts with chemical synthesis of a DNA library. Taking advantage of solid phase synthesis, stretches of random sequence are generated. Repeating the random cycles for n positions can produce a stochastic collection of up to $4n$ variant sequences. The total size of the average library is in the order of 10^{15} to 10^{16} unique molecules. RNA is generated from this DNA library through in vitro transcription. The pool of variant RNA molecules is then subjected to selection for a physical property of interest, such as binding to a target structure. RNA molecules that have steric and electronic complementarity to the target structure can be partitioned from nonfunctional variants by their binding properties. The selected variants are amplified by reverse transcription, polymerase chain reaction (PCR) and transcription. This process of selection and amplification is then repeated until the sequence complexity of the pool is sufficiently reduced. The specific individual

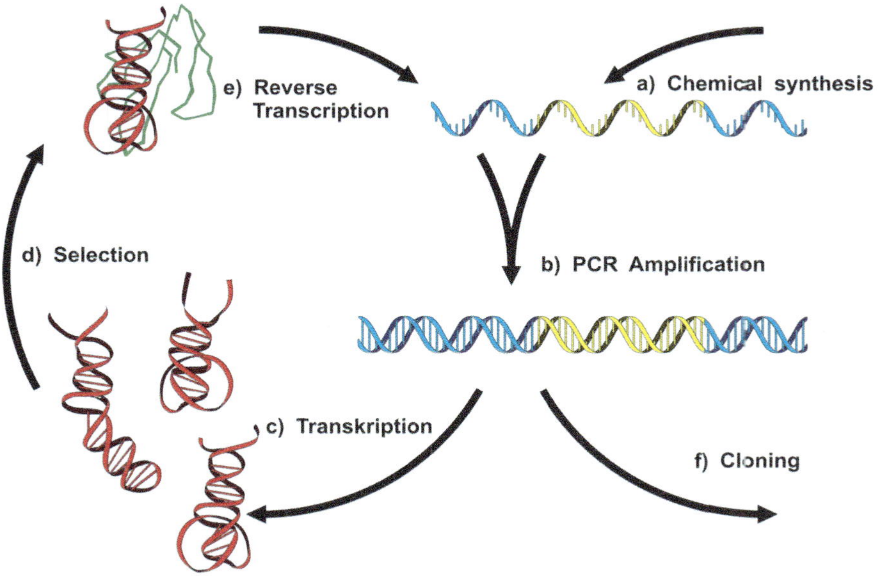

Fig. 3.30 Selection of RNA aptamers. A combinatorial library of DNA molecules is produced via chemical synthesis. The central portion contains random sequences (*marked in yellow*). The lateral areas (*marked in blue*) possess complementary sequences to align primers for amplifying the library through Polymerase Chain Reaction (PCR, step b). The DNA library is then transcribed into corresponding RNA molecules (*marked in red*) that adopt a large variety of shapes due to their different sequences (step c). The RNA library is subsequently brought into contact with the target molecule (step d). Affine RNA molecules bind to the target (*marked in green*). After separating the unbound molecules, the binders are reverse transcribed into DNA and then amplified via enzymatic reactions (steps e and b). This procedure is continuously repeated until the pool is sufficiently enriched with specific binders for the target. The individual variants are separated and identified through molecular cloning and sequencing of the PCR products (step f). This process identifies an initial binder along with dozens of other candidates. If the initial candidate falls short of expectations in screening assays, it can be quickly replaced with one of the many other candidates.

sequences of functional RNA molecules from the enriched pool are finally obtained through molecular cloning and sequencing of their DNA copies.

There are several advantages of SELEX compared to other strategies employing combinatorial libraries. First, the dual nature of nucleic acids, having a defined sequence and shape, allows to identify functional molecules from a large pool of variants. Through repeated amplification and selection the binding variants are efficiently enriched and the synthetic history of the process is simply determined by sequencing the DNA copies. This renders procedures like tagging of the library obsolete that might interfere with the binding properties of variants. Second, unlike short peptides, small oligonucleotides form stable intramolecular structures (Gold et al. [GPUY95]). The stability of short oligonucleotides is likely due to complex interactions of the standard nucleotides, such as Watson-Crick- and non-Watson-Crick-basepairs, bulges, hairpin loops, pseudoknots and G quartets. Functional oligonucleotides obtained by SELEX are often as small as 25 nucleotides and have binding affinities that rival those of large proteins. Finally, the diversity of sequences can be expanded throughout the process. Through error-prone amplification in the individual cycles, new variants from the enriched sequences can be obtained. Therefore, new RNA molecules with the desired properties are isolated through repeated cycles of variation, selection and replication, much like the process of natural evolution.

The obvious limitation associated with the application of aptamers in biotechnology or medicine is their instability in the presence of biological media. This problem has been addressed by using nucleotide analogs that render the ligands more stable against nuclease degradation. The studies focused on substitutions that are compatible with the enzymatic steps of in vitro selection. Employing 2'-aminopyrimidine ribonucleoside triphophosphates, modified RNA ligands were selected that confer similar affinities and specificities as unmodified RNA ligands(Jellinek et al. [JGBJ94]).

RNA aptamers have been selected which are specific for a variety of small molecules. The target structures consist of organic dyes (Ellington and Szostak [ES90a]), nucleotides, such as ATP (Sassanfar and Szostak [SS93]) and GDP/GMP (Connell and Yarus [CY94]), and amino acids, such as arginine (Famulok [Fam94]) and tryptophan (Famulok and Szostak [FS92]). The tryptophan-specific aptamer demonstrated stereo-specificity of the interaction. While D-tryptophan-agarose is recognized with a dissociation constant (K_d) of 18 μM, L-tryptophan-agarose is not recognized within detection limits. Lauhon and Szostak [LS95] have successfully isolated an aptamer that is specific for NMN and NAD. The aptamer discriminates between different redox states of the cofactor, having a K_d of 2.5 μM for NAD but recognizing NADH with a K_d of about 40 μM. Furthermore, SELEX has been applied to small molecules of pharmacological value, like vitamin B12 with a K_d of 90 nM, and the beta-2 agonist theophylline, with a K_d of 0.1 μM (Jenison et al. [JGPP94b]). In their selection of theophylline-specific aptamers, they applied a new method. The specificity of the oligonucleotide ligand was improved by counter selection to avoid cross-reaction with caffeine. The selection was performed by partitioning the RNA library using immobilized theophylline. Prior to eluting with theophylline, cross-reacting RNAs were removed by elution with caffeine. The resulting aptamers were able to discriminate between the two xanthine derivatives with 10-fold higher specificity than monoclonal antibodies against theophylline.

The first target of SELEX was bacteriophage T4 DNA polymerase (Tuerk and Gold [TG90a]). Since this initial application, a variety of protein-specific aptamers have been described. The targets include proteins that naturally interact with RNA. For example, Schneider et al. [STG92] raised aptamers to the bacteriophage R17 coat protein, Ringquist et al. [RJS$^+$95] obtained aptamers for ribosomal protein S1, and Gold et al. [GPUY95] successfully isolated aptamers for bacteriophage Q_β replicase. The method was applied to mammalian RNA-specific proteins. These studies primarily focused on projects with the potential to lead to the development of new therapeutics. They include the HIV-1 Rev-protein (Bartel et al. [BGS91]), the HIV-1 tat protein (Tuerk and MacDougal-Waugh [TMW93]) and the reverse transcriptase of HIV (Tuerk et al. [TMG92]). The binding constants determined for these protein specific-aptamers rivaled or exceeded those obtained for the natural RNA-protein interaction (K_d of 1–5 nM).

Aptamers have also been raised to proteins that are not known to bind naturally to RNA. The binding constants obtained were in the low nanomolar range for thrombin (Kubik et al. 1994), the IgE class of antibodies and a mouse monoclonal antibody raised against the human insulin receptor (Doudna et al. [DCS95]). Surprisingly, Jellinek et al. [JLRJ93, JGBJ94] described binding constants for basic fibroblast growth factor (bFGF) and vascular endothelial growth factor (VEGF) in the picomolar range, thus far exceeding those obtained for known nucleic acid-binding proteins.

The advent of SELEX in 1990 and the successful selection of many RNA aptamers in the following years led to a scientific enigma for over a decade. If RNA molecules carry the inherent ability to recognize virtually any molecular target, why then is nature not using this ability? Starting with the discovery of Winkler and Breaker [WNB02], however, a multitude of so called "riboswitches" have been identified in natural systems. These riboswitches comprise an aptamer domain that is responsible for binding a metabolite and an expression region that transmits the ligand-binding status of the aptamer domain through a conformational change. They can regulate gene expression either at the level of transcription or translation, and it has been proposed that riboswitch mechanisms are even used to regulate the processing of mRNA. Riboswitches are common to many different genes and organisms. Taking the knowledge of the Human Genome Project into account that less than 2% of our genetic information is translated into proteins but more than 60% is transcribed into RNA, it seems likely that yet more surprises may be just around the corner.

3.5.2 DNA Aptamers

Specific ligands can also be derived from DNA libraries, since DNA molecules are also capable to fold into the necessary complex structures (Ellington and Szostak [ES92], Bock et al. [BGL$^+$92]). The in vitro selection of DNA aptamers is shown in Fig. 3.31. The binding constants obtained for DNA ligands were in the same range as those achieved for RNA ligands. The selection for DNA aptamers is simpler, since PCR amplification is the sole enzymatic step used for the selection of functional ligands. The main difficulty arises from producing single-stranded DNA during or after PCR amplification. This has been overcome by asymmetric PCR or removing a biotinylated strand by base denaturation and subsequent separation on an avidin-agarose column. DNA aptamers might be more suitable for potential applications than RNA molecules because of the greater stability of DNA.

Ligand-binding DNAs were selected for small dye molecules (Ellington and Szostak [ES92]), ATP (Huizenga and Szostak [HS95]) and proteins, like HIV-1 reverse transcriptase, elastase and human thrombin (Bock et al. [BGL$^+$92]). The

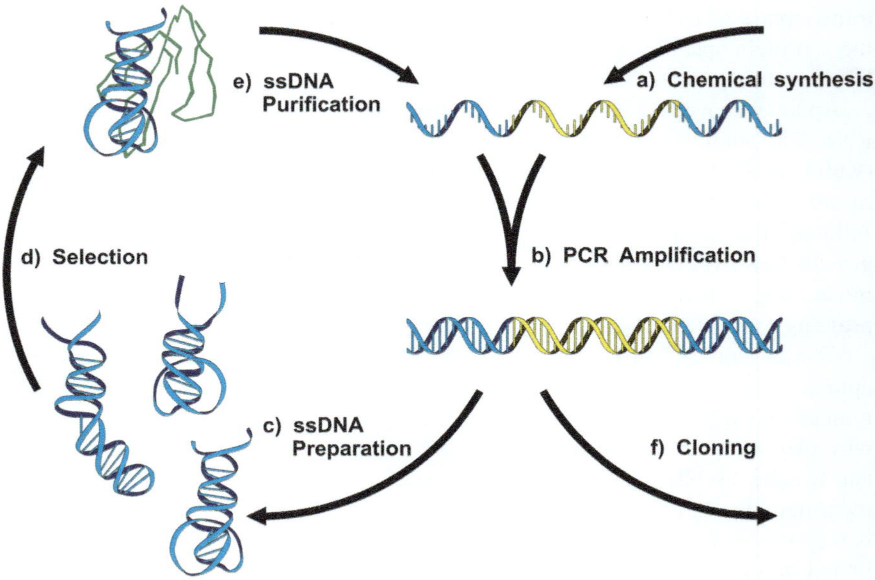

Fig. 3.31 Selection of DNA aptamers. The design of DNA aptamers resembles that of RNA aptamers (see Fig. 3.29). Again, a chemically synthesized DNA library with random sequences (marked in yellow) is amplified by PCR (steps a and b). The double-stranded library is then converted into single-stranded DNA molecules (ssDNA) that adopt different shapes due to their sequence composition, just like the RNA molecules in Fig. 3.29 (step c). After binding to the target molecule (step d), the selected variants are purified (step e) and again amplified via PCR. After several rounds of enrichment, the DNA aptamers are identified through molecular cloning and sequencing of the PCR products (step f).

thrombin-specific ligands have anticlotting activity and are thus candidates for new lead compounds. The structure of this aptamer was solved by NMR, revealing a central binding motif composed of an intramolecular G quartet. The interactions of thrombin and aptamer were also evaluated by applying surface plasmon resonance, either by attaching a biotinylated DNA aptamer to a sensor chip and measuring thrombin binding or by immobilizing thrombin on the sensor surface and monitoring the binding of aptamers.

3.5.3 Spiegelmers

The limitation associated with the diagnostic or therapeutic application of oligonucleotides is their instability in the presence of biological media. This problem has been addressed by modifications that render the RNA ligands more stable against degradation. The strategies focused on modifications that are compatible with the enzymatic steps of in vitro selection (pre-SELEX). Employing 2'-aminopyrimidine ribonucleoside triphosphates, modified RNA ligands were selected, which confer similar affinities and specificities as unmodified RNA ligands, but show increased serum stabilities. In addition, oligonucleotide ligands can be stabilized after the selection experiment by introducing chemical modifications (post-SELEX). The combination of both approaches led to aptamers with substantially improved stabilities.

We have developed a new method for the design of highly stable mirror-image ligands (Fig. 3.32, Klussmann et al. [KNB+96], Nolte et al. [NKB+96]). The approach is based on the observations of Pasteur (1861), that enantiomers display identical molecular properties. The "key and lock" hypothesis of Fischer (1894) predict that two chiral partners interact through shape complementarity and are thus stereospecific. The majority of aptamers identified by SELEX bind to chiral targets. Following the principles of stereochemistry, the mirror-image ligands (l-RNA) should recognize the enantiomers of those targets. Due to the homochirality of most biochemical compounds, such mirror-image RNA ligands would be of limited practical importance. There is, however, the option to raise an aptamer for the mirror-image of an intended target molecule. Applying the rules of symmetry, the mirror-image RNA ligands do consequently bind to the natural target molecule.

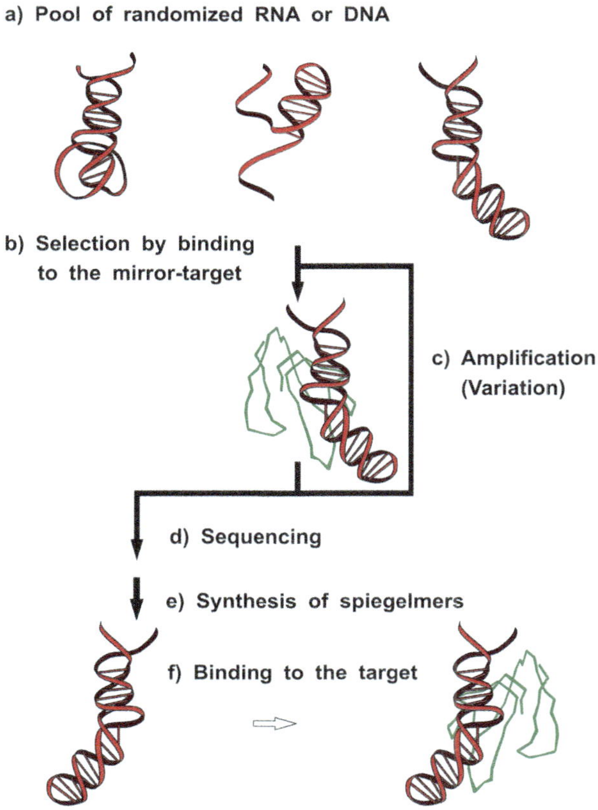

Fig. 3.32 Design of Spiegelmers. Oligonucleotides (*red*) with high affinities for the mirror image of a target molecule (*green*) are selected from a combinatorial library. The mirror image oligonucleotide (spiegelmer) is then able to bind to the target molecule in its natural handedness.

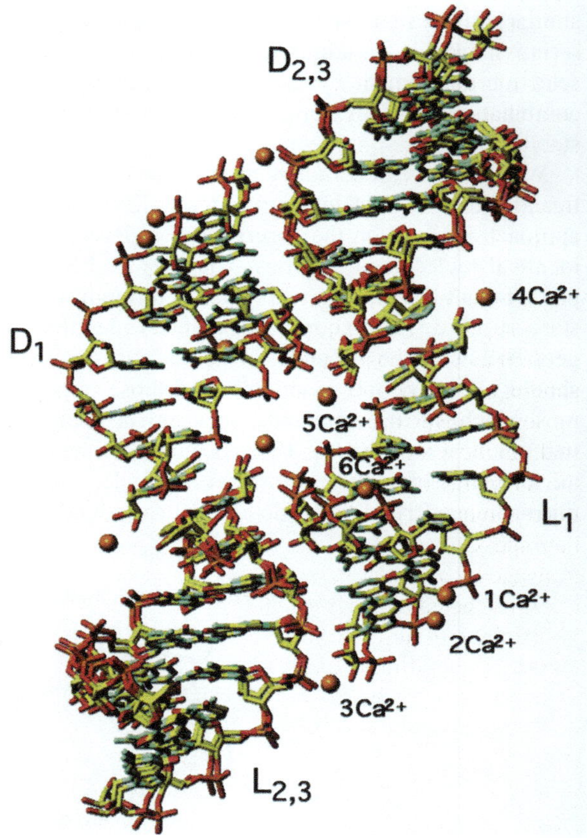

Fig. 3.33 First Spiegelmer Structure. The figure shows a natural RNA with its mirror image at high resolution (1.3 Å), formed by the racemate of the RNA duplex r(CUGGGGCGG). r(CCGCCUGG) from Thermus flavus 5S rRNA.

Spiegelmers and their building blocks, the mirror-image nucleotides, have never been found in nature. They are thus not cleaved by nucleases. In addition, since they are chemically similar to natural nucleic acids, they are non-toxic and non-immunogenic. These properties render spiegelmers as promising candidates for future pharmaceutical developments. In order to understand the structures and therefore the functions, we have crystallized the first spiegelmer as shown in Fig. 3.33 (Rypniewski et al. [RVP$^+$06]).

3.5.4 New Ribozymes

Ribozymes are catalytic RNAs. The name derived from a combination of RIBOnucleic acids and enZYMES. Catalysis by biopolymers was originally considered to be only performed by proteins. Even though early proposals on the catalytic function of RNA date back to 1964 (Rich [Ric62]) it was not suspected until the findings of T. Cech and S. Altman that catalytic RNAs still exist as functional molecules in present days biosphere (Kruger et al. [KGZ+82], Guerrier-Takada et al. [GTGM+83]). Even though there are indications for a possible catalytic involvement of ribosomal RNA in peptide bond formation (Noller et al. [NHZ92]), the well characterized activities of natural ribozymes are restricted to splicing, transesterification and cleavage of phosphodiester bonds. It has been shown that metal ions are essential for their activities. In particular, the activities of the group I intron from Tetrahymena, the Rnase P and several small ribozymes, like hammerhead and hairpin ribozyme, are completely dependent upon the presence of divalent metal ions. Yarus [Yar93] proposed in his conjecture that RNAs and proteins can fix metal ions in space with similar versatility and that RNAs should therefore be able to perform any reaction also carried out by protein metalloenzymes. Ribonucleic acids might thus be able to function in such diverse mechanisms as electrophilic catalysis, general acid/base catalysis or the catalysis of redox reactions.

Several investigators have successfully modified the activity of natural ribozymes, such as the group I intron from Tetrahymena (Robertson and Joyce [RJ90]), and the hairpin ribozyme (Berzal-Herranz [BHJB92]). Furthermore, the combinatorial library approach has been applied to select new ribozymes from random libraries. Pan and Uhlenbeck (1992) identified RNAs that catalyzed Pb2+-dependent self-cleavage by a similar mechanism as described for ribonucleases. Bartel and Szostak (1993) selected an RNA catalyst capable of ligating a nucleoside triphosphate to the 3'-terminus of an oligonucleotide, thus mimiking the basic mechanism of a polymerase. By modifying an ATP-specific aptamer, Lorsch and Szostak (1994) isolated a new ribozyme with polynucleotide kinase activity. In addition, RNAs were identified that catalyze aminoacylation (Illangsekhare et al., 1994) or the cleavage of an amide bond (Dai et al. [DMJ95]). The catalytic potential of nucleic acids, however, is not restricted to RNA. Breaker and Joyce [BJ94] described a new deoxyribozyme, capable of cleaving the phosphodiester bond of an oligonucleotide. Most recently, Cuenoud and Szostak [CS95] reported the isolation of a DNA metalloenzyme with even DNA ligase activity. Since new ribozymes can be selected for arbitrarily chosen reactions, they might supplement the set of available catalysts for meaningful applications in the near future.

3.5.5 Ribosome Display

Another in vitro evolution methodology is referred to as the ribosome display (Lamla et al. [LME01]). This technology is based upon the in vitro ribosomal synthesis of proteins to which the element of in vitro evolution has been added. The general principle of this method is that the mRNA coding for a specific protein is statistically varied in any part of the protein coding sequence, so that the characteristics of the proteins can be altered. The changes can be many fold and they can include functional changes of an enzyme, or the development of new enzymatic or structural functions, or even affinities to just about any target imaginable.

One of the major advantages of the in vitro protein synthesis is the fact, that just about any protein can be synthesised in vitro and that the in vitro synthesis of proteins permits also the synthesis of toxic proteins, proteins selectively labelled with isotopes, and the introduction of unnatural amino acids at specific positions in the protein (Lamla et al. [LSE02], Lamla and Erdmann [LE03, LE04]). As an example for the large spectrum of possibilities in which the in vitro protein biosynthesis can be used, we would like to show its potential in the ribosome display experiment in which the elements of molecular evolution are added, so that a protein gains, for example, a high specificity and affinity for streptavidin (Lamla et al. [LSE02], Lamla and Erdmann [LE03]). In nature streptavidin is interacting with a small compound called biotin. This interaction is so specific and stable that it is widely used in molecular biology and biotechnology as an affinity system for the isolation of proteins and other molecules. As a matter of fact these interactions between biotin and streptavidin have also been employed in different types of biochips.

In order to fully comprehend the potentials of the ribosome display method a brief description of protein synthesis has to be made. As can be seen in Fig. 3.34 protein synthesis requires that the information for the sequence of a protein encoded in the DNA is transcribed into a messenger RNA (mRNA). Theses mRNAs are divided, according to the genetic code, into so called triplets, each one representing one of the 20 different amino acids, the building blocks of proteins. The triplets are made up of three of the four possible nucleosides (building blocks of RNA molecules), which are abbreviated with the large letters A, C, G, and U (Fig. 3.35). For example as can be seen in this figure the triplet UUU will code for the amino acid phenylalanine and CGA for arginine.

It is the role of another type of RNA molecules, the transfer RNAs (tRNAs), to bind specifically one of the 20 amino acids and to deliver this amino acid to the ribosome. The ribosome is the very complex organelle consisting in bacteria of three

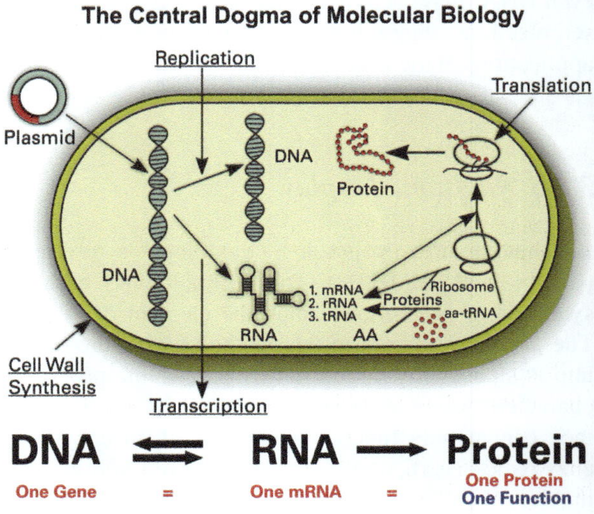

Fig. 3.34 The Central Dogma of Molecular Biology illustrated in a bacterial cell. During the growth phase of the cell three important reaction phases take place, which are the replication phase in which the DNA is duplicated, the transcription phase in which the DNA sequence is translated into mRNA molecules and in the translation phase in which these mRNAs are translated to proteins with the aid of the ribosomes.

The Genetic Code

First Position (5' end)	Second Position U	Second Position C	Second Position A	Second Position G	Third Position (3' end)
U	Phenylalanine	Serine	Tyrosine	Cysteine	U
U	Phenylalanine	Serine	Tyrosine	Cysteine	C
U	Leucine	Serine	Stop	Stop	A
U	Leucine	Serine	Stop	Tryptophan	G
C	Leucine	Proline	Histidine	Arginine	U
C	Leucine	Proline	Histidine	Arginine	C
C	Leucine	Proline	Glutamine	Arginine	A
C	Leucine	Proline	Glutamine	Arginine	G
A	Isoleucine	Threonine	Asparagine	Serine	U
A	Isoleucine	Threonine	Asparagine	Serine	C
A	Isoleucine	Threonine	Lysine	Arginine	A
A	Methionine	Threonine	Lysine	Arginine	G
G	Valine	Alanine	Aspartic Acid	Glycine	U
G	Valine	Alanine	Aspartic Acid	Glycine	C
G	Valine	Alanine	Glutamic Acid	Glycine	A
G	Valine	Alanine	Glutamic Acid	Glycine	G

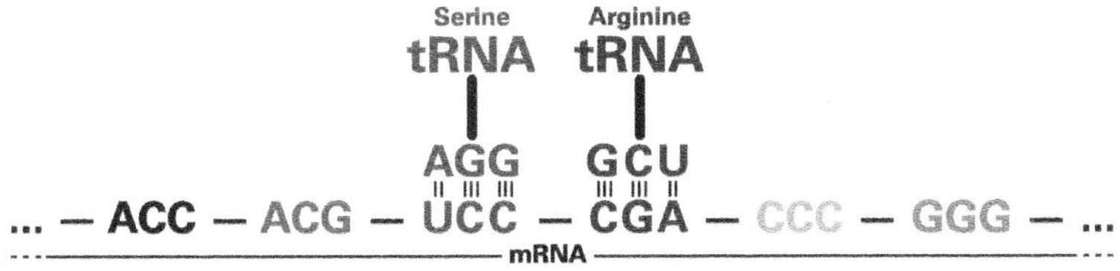

Fig. 3.35 The genetic code. The synthesis of proteins requires the translation of the mRNA sequence into a protein sequence. In order to accomplish this, the mRNA is divided into triplets (sequences of three nucleosides). Each triplet can be assigned, according to the genetic code, to one of the 20 amino acids, the building blocks of proteins. As can be seen in the table of the figure the triplet CUG corresponds to the amino acid leucine and AAC to the amino acidasparagine. On the bottom of the figure the process is shown by which the ribosome decodes the mRNA with the help of aminoacyl tRNAs, which carry their specific amino acid to the ribosome. As indicated in the lower part of the figure, the tRNA carries at one end the amino acid and at the other end also a triplet sequence, which is complementary to a triplet sequence on the mRNA. Thus a highly specific interaction by Watson-Crick base pairing can take place, between the triplets of the mRNA and the triplets of the tRNAs. An enzyme called the peptidyl transferase (also a ribosomal RNA, and a ribozyme), has the function to covalently link one amino acid with the next one, so that the synthesis of proteins is insured.

different ribosomal RNAs (rRNAs) and 35 different ribosomal proteins. The total molecular weight of this RNA-protein complex is approximately 3 million Daltons. The ribosome is responsible for synthesising proteins in the cell. That the correct protein sequences are synthesised is guaranteed by the genetic code in which the triplets from the mRNAs are interacting, according to Watson-Crick base pairing rules, with a complementary sequence in the "anticodon loop" of the corresponding tRNA. The so called Watson-Crick base pairing rules require that the nucleotide A can base pair by hydrogen bonding with U and G with C (Fig. 3.35). For example the triplet UUU, coding for the amino acid phenylalanine, will base pair with a AAA sequence belonging to the phenylalanine specific tRNA. The synthesis of a protein is started by a special initiator tRNA (using the AUG triplet) and it is terminated by so called stop codons (UAA, UAG, and UGA) for which no corresponding tRNAs exist.

If the ribosome display is to be used to develop a new high affinity tag at the so called N-terminus of a protein, the following has to be done, as shown in this case in Fig. 3.36. As can be seen the normal mRNA, which codes for the Fatty Acid Binding Protein (FABP), will be extended by 45 nucleotides corresponding to 15 additional triplets at the 5' end of the coding sequence of the mRNA. Since the nucleotide sequence is in principle generated by the SELEX method, see above as illustrated for aptamers, statistically 1014 different mRNAs are made, each varying only in the variable region from one another.

After the different mRNAs have been translated, the protein synthesising reaction is stopped in such a way that the complexes are retained consisting of the ribosomes, the 1014 proteins synthesised, and the corresponding mRNAs (Fig. 3.37). These complexes are then exposed to an affinity column or magnetic beads containing the

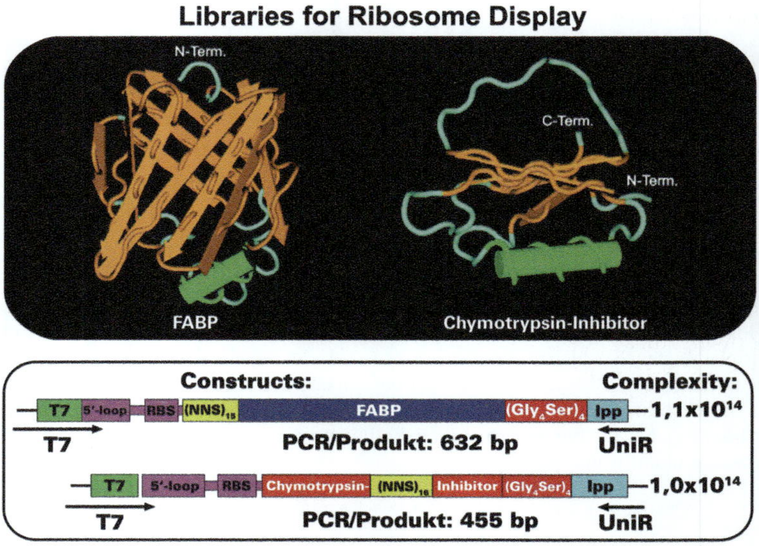

Fig. 3.36 Design of protein libraries for ribosome display. Development of protein libraries to add peptide sequences at the N-terminus (fatty acid binding protein) or internally (chymotrypsin inhibitor) of a proteins in order to develop proteins by molecular evolution with new structural and functional characteristics. The construct for FABP with a complexity of 1 × 1014 was used in the experiments described in the text and the following figures.

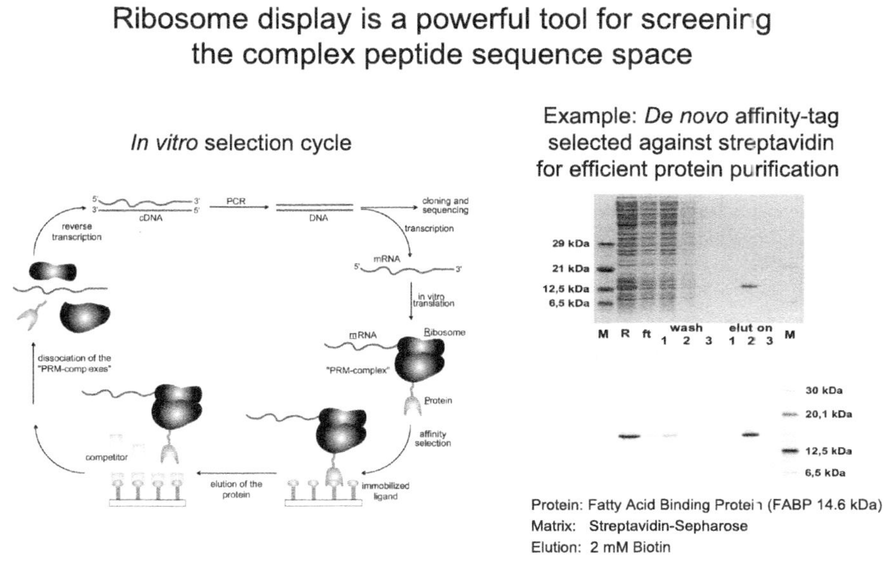

Fig. 3.37 NANOTAG development by ribosome display. On the left side of the picture protein synthesis is shown from the mRNA library generated by molecular evolution. The NANOTAG attached to FABP being part of the FABP-mRNA-ribosome complex is isolated by affinity selection towards streptavidine bound to magnetic beads or an affinity column. The right side of the figure shows a polyacrylamide gel electrophoresis and the high potentials of using the NANOTAG to purify or isolate a protein, in this case the fatty acid binding protein.

covalently attached target. In our case the target chosen was the protein streptavidine. Since most of the proteins synthesised by the ribosomes will be exposed on their surfaces, specific interactions between some of these proteins and the target will be possible. Therefore the ribosome-protein-mRNA complexes can be isolated. Next the mRNAs from these complexes are isolated and retranscribed into DNA. After amplification of the DNA by PCR (see aptamer section) the DNA can be sequenced and from that sequence the sequence of the protein can be deduced applying the rules of the genetic code (Figs. 3.35 and 3.36).

Following these procedures, we have isolated a fatty acid binding protein to which at its N-terminal sequence 15 amino acids where attached with very high binding affinities towards streptavidin. Our best tag, called NANOTAG exhibited a dissociation constant of 10-9 M. In Fig. 3.37 the results are shown, when this tag is used to isolate or purify a protein, in this case the fatty acid binding protein.

In order to understand, on the molecular level, the interactions between the NANOTAG and streptavidin, we attempted to crystallize these complexes in order to determine their three dimensional structures by x-ray analysis. As illustrated in Fig. 3.38 the x-ray analysis permitted a structural characterisation at 1,15 Å resolution. At this very high resolution the atomic interactions of a seven amino acid long peptide, corresponding to the first 7 amino acids of the NANOTAG, were shown to bind at the same binding pocket at which the natural biotin-streptavidin interaction

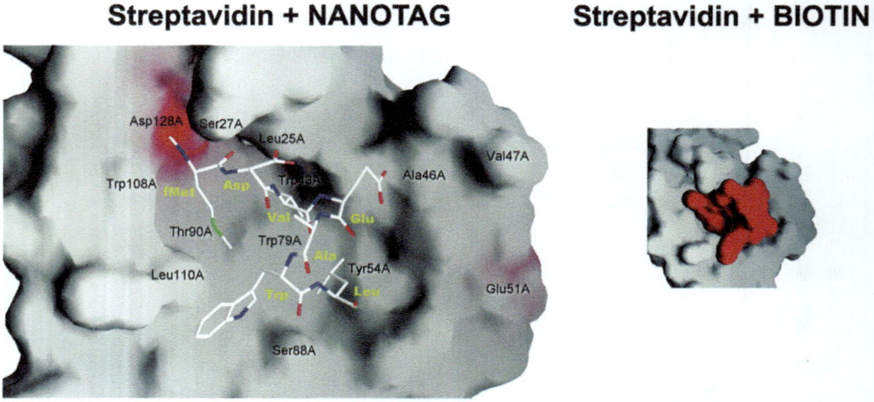

Fig. 3.38 NANOTAG-streptavidin interaction at atomic resolution. The high affinity of the NANOTAG peptide sequence towards streptavidin is also demonstrated by the ability to crystallise the NANOTAG-streptavidine complex at very good quality so that its atomic structure could be determined by x-ray analysis at very high resolution (1.1 Å). In comparison the crystal structure of the complex between biotin and streptavidin is also shown and it demonstrates that biotin and the NANOTAG compete for the same binding pocket on streptavidin.

is taking place. The knowledge of these interactions combined with methods of bioinformatics is now applied to design peptide tags with even higher affinities to the streptavidin protein. The information gained by these studies will also make a step forward to develop the techniques to learn how to design protein structures with specific biological functions.

3.5.6 Summary and Outlook

As presented in this chapter the in vitro evolution represents a powerful tool to develop new types of molecules for the benefit of biotechnology and medicine. This has been documented by the examples of aptamer and spiegelmer development and the employment of ribosome display for the design of new proteins. It is of interest to point out that the in vitro evolution experiments described here follow a Darwinian principle, in which the pressure for a certain selection is governed by the experimenter in the laboratory.

As discussed, the SELEX selection permits the in vitro development of 1015 different high affinity nucleic acid molecules and that these molecules can be successfully directed just about against any target, which may range from small molecules like amino acids, antibiotics or even TNT, to proteins, or living cells, and even living cells, which may be infected by a virus. Once the in vitro developed RNA or DNA molecules have been characterised they can be chemically synthesised by an automated nucleic acid synthesizer. The stability problems of nucleic acids in human serum have been solved by the development of Spiegelmers, which are mirror images of the common nucleic acids. Since these Spiegelmers are not made in nature, there have been no enzymes developed in evolution to degrade these molecules. These high affinity nucleic acids are very similar to the well known antibodies, but they have several advantages, which can not be matched by the antibodies. To these advantages belongs the fact that nucleic acids are not immunogenic, that they can be reproducibly synthesised in any laboratory around the world and that they can be extremely stable in the living cell if they are chosen in the form of Spiegelmers.

Following the basic principles of the in vitro selection of high affinity nucleic acids described above a method has been developed, which has been termed as ribosome display. In this method mRNAs are statistically varied, for example in 15 codons, so that a pool of 1014 different mRNAs will be developed, which after translation in an in vitro protein synthesising system will nearly yield as many different proteins. Once these proteins have been synthesised they can be selected for unique structural or functional features. In this chapter we described as an example the isolation of a high affinity peptide, called NANOTAG, which when attached to the fatty acid binding protein could specifically interact with streptavidin with dissociation constants of 4 nM, which equals those observed for biotin a natural molecule binding to streptavidin.

Thus, the chapter summarised the newest techniques to develop high affinity RNA, DNA and protein molecules. Clearly these methods are urgently needed in molecular medicine, for the development of new diagnostic and therapeutic tools, because the current methodologies have been proven to be as not satisfactory to solve the medical problems of today. To look into the long term future, we foresee that the accumulation of more structural and functional knowledge generated by these powerful in vitro evolution methods will finally lead to a situation, in which the molecular design of any molecule with desired structures and functions will be accomplished by bioinformatics (see Sect. 3.4).

3.6 Optimal Structures of Residential Heating Systems

Peter Roosen

3.6.1 Problem Definition and Scope

Developing innovative space heating concepts necessitates consideration of various factors influencing the pragmatic target function "warm, comfortable housing space". Additional goals of the individual user must be considered, e.g. the accepted energy consumption, the costs conjoined with that, the flexibility of utilization, and many more. Besides the purely technical development of a respective heating system it is thus necessary to identify a suitable application niche relative to already established systems: It is not to be expected that a new design will outperform existing ones in every respect. There are many application profiles and boundary conditions for the requested space heating demands. Prior to an extensive development of a novel solution the primary advantages need to be identified.

For every heating system there are several configuration parameters (such as area proportions, insulation thicknesses, temperature levels, operation control concepts) exhibiting specific effects on the usually contradicting pragmatic target functions, such as overall costs, cosiness, primary energy consumption etc. The targets are affected differently, depending on the foreseen usage scenarios (such as constant vs. episodic usage of a flat, modulating usage of a utility building vs. stochastic usage of an event hall), clothing habits (strict business dress code vs. casual wear) and unchangeable parameters for a given building such as the window area ratio. Therefore the optimal system configuration has to be determined individually with respect to the potential relative target function contributions.

Varying the relative target function weighting and monitoring resulting major system parameter changes, such as a switching of the basic heating system type, yields a notion on *decision stability*. A stable decision in this sense is a constant selection of one certain system type even in the light of slightly changing subjective target function weightings.

The method of choice for identifying dominant and stable ranges of a heating system is the pareto optimization. The set of optimal solutions for two or more partial targets, the pareto set, is determined, covering all possible relative weightings of the considered partial targets. Due to the fact that the individual target function value is calculated independently from any other and not traded against them there is no need to fix an a-priori weighting. Instead, the targets are pursued simultaneously and independently.

In the scope of this exemplary study the following *target values*:

- a comfort measure
- the primary energy consumption

for the *simulated space heating variants*:

- gas fired floor heating
- gas fired wall radiator heating
- dynamic electric wall heating (EDDH, electrodynamic direct heater)

are considered. The analysis is carried out for three *usage scenarios* and a floor area of 78 m^2:

- Continuously used flat (living room, sleeping room, kitchen, hall) with differentiated usage profiles and respective heat gains in the different rooms,
- same flat with same relative usage profiles, but pure weekend usage (Friday from 18:00 hours and Sunday 18:00 hours),
- open-plan office with same ground area, 15 computer work places and flexitime usage (workstart in the interval 7:30–8:30 hours, combined with respective distributed work end in the afternoon).

The diversity of usages and the respective inner gains with their varying amounts and temporal distributions will show the differences in heating systems performance and their selective advantages for those tasks.

3.6.2 Simulation and Optimization

The aim of the reported work was to identify effects of specific usage scenarios and boundary conditions on the achievable optimality of a heating system with respect to the target functions "energy consumption" and "comfort". The desire to obtain significance for practical purposes led to the conclusion that the underlying system simulation should be as realistic as possible. Several energetic simulation systems, each of them cumulating several ten years of development times, are generally suitable for this purpose. A re-implementation of such a simulation core is usually prohibitive due to the complexity of such a task. Taking an existing complete simulation package usually leads to the fact, though, that there is no possibility to interact with inner structures of the simulative calculations from the outside, via the user interface. It is rather necessary to treat the simulation package as a whole as a kind of "black-box" performer that may respond to parameterized simulation task definitions with its specific way of result response. The response, typically made available in the form of files, has to be scanned and parsed for significant target function information by the optimization procedure.

The simulation system ENERGYPLUS (e^+) [Dep03b], provided by the US Department of Energy, seems especially suitable for optimizing the energetic and comfort behaviour of residential space heating systems. It combines a very great detailedness of simulatable effects with a relatively simple system specification language in input files that may be artificially created and/or modified by a custom optimization module. A simulation task may be invoked by a command line call, referencing a respective pure ASCII input file of some 200–300 kB in size. The input file contains the individual settings of (a part of) the building to be simulated, such as geometries, geographic orientation, wall structures, usage profiles, efficiency values for heating devices and such. They are defined by comma-separated lists depending in their structures on precursory keywords, relating to each other. Simulations are performed for real weather data sets defined with respect to temperature, humidity, and irradiation data for numerous cities of the world. Consequently, solar irradiation gains through windows are considered in detail, with their geographic orientation being taken into account. Here, a weather data set of Düsseldorf/Germany is chosen for the optimization calculations documented below.

Some configuration parameters, while being just simple numbers to the optimizer module, require a somewhat detailed intermediate treatment for their transfer into the simulation module. As an example, the shift in an onset time of a heater needs to be reformulated into a completely rewritten 24 hours scheme in order to make e^+ recognize the change. For this task the macro preprocessor m4 [Sei00] was used. It creates the actual simulation input file with valid 24 hours schemata by changing a template containing respective macros.

Upon request e^+ creates very detailed load profiles, energy flux reports etc., and also time-resolved comfort calculations according to the Fanger model [Dep03a] that are written into respective output files. In order to provide the optimization module with the necessary scalar target function values for each previously defined system setting the output files have to be pre-processed in an appropriate fashion. For this task a Python script was put to work.

3.6.2.1 The Fanger Comfort Model and its Adaptation for Optimization Purposes

The Fanger comfort definition, being one of the most commonly adopted, is based on the six factors air temperature, humidity, air speed, mean radiant temperature (MRT), metabolic rate and clothing levels. (For an elaborate, in-depth discussion on Fanger's and other comfort models see [RdD97]) Fanger's model is based upon an energy analysis that takes into account all the modes of energy loss from the body, including: the convection and radiant heat loss from the outer surface of the clothing, the heat loss by water vapor diffusion through the skin, the heat loss by evaporation of sweat from the skin surface, the latent and dry respiration heat loss and the heat transfer from the skin to the outer surface of the clothing. The model assumes that the person is thermally at steady state with his environment. It was derived from experiments with American college age persons exposed to a uniform environment. The comfort equation establishes the relationship among the mentioned environment variables, clothing type and activity levels.

However, the equation only gives information on how to reach the optimal thermal comfort by combining the variables involved. Therefore, it is not directly suitable to ascertain the thermal sensation (in terms of discomfort) of a certain person in an arbitrary climate where the variables do not satisfy the equation. Fanger used the heat balance equation to predict a value for the degree of sensation using his own experimental data and other published data for any combination of activity level, clothing value and the four thermal environmental parameters.

Even though the comfort model is based on objective thermodynamical interdependencies there is a substantial influence of subjectivity to the experimentally observed results (see also Sect. 2.5.4). Accordingly test persons disagree to a limited extend on the experienced comfort, necessitating a statistical approach to represent their sensation as an averaged value. Therefore the Predicted Mean Vote (PMV) was defined that is expected to arise from averaging the thermal sensation vote of a large group of people in a given environment. The PMV is a complex mathematical expression involving activity, clothing and the four environmental parameters (see [Dep03a]).

The thermal sensation/discomfort is expressed as an index value, derived from the commonly used seven point psycho-physical ASHRAE scale, as summarized in Table 3.2. A PMV between -1 and 1 is regarded as the comfort zone.

The interpretation of the Fanger PMV comfort value provided by e^+ needs some additional comments. In contrast to the discrete model originally devised by Fanger the comfort value returned by the simulator is a continuous one. But the general

Table 3.2 Thermal sensation PMV scale after Fanger, using a seven point psycho-physical ASHRAE scale

Sensation	Hot	Warm	Slightly warm	Neutral	Slightly cool	Cool	Cold
ASHRAE scaling	1	2	3	4	5	6	7
Fanger PMV values	−3	−2	−1	0	1	2	3

idea is preserved by modeling coldness discomfort notions as negative, hotness discomforts as positive values. In order to attribute comfort values to heating system configurations the hour-wise and room-resolved *absolute* Fanger values – in the sense of a directionless deviation from the optimal sensation – are averaged after weighting them with the number of inhabitants present in each one-hour period. Hence the goal of the optimization process is to strive for a positive bound value of zero as comfort measure. Unavoidable periods of overheating during summertime limit the lower bound of it to values well above zero, though, as the investigated spaces are only equipped with heating, not climatization devices. Underheating and overheating periods in unpopulated rooms do not contribute to discomfort as there is no person present to notice it.

The inhabitants' comfort assessment is closely coupled with their (simulated) clothing thickness. Changes in the range from light summer casual wear to thick winter pullovers yield significant differences in valuation of otherwise identical thermal environment conditions. For practical purposes it is therefore not sensible to assume constant clothing conditions: It is definitely more realistic to assume the typical dynamical behaviour of the inhabitants to partially undress or add pieces of clothing within the respective socially controlled limits if slight notions of discomfort are experienced. Such an adaptive clothing schema is not feasible in e^+, though. Here a fixed hourly schema for a clothing thickness is to be provided.

To achieve a practically realistic statement a calculational trick has been implemented. The number of inhabitants of every room is halved, with one half being clad as light as possible and the other half as thick as possible, referring to the socially accepted clothing rules under the given circumstances. As the ideal comfort value is zero, a change in the sign of the Fanger value for the two halves is taken as an indicator that there exists a clothing thickness in the foreseen range providing maximum comfort. Accordingly, for that interval the practical comfort target function value is set to zero, designating the optimal one in the positive definite range of comfort assessment absolute values. If the signs of both halves are same, either the thinnest clothing is still to warm or the thickest is still not warm enough to obtain the maximum comfort. In that case the lowest absolute value will be taken as target value, assuming that the inhabitants will adopt their clothing to the nearest socially acceptable level. The latter changes significantly in context with the simulated scenario: The variation span for private living conditions is definitely larger than the one for the bureau business condition.

3.6.2.2 Optimizer Adaptation

Since the simulator's internal organisation and the functional structure of the target function in its dependence on the available configuration parameters remain hidden

Fig. 3.39 CPU load curves of four of the cluster of seven computers working in parallel. The dead times of individual units appear statistically during the closure runs of each generation, recognizable as small dips in the respective activity displays.

to the user's scrutiny, the optimization module that is to cooperate with the simulator should react as insensitively as possible against non-differentiabilities and unsteadiness. Therefore a custom developed evolutionary optimization system was put to work [Roo99]. It provides a dynamic list population management, a Pareto set oriented multicriterial assessment scheme and could quite simply be parallelized with respect to the invoked simulator runs by using the PVM system [PVM, GS92]. This simple parallel execution of up to seven simulation runs on separate TCP/IP network coupled computers was sufficient in the context of the given task as is shown in Fig. 3.39. Without changing the relatively inflexible generation paradigm the dead times imposed by idling processors at the end of a generation's simulation runs do not add up and reduce the overall calculational speed too much: The ratio $1/2 \cdot CPU\ count/individual\ per\ generation$, representing the statistical idle time per generation, is simply too small. The calculation time of one simulation run is typically about 40 sec (simulator machines' performance category: Celeron 1 GHz). Accordingly the speed loss due to necessary information transfer times was negligible for the less than 1 kB communication requirements per simulation invocation. Even when using more powerful processors the described concept will still be favourable and applicable without larger performance losses.

3.6.3 Heating System Comparison for Different Usage Scenarios

All scenarios were optimized for the same basic boundary conditions of floor area and geographic orientation. Furtheron it is assumed that identical usage profiles apply for adjacent units above and below the calculated one, so no net heat transfer is calculated via floor and ceiling. The residential units are simulated as four zone models (see Fig. 3.40: living room, sleeping room, kitchen, hallway without heating). The bureau is rendered as one zone (open-plan office).

According to the different usage profiles some hourly scenarios were introduced: The forced venting by outside air (infiltration) was significantly reduced in times of absence. The temperature level was accordingly adapted for bureau (Mo – Fr) and

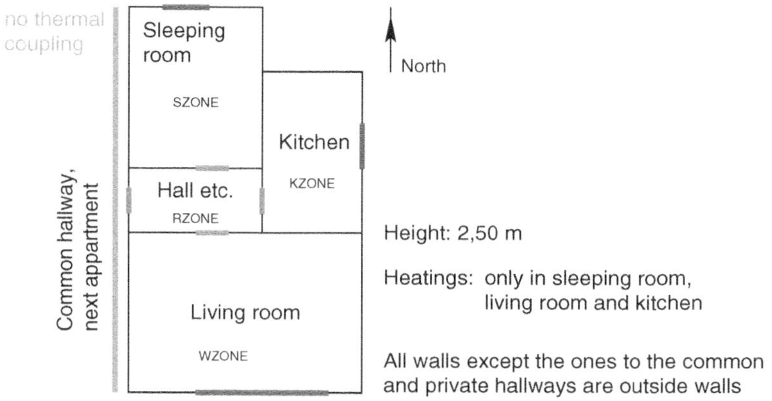

Fig. 3.40 Structure of the residential units, both continuous usage and weekend use scenario. There is no heat exchange via floor, ceiling, and adiabatic walls to next flat on same floor.

weekend flat (Fr – Su) usage. The temperature settings for nightly drawdowns were applied during the other times.

Since the different heating systems exhibit very different heating dynamics – with the floor based heating being known as exceptionally slow – different pre-heating times have been introduced as optimizable parameters. So competitive comfort values can be reached, possibly conjoined with significantly increased energetic inputs, though.

The following configuration parameters were set free to be changed by the optimizer within reasonable limits:

3.6.3.1 Technical Section

Thickness of inner walls. This value describes the thickness of the walls between adjacent rooms inside the calculation region. Since massive walls act as heat buffers this value has direct influence on the temperature characteristics resulting from short-term stochastic heat flux changes (solar irradiation through windows, ventilation events).

Thickness of wall insulation. The outer face of the facade is equipped with a respective heat insulation, interpreted as a polyurethane layer.

Maximum heating power. This value limits the energy output of the heating system, limiting the energy consumption as well influencing the comfort.

Heat transfer coefficient of the radiators. This value only exists for hydraulically operated heating systems and describes the heat transfer coefficients for the individually simulated rooms. In the case of floor heating the numerical value describes the total length of heating tubes which, for a specified width of the tubing, is a direct representative of the heat transfer coefficient.

3.6.3.2 Performance Relating Section

Temperature settings (regarded as constant heating control system targets over the whole simulated period)

Daytime temperatures in individual rooms. Temperature targets of living room and kitchen (together) and sleeping room (separate) are specified. The hallway is only passively heated by its energetic coupling to the other rooms. Living room/kitchen presets are interpreted as setting for the bureau one zone modelling.

Nighttime temperatures in individual rooms. Same specifications as for daytime temperatures, but given for the drawdown times. Same treatment in the one zone model, like with daytime temperatures.

Begin of daytime operation. This is, like the other temperature settings, interpreted as time of day at which the heating is switched to daytime operation; living room and kitchen.

Stop of daytime operation. Time of day when nighttime operation starts.

Preheating period. This period denominates the advance switch-on interval for the weekend usage simulation, to achieve a sufficient comfort level at the time of personal presences.

Absence temperature. Prevention of freezing and the required preheating period upon arrival is influenced by this temperature being set during times of absence.

So all scenarios contain at least 10 configuration parameters that are subject to optimization.

Continuously inhabited flat. The continuously inhabited flat accomodates a varying number of people. Following the assumption that the sleeping room is practically uninhabited during daytime, while the kitchen and the living room is not used during night time, there are periods of potential temperature lowering and hence energy consumption reduction without loss of comfort. Depending on the characteristics of the heating supply the rooms react differently if one tries to put this saving potential to work. The simulation periods contain two weeks per season.

First of all, Fig. 3.41 shows clearly differing Pareto fronts for the individual systems. Each colored "curve" in reality consists of numerous individual points, each representing a certain parameterized set of design variables. A same coloring designates a same basic heating type, with no further indication of the settings of the other configuration parameters. Along a front they change quasi-continuously, producing a visual envelope of the relative tradeoff of comfort vs. primary energy consumption.

To aid curve interpretation the EDDH curve (red) shall be discussed in short. On it, we may observe a point with a comfort value (Fanger PMV) of 0.2 and a energy consumption of approx. 8000 kWh/a. According to the optimization results there is no attainable EDDH configuration within the bounds of the configuration parameters that would yield both a better Fanger value and a smaller energy consumption. If we wish to reduce the energy input we must accept an inferior (and hence higher) Fanger value, leading us upwards to the left to the next red point with

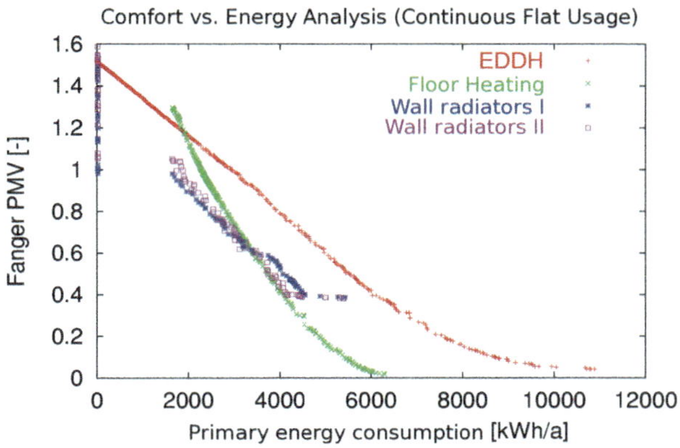

Fig. 3.41 Comparison of the Pareto fronts of three space heating systems with respect to comfort and primary energy consumption, under the condition of a continuous but differentiated usage of the available rooms. For discussion see text.

it's respective configuration parameter settings as potential candidate. Accordingly, there is a tradeoff curve for each basic heating design. Due to the different dynamic responses of the basic designs their curves may well penetrate each other, indicating that at the intersection point another design takes the lead in the global, combined Pareto front.

While floor heating and EDDH develop continuously bent curves the radiator heating seems to exhibit only piecewise steady ones that strongly jump at a Fanger value of about 1. This jump coincides with error messages of the simulator that indicate a stability problem of the respective simulator runs. Those cannot be detected by the optimizer evaluator, though. The very small energy inputs observed in that range of the Pareto front indicate a non-converged mode of the simulation. Therefore the wall radiator results should only be considered at Fanger values less than one. The practically wrong results have been deliberately retained in the result display as they should be taken as a pointer to a rather important issue in simulation-based numerical optimization: The optimization process itself may drive the simulator into parametric regions where it defects and delivers unreliable results without the tutoring human to notice it. So, in complex and unclear situations additional inspection facilities beyond the pure target function values should be provided to help the supervisor classify the conciseness of the calculated results.

Furtheron the figure shows two separate wall radiator optimization run result sets to highlight the statistical behaviour of results. Both runs yield the "ideal" Pareto front in varying parts and stay inferior in others. Here the phenomenon of optimizational stagnation is evident: with evolutionary optimization there is no proof that the optimum really has been approximated. Repeated optimizations are mandatory, and even then there is no security on how far the real optimum frontier is still away, even if in most cases some asymptotic behaviour is likely.

Apart from the technical aspects of optimizing simulation, the comparative results yield also practical insights: In the continuously inhabited flat the unpopulated intervals in the different rooms seem to be too short to introduce an observable primary energy consumption effect from the rather rapid and dynamically operating EDDH, or the wall radiation system in a reduced amount, relative to slow floor heating appliance. Even though the EDDH end energy input (not shown in the diagram) is less than those of the two competing systems, it is dominated by the more conventional heating systems over the whole investigated range due to the conversion losses in the production of the electrical energy (assumed effectivity of 40%). The flat seems best to be kept on a kind of quasi-static temperature level to reach the best possible tradeoff curve. The general, mean heat losses through the outside walls and windows are not evened out by inner gains and dominate the summed-up consumption picture.

Flat with weekend usage. The partitioning of the weekend flat is identical to the continuously inhabited one. During its times of usage the same inner gains are assumed, but only from Friday 18:00 h to Sunday 18:00 h. During the rest of the week all temperatures are set to the nighttime operation level, the forced ventilation is set to a very low rate, and no inner gains are assumed. The comfort rating is performed as usual: The absolute Fanger value is weighted with the number of inhabitants. This leads in effect to a non-rating over the working-days part of the week. As the flat may severely chill out just before the weekend inhabitance starts again a freely choosable preheat time is provided to achieve sensible comfort values at the onset of each inhabitance period. This preheat time is a parameter in the performed optimization, leading to both increased cosiness and increased energy consumption and thus to a conflict in goals fulfillment.

Since the few usage periods in the simulated observation times of the continuously used room systems would presumably lead to erroneous results the complete simulation time was raised to 200 days per year, again assuming the weather reference data of Düsseldorf, Germany, provided by the e^+ system. This adaptation yields a number of comfort-assessed days comparable with the continuous usage

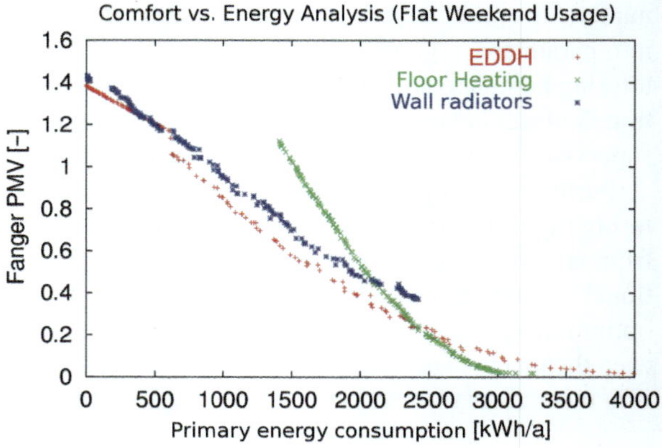

Fig. 3.42 Comparison of the Pareto fronts of three space heating systems with respect to comfort and primary energy consumption, under the condition of a weekend usage. See text for discussion.

situations. Nevertheless it is important to consider the intermediate, unpopulated days as well: Their settings influence both primary energy consumption and achievable comfort levels during inhabited periods.

In the weekend usage mode the electro-dynamical direct space heating concept can show its advantages: It takes a leading position in most parts of the Pareto tradeoff curve and surpasses the radiator solution in any case (Fig. 3.42). Only in the high-comfort (but accordingly also high energy input) domain the floor heating concept is slightly superior. In the Pareto range around 1500 kW/a most EDDH parameter sets show a very short preheat time prior to the 18 hours arrival time of the inhabitants (17 hours and later). This suffices due to the small thermic masses that need to be warmed up to obtain reasonable comfort values.

Bureau situation. Relatively large inner gains are present in the bureau situation, assumed with the same floor area as the residential usages: An occupation density of 15 computer workplaces with the system switched on during normal bureau times is assumed. Accordingly a remarkable (dis-)comfort baseline is observed in the Pareto sets of all heating concepts (Fig. 3.43). The heat produced by inner gains cannot be eliminated during summer days since no climatization, but only ventilation is assumed and simulated to preserve a certain level of comparativity. In addition, in the bureau environment the clothing order is somewhat more restricted with respect to very light clothing, leading also to a raised level of discomfort during hot summer days.

The relative positioning of the heating systems is somewhat similar to the results for the weekend used flat: While the floor heating concept dominates the range of high comfort (and high energy consumption) the low energy input (and combined with that, the low comfort) range is the domain of the EDDH. The radiator wall heating concepts exhibits an interesting distribution: Its Pareto set mostly coincides with the dominated parts of the other two concepts, without ever being a leader itself.

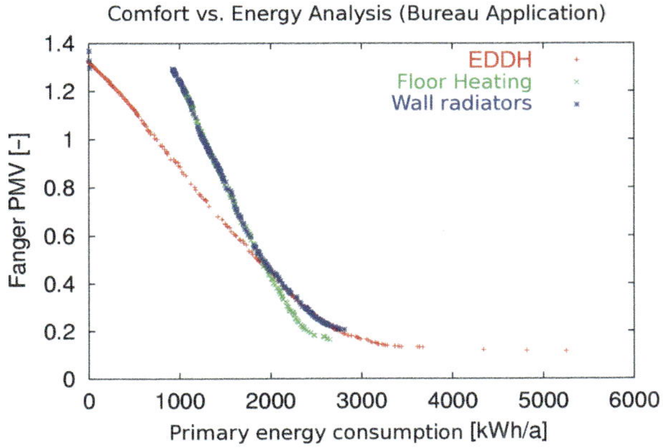

Fig. 3.43 Comparison of the Pareto fronts of three space heating systems with respect to comfort and primary energy consumption, under the condition of a bureau usage. See text for discussion.

3.6.4 Interpretation of Results

The results allow some insights into the potential, but also the limitations, of space heating system optimization. First of all, the anticipated outcome is confirmed. "There is no such thing as a free lunch" – or in other words: For each system in question one has to balance comfort requirements with an appropriate (primary) energy input, although the amount of the latter may be systematically different across heat provision system types. Each heating system produces its own Pareto front in the comfort vs. energy input plots.

There is no general leader system completely dominating other existing concepts. Depending on the structure of the heating request – mainly continuous, intermediate with relatively long on/off periods, or short periodic changes as three example situations – the advantages and disadvantages of the supply systems become visible.

In a continuously populated residence the determining factor for primary energy consumption is the heat loss through walls and windows that has to be balanced by almost steady heat input during phases of cold weather. Here the overall conversion efficiency of the heat provision system plays the most important role for the primary energy input calculation. Since the production of the very high energy carrier medium "electricity" from primary (fossil) energy coincides with relatively high conversion losses, this in turn leads to a general energetical inferiority of ohmic heat generation when the same levels of comfort are to be reached. In the continuously used flat there are no major dynamical effects the highly adaptable electric heat production can utilize to lessen this general disadvantage.

The weekend use of a flat is almost a complete antithesis. Long chillout periods during the absence of the residents lead to the necessity to heat up relatively large volumes of thermically capacious, heat absorbing material. Here the established systems of radiators or floor heating need long preheat times in order to get the living space itself sufficiently cozy. After the inhabitants once again leave the flat, the heat energy mainly introduced into the building's walls slowly dissipates into the environment without any further benefit. Contrasting to this behaviour the EDDH system mainly heats the space while leaving the surrounding walls mainly in their cold states during the rather short usage periods. So the total input heat energy per weekend is significantly smaller than with the other systems. This effect overbalances the conversion losses.

The slow start profile of the scenario, due to the slow crowding of the workplace according to the simulated flexitime usage, once again seems to promote the relatively slow floor heating system, at least for sufficiently high comfort levels. The dynamic effect that would presumingly promote the EDDH seems not large enough to overcompensate the general electrical conversion loss on the primary energy target function.

Although these results are quite convincing and can well be interpreted in terms of general scientific concepts, they rely strongly on the modeling input flexibility of providing the simulation model with practically relevant detailedness – and therefore on the effort in setting up the model. Some of those limitations shall be scrutinized in more detail below.

One major aspect of an EDDH application is the *very* fast response of the comfort value on any control action performed to the actual heat input into the room. This response is in the order of minutes, while the simulation system will only yield reasonable comfort values on the basis of 10 minute steps. A faint approximation is the assumption that each change in population of the simulated space, and any change of inner gains, just coincides with full hours for which the time schedules must be defined for the e^+ simulator. A more detailed simulation would require both many more time steps and a simulation system explicitly taking into account dynamic heat distribution processes in the surrounding walls. This in turn would multiply the required simulation time which was already at its limit for the available hardware.

e^+ provides some automatisms for adapting the input heating energy into the room, as well as for the venting in case of overheating etc. Although these procedures are already quite elaborate with respect to realistic scenario modeling, it does not suffice to describe pragmatic personal actions. Just to give some examples, what is *not* sufficiently described: (i) When people populate a flat the doors between rooms tend to be left open in case of rather similar temperature distributions while they are usually closed more rapidly when temperature gradients occur. (ii) The comfort feeling in a surrounding with a temperature gradient – both in time and space – is not covered, as the Fanger model does not support this boundary condition. Even if people feel slightly uncomfortable due to a slightly too cold surrounding this feeling is somewhat compensated if the surrounding tends to get warmer. (iii) If there are different temperatures in different rooms a person wandering frequently between them will have a different sensation compared to his or her stay in just one of them. If a room is used only for short times (e.g. some minutes in each case), but this rather often, a discomfort value attributed to this room will in practice be less perceived in comparison to a person staying in that environment for a longer time.

There is one additional aspect that cannot be covered even by the best possible situation-aware modeling: The potentially changing room assignment in a given complete space. Exchanging a bedroom against a living room, or splitting a large living room into smaller sections that become somewhat autonomous in their usages, changing cooking behaviour – such as switching from conventional cooking with large inner heat gains to microwave ovens with almost no external heat production, or exchange of slow bureau computers to faster but more energy demanding ones: Those changes in usage cannot be foreseen, so another pragmatic optimization target should be kept in mind as very important for most cases: the usage flexibility.

Trying to integrate this aspect leads to the frontier of modeling complexity and the possibility to interpret the obtained results, as there is no objective method presently in sight to economically (with respect to modeling setup) take this into consideration. The only resort that may address this problem might be the qualitative comparison of the magnitude of calculated scenario differences and their attribution to the probability of happening scenario switches.

3.7 Biochemical Engineering: Knowledge Based Process Design

G. Schembecker, K. Bauer

3.7.1 Introduction

Biochemical processes become more and more important. Especially in the pharmaceutical industry, fermentation products are a valuable source for new therapeutics. Also in the food and feed industry, the production of alcohols, amino acids and such products is more and more based on biochemical reaction steps.

Designing those industrial biochemical production processes is a typical task for a chemical or biochemical engineer. Interestingly, the decision making process during the development of the production plant is quite similar to the one a doctor faces choosing the right treatment for a patient. A decision has to be made based on few hard facts only (e.g. fever, blood pressure) as the patient usually provides qualitative information only (I feel sick) and a full analysis or examination would cost too much time and effort.

Process designers face the same situation. They have to take decisions with tremendous impact on the economic performance of the future production plant based on vague information just applying their experience. Usually, the starting point is an idea generated and tested in laboratory, and some experimental data on the reaction is available. In order that the idea becomes an operating industrial-scale plant, many decisions have to be taken within a comparatively short time. These cover the selection of an appropriate reactor and the selection of the separation steps including the definition of favorable operation parameters (temperature, pressure) for every operation. As even within a relatively simple process there are many different unit operations, constructions, possible operating points, and interconnections, it is easy to see that thousands of variables have to be dealt with. There are discrete and continuous variables both dependent and independent. As there is usually more than one possibility to affect the economics of a technical process, the result is a complex optimization problem, which can not be solved rigorously. Even the simple question whether it is better to have higher conversion or selectivity in a reactor is extremely difficult to answer and may even lead to contradictory solutions.

In addition, at the beginning of a process development the knowledge about the system is quite limited and mainly of qualitative nature. Even if a laboratory process already exists, a lot of information needed for designing and operating an industrial plant is still missing such as property parameters of the components, reaction kinetics, color, odor and stability of the products, or the effect components present in traces only have on the process performance. However, the decisions, which have to be made in the beginning, narrow down the degrees of freedom for the later process design. Thus, they strongly influence the efficiency of the final process. The relationship between process efficiency (here cost and environmental impact) and knowledge is illustrated in Fig. 3.44. If it were possible to enhance the process knowledge available during early development ($A_1 \longrightarrow A_2$), the impact on the performance of the resulting production process would be large ($B_1 \longrightarrow B_2$).

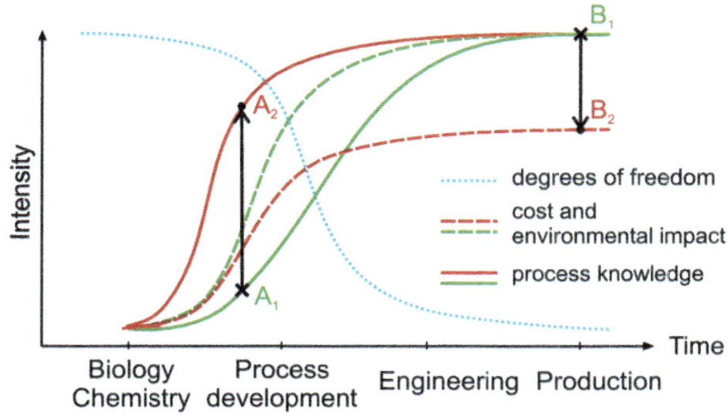

Fig. 3.44 Process knowledge and efficiency; Heinzle and Biwer [HB07], modified

3.7.2 Design Methodology

The design of a chemical or biochemical process presents a complex task. A key challenge is to deal with a number of uncertainties on the one hand but to come up with a technically feasible and economically viable process concept on the other hand. Among other sources, such uncertainties may relate to

- economical issues such as uncertain prediction of future sales or unknown business plans of the competition or
- technical issues such as incomplete knowledge about the reaction (e.g. by-product portfolio), unknown physical properties especially for complex mixtures or unknown impact of process parameters on product properties.

Another key challenge is presented by the nearly infinite number of potential process flowsheet options. Therefore, there is a need for a design methodology which

1. constrains the solution space as much as possible,
2. quickly comes up with good process options and
3. excludes non-feasible or non-economic solutions as soon as possible.

One possibility to cope with these requirements is to use heuristic rules (experience expressed as rules of thumb) combined with a well structured decomposition approach. Such a combination of several design methods allows finding reasonable process concepts in reasonable time. The methodology tool box and its implementation in so-called expert systems will be described in the following chapter.

3.7.2.1 Decomposition

According to systems theory a complex design problem can be decomposed into sub-problems of less complexity in order to reach a manageable problem size. For the conceptual design of (bio)chemical processes such a decomposition is described in Fig. 3.45. On the first level, the whole process is regarded as a black box, which converts some input streams into the desired products and separates the contaminants from the product stream (output streams). The performance of the whole

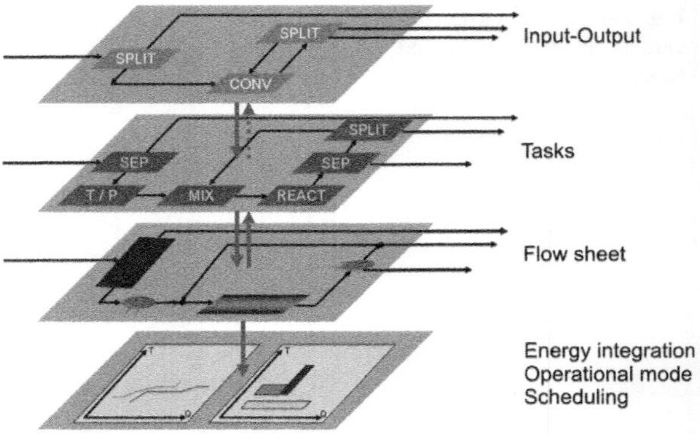

Fig. 3.45 Decomposition; Schembecker [Sch06a], modified

process is set and constraints concerning the products are defined, such as the purity that has to be achieved. Knowing the product prize, it is also possible to calculate in which range the production and investment cost may lie, which has an influence on the later choice of separation steps.

On the second level, the tasks level, the whole process is divided into subfunctions that are necessary to achieve the performance set on the previous level. Normally, there is at least a reaction (or in biochemical processes a fermentation) step and as many separation steps as needed to get the desired product in desired quality.

These tasks, which have to be carried out, are further specified on the flowsheet level. Based on property differences between the components and streams, unit operations are selected and combined resulting in a flowsheet structure.

Finally the process is further optimized with regard to energy consumption, operational mode of the unit operations (batch, semi-batch, continuous) and, if some parts of the process are operated in batch mode, scheduling.

3.7.2.2 Heuristic Rules

On each level of the decomposed problem according to Fig. 3.45, decisions have to be taken based on vague information. For this task heuristic rules are widely used in the development of chemical production processes. In those rules of thumb, knowledge of experts in a certain area is combined and abstracted so that it can be transferred and applied to new problems. As most of the knowledge is of rather qualitative nature, the rules normally give hints and recommendations on which unit operations to choose in which sequence, while the results in terms of mass or energy balance can not be predicted quantitatively. A heuristic approach always results in numeric or experimental comparative studies of the process alternatives achieved.

Heuristic rules carry most often uncertainties with them, may be contradictory and are valid depending on the context only. Nevertheless, it is possible to find a solution even if partial information on the characteristics of the system is available only. Such rules can be used to define the structure of a chemical process. An example may give an idea about the decision making procedure. Take the following two (simplified) rules:

1. Condensation seems to be an economic option for the separation of a process stream if the normal boiling point of some components in the mixture is substantially higher than cooling water temperature and if the normal boiling point of some other components in the mixture is substantially lower than cooling water temperature.
2. Absorption seems to be an economic option for the separation of a process stream if the normal boiling point of the components with low concentration in the stream is substantially lower than of the other components in the mixture.

These rules allow to decide on the selection of a separation process based on qualitative information only. The most important reason leading to the decision is a significant difference in boiling temperature. Moreover, the rules express that condensation and absorption are competitive unit operations. They also show the difference of the two unit operations. Condensation has an advantage over absorption if a significant separation can be achieved by using cooling water. Absorption might be a good alternative in case an impurity must be separated from a stream. In some cases both rules might be fulfilled and present two alternatives which cannot be ranked just using heuristics. Nevertheless, it becomes clear that heuristic rules for the structural design of process flowsheet implicitly cover economic relationships.

3.7.2.3 Expert Systems

Expert systems, i.e. computer programs, in which human problem solving methods are modeled, are using heuristic rules which are integrated in a problem decomposition environment. An expert will narrow down the search by recognizing patterns and solve the process synthesis problem based on experience, common sense and intuition. In an expert system, that knowledge is stored in heuristic rules in an "if – then" format very similar to the rules shown above.

For the development of chemical production processes, several expert systems exist. As an expert system alone is not sufficient to find an optimal solution, they are often combined with numerical calculations as in the READPERT system, which is used to design and select chemical reactors [SDWS95]. READPERT needs information about the kinetics of a chemical reaction. Such information may be available as numeric expressions. However, in most cases, qualitative data will be available only. Instead of quantitative reaction constants, reaction orders etc. the chemists know about the qualitative impact of temperature and pressure (e.g. on selectivity) and have an idea about the reaction system (e.g. consecutive or side reactions, equilibrium data etc.). By applying rules like "IF consecutive reactions exist, avoid back-mixing" the software guides systematically through the design of a chemical reactor. It follows a fixed sequence for decision making. Starting with some fundamental decisions on optimal conditions at the place were the reaction takes place (e.g. the surface of the catalyst particles) some general operating conditions like temperature and pressure profile are worked out. Finally a technical reactor type is selected and heat transfer is handled.

Another example for a computer aided decision support tool is an expert system for the selection and synthesis of large-scale protein purification processes developed by Asenjo et al. [AHB89, Ase90, LA92]. Within this expert system, the downstream process is divided into two subprocesses: recovery and purification. A selection of unit operations is dedicated to both subprocesses. The recovery subprocess comprises processing the fermentation broth until a cell free solution is obtained and the total protein concentration is at a preassigned value. After that, the protein solution is further treated in the purification subprocess until the required product purity is achieved. Starting the synthesis of a process, the final product and its application, which determines the final purity and tolerable impurities, has to be defined and the starting material must be characterized. Based on this information a suitable separation sequence is selected applying heuristic rules. There are main heuristic rules (see below [Ase90]), which provide general advice in how a sequence should be designed and several special ones, which are based on industrial expert's knowledge and deal with specific purification tasks or unit operations.

1. Choose separation processes based on different physical, chemical or biochemical properties.
2. Separate the most plentiful impurities first.
3. Choose those processes that will exploit the differences in the physicochemical properties of the product and impurities in the most efficient manner.
4. Use a high-resolution step as soon as possible.
5. Do the most arduous step last.

Due to the division of the problem (process) into two subproblems (subprocesses) the possible alternatives for the downstream sequences are narrowed down very early, which simplifies the decision making process. Furthermore, fixed constraints are preset in the program such as the concentration, which has to be met after the recovery subprocess. Hence, the solution space is limited so that it is doubtful that the optimal solution will necessarily be achieved using an expert system like this. However, it shows that good solutions for a process structure can be found very quickly by applying heuristic rules on a very generic level.

As it is not possible to evaluate the solutions obtained by a pure heuristic approach and a solely mathematical optimization is not possible because the problems can not be completely described in mathematical equations, Prokopakis suggests the combination of heuristic and simulative methods [PA90]. The heuristically found sequence can be further optimized by using evolutionary algorithms and simple models for the simulation of the process performance.

Lienqueo and co-workers extended parts of the expert system of Asenjo et. al. to a hybrid expert system, which combines rules stated by experts and mathematical relationships using the properties of the proteins from a database to design the purification subprocess [LA00, LLA96, LSA99]. As there are usually many different possible separation sequences, two criteria for selection of the best solution are implemented. Both criteria represent a mathematical description for the third heuristic rule shown above. It is tried to express how efficiently a certain separation step utilizes the property differences between product and protein contaminants. Target is to design a separation sequence with as few steps as possible to meet the required purities. As much data must be available for the calculation of the criteria for each step and the possible composition afterwards, an enormous database on contaminants expressed by typical fermentation organisms must be available meaning large time and effort for an experimental exploration.

Based on the description of the property differences in the criteria, mathematical optimization models were developed to design the purification subprocess [VALP01]. For a purity given, the sequences with a minimal number of steps are selected. Afterwards, the minimal number of steps is set as total number and a sequence with a maximal purity is generated. As just the purity achieved in and/or the number of steps of the purification subprocess is used as optimization targets and not the performance of the whole process including fermentation and recovery, the economic optimum may not be reached by this approach.

3.7.3 Process Design for Biochemical Processes

In the field of chemical process synthesis, it is possible to design new processes quickly based on mostly qualitative information (see [GSD06]) applying existing heuristic rules. When designing the structure of a complete biotechnical process no such approach exists. However, it should be possible to transfer at least parts of the methodology to biotechnical processes and to use the experience from the chemical industry for the biochemical industry.

Like for process synthesis of chemical processes the decomposition method provides a hierarchical approach, in which the level of detail is increased systematically. As the amount of data available increases during process synthesis due to further research, the decomposition method presented in Fig. 3.45 should suit for biochemical process designs as well.

On the input-output level, the basic constraints and specifications of the product are defined, as well as the production source (bacteria, yeast, mammalian cells etc.) and possible contaminants. The starting point for the process synthesis is set. A rough structure of the process is developed on the tasks level. For biotechnical processes there is usually one step to recover the product from the fermentation broth. If the product is located in the cells, cell disruption, steps to separate the contaminants, as well as steps to purify and polish the desired product become necessary.

The structure generated is the starting point for the development of the process flowsheet. Based on differences in the properties of the components (charge, size, solubility etc.) and streams (phase, concentration etc.) the unit operations used are defined as well as their sequence in the process. It is necessary to know at least some properties of the main components. On this level, the heuristic approach can demonstrate its potential for efficient process design. Normally, the rules lead to several alternative flowsheets. As thermodynamic data and mathematical models are usually not available for biotechnical processes and products, it is not possible to use rigorous methods to develop new downstream processes. Therefore, empiric models are used to calculate the efficiency of the different process alternatives to find the most promising ones, which are further developed and optimized on the laboratory scale.

Especially between the tasks and the flowsheet level, there are interactions. Most often the results of the tasks level must be revised, when further information on the process and the components involved become available. This is repeated, until a process is developed which is as optimal as necessary.

On the last level, the whole process is further optimized. It may be possible to exchange heat between certain stages (energy integration) so that the total energy consumption decreases. Many unit operations can be operated batchwise or continuously. A final decision on the operation mode has to be made and, when dealing with batch operations, the schedule has to be optimized leading to a production plan.

The decision, which structural alternatives generated will be good enough for further investigation is taken on each level based on the economics. If it does not seem to be possible to find a process within the cost range defined on the first level, any further investigation will be stopped.

3.7.4 Process Design Example

Separation and purification of the products from a fermentation broth are most often responsible for the major manufacturing cost in biochemical processes. Therefore, the following example focuses on the steps necessary to isolate and purify a fermentation product from the broth, usually done in a sequence combining several different unit operations. It demonstrates the power of knowledge based decision making for the structural design of biochemical processes.

Target is to develop alternative process concepts for the recovery and purification of fermentatively produced L-alanine (the smallest amino acid) [BS08]. As a continuous fermentation process with immobilized cells is assumed, the starting point for process design is the fermentation broth. It contains mainly water, some salts, unconverted glucose, the desired product L-alanine, some side products, and possibly cell debris, if some cells have been disintegrated during the fermentation process. Therefore, referring to Fig. 3.45, we are located on the flowsheet level just having taken the decision on the next task: separation of the process stream.

The properties of the components, which can be exploited for a separation, are summarized in Table 3.3. The different colors indicate in which fraction a component would be applying a separation principle exploiting differences in properties listed in the first column. For example, if separated based on size, glucose and by-products cannot be separated from L-alanine (blue fraction). The small molecules water and the salts form another fraction (green fraction), large sized cell debris can be separated in yet another fraction (red fraction).

Table 3.3 Properties of the fermentation broth components

	L-alanine	By-products	Water	Na_2HPO_4 NaH_2PO_4	$(NH_4)_2SO_4$	Glucose	Cell debris
Size [μm]	small	small	very small	very small	very small	$8,6 \cdot 10^{-4}$ small	0,2–1
Molecular weight [g/mol]	89,1	68–112	18	142/120	132	180	–
Charge	pH dependent pI = 6,1	salts: charged acetoin: uncharged	–	charged	charged	uncharged	uncharged
Solubility [g/l_{H_2O}]	165	550–1000	–	77/850	754	470	–
log P_{OW}						$-3,2$	–

Based on this classification, the heuristic rules given on page 158 are used to develop alternatives in which sequence the differences in properties shall be exploited. Afterwards, suitable unit operations are assigned to the property differences giving the process structure. For example, if the size is the property exploited, possible unit operations would be filtration, sedimentation or centrifugation. Which unit operation suits in a certain process synthesis problem depends on constraints such as product stability and operation range of the unit operation. This leads to a kind of alternatives tree as shown in Fig. 3.46.

The alternatives tree includes the standard process to separate amino acids comprising an electrodialysis to get rid of the salts, a capture (usually ion exchange), another electrodialysis, evaporation and drying of the final product. In addition, the tree shows several other alternative process structures which have been selected using heuristic rules. For example, an ultrafiltration unit could be an economic option as first processing step in the sequence utilizing the difference in size of the molecules to be separated.

Furthermore, the example demonstrates how heuristics help to quickly exclude non-feasible and non-economic unit operations. Typical laboratory separation technologies such as gel electrophoresis applying differences in charge and molecular weight have been excluded due to their low product yields combined with high cost.

In order to get a first idea about the quantitative performance of the process alternatives generated rough mass balances are calculated and the total yield of each process is estimated, Fig. 3.47. Again this estimation is based on experience expressed in heuristic rules. The calculation uses typical performance data which could be achieved in the past using similar process units. The above mentioned standard process (alternative 6) is ranked at medium level. Interestingly, the alternatives with the highest and lowest yield both start with a crystallisation step, which means it can not be generally stated that a specific unit operation should or should not be the first step of a separation sequence. A limited number of good process

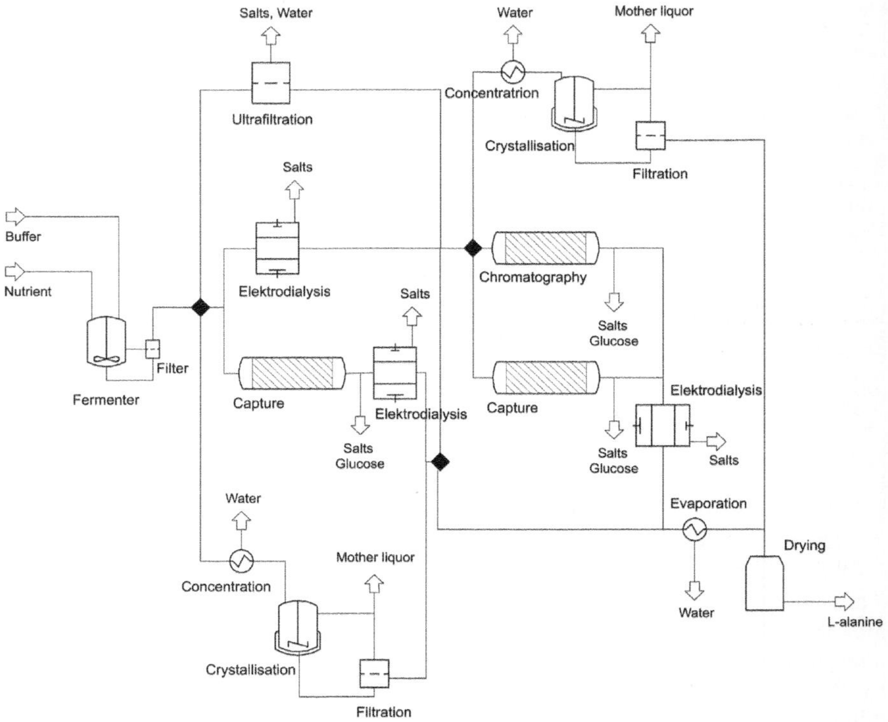

Fig. 3.46 Downstream process alternatives for the production of L-alanine; Bauer and Schembecker [BS08]

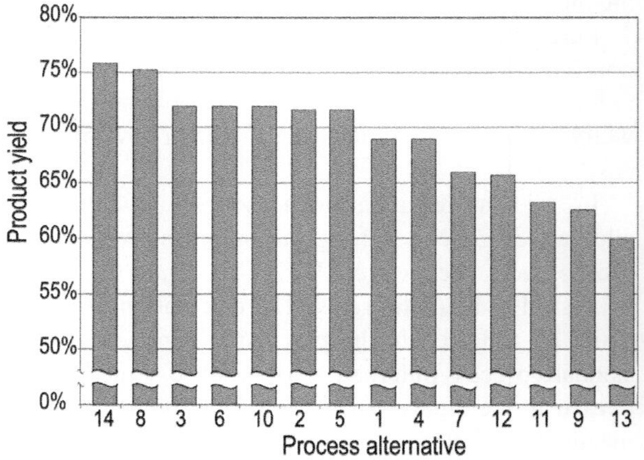

Fig. 3.47 Product yields of the different alternatives

alternatives must now be fed to further development in more detail using improved simulation models and laboratory experiments.

3.7.5 *Conclusion*

The development of new industrial scale production processes is a fuzzy reasoning problem, as an engineer has to take decisions based on mainly qualitative information while dealing with a vast quantity of variables. For chemical process synthesis, there exist different methods such as decomposition and heuristic rules. It has been shown that a combination of heuristic rules and numeric estimation quickly leads to reasonable results.

The transfer and application of this approach to biochemical process development leads to promising results as exemplified in Sect. 3.7.4. However, instruments to estimate the process performance numerically are incomplete, as there is lack of models describing the performance of unit operations and the properties of components and streams.

This discussion of the potential of expert systems applied to chemical and biochemical process designs only even touched the purely technical aspects of the problem. As soon as additional commercially relevant, non-technical issues are taken into consideration the fuzziness and the number of variables increases even more. Aspects like reutilisation of available plant parts, operability of strongly interconnected flow schemas, danger of patent conflicts by choosing certain processing paths seen in relation to their promise of raised payoffs, and much more influence the final decision on a technical realisation of the process in question significantly as well. As of today there is no integrative framework that combines these aspects into a more global decision support tool, rendering the underlying system of an also commercially paying engineering design as a strongly evolutionary process with a high amount of trial and error.

3.8 Structural Design and Process Shaping with Uncertainty

Dietrich Hartmann

3.8.1 *Introduction*

Structural design of contemporary systems in civil engineering, as well as shaping of processes associated with such systems, is characterized by an increasing complexity. This constitutes a big challenge to engineers which can be mastered only by means of powerful computer-aided methods. To obtain practicable results, the computer models established must represent realistic real world scenarios. Intensive simulations are indispensable, as well with regard to the number of simulations as also for the purpose of verification and validation. The high complexity additionally calls for multidisciplinary solutions and necessitates multi-level, multi-scale and even multi-paradigm models leading to micro-, meso-, macro- and/or super-models interwoven to each other.

In the following, three different examples are dealt with which demonstrate how the handling of uncertainty in complex structural engineering problems can be accomplished, according to the general discussion in Sect. 2.4.4. The first example is concerned with the lifespan-oriented design of large scale structures subject to time-variant stochastic loading taking into account deteriorations. In the second example, the computer-aided destruction and collapse of complex structural systems using controlled explosives is discussed with respect to the realization of fuzzy randomness. Finally, the third example addresses the structural optimization considering stochastic imperfection in the geometry introduced during the erection process.

3.8.2 Lifespan-Oriented Design

As an example for the lifespan-oriented design of structural systems in civil engineering we consider the optimum design of steel structures accounting for deteriorations and damages induced during the utilization of a structural system. The design problem is thus transformed into an equivalent structural optimization problem.

In the case of steel structures, the design against failures due to fatigue phenomena often becomes crucial. To give an example, industry halls composed of frames (Fig. 3.48) and used as a first reference system may fail under repetitive, stochastically multiply correlated as well as instationarity wind actions. A further example are steel arched bridges (Langer's beams) which are considered as the second type of a reference system (Fig. 3.49).

Surprisingly, it has been detected that even newly erected bridges show failures in the hanger connection plates (Fig. 3.50) linking the vertical arch hangers which are attached to the main girder of the bridge.

Extensive research and testing have identified more often than not that mostly transverse-induced across vibrations are the cause of the damage (cracks in the plate), shortly after the erection. Again, the randomness of the wind actions, even if stationarity can be assumed, affects the damage-sensitive vibrations within a so-called lock-in domain (Fig. 3.51).

From the above remarks, it can be concluded that the lifespan-oriented structural analysis and design is substantially governed through a plenitude of uncertainties in the stochastic load processes (actions) with regard to appearance, duration and intensities. Accordingly, the structural response in terms of displacements, vibrations, strains and stresses have to be regarded as time-variant stochastic processes themselves, yielding stochastic deteriorations and a temporal degradation in the total system. Of course, the relevant structural data have also to be treated stochastically in terms of basic variables (simple stochastic variables). To elucidate, in principle, how randomness as one of the categories of uncertainty is incorporated into

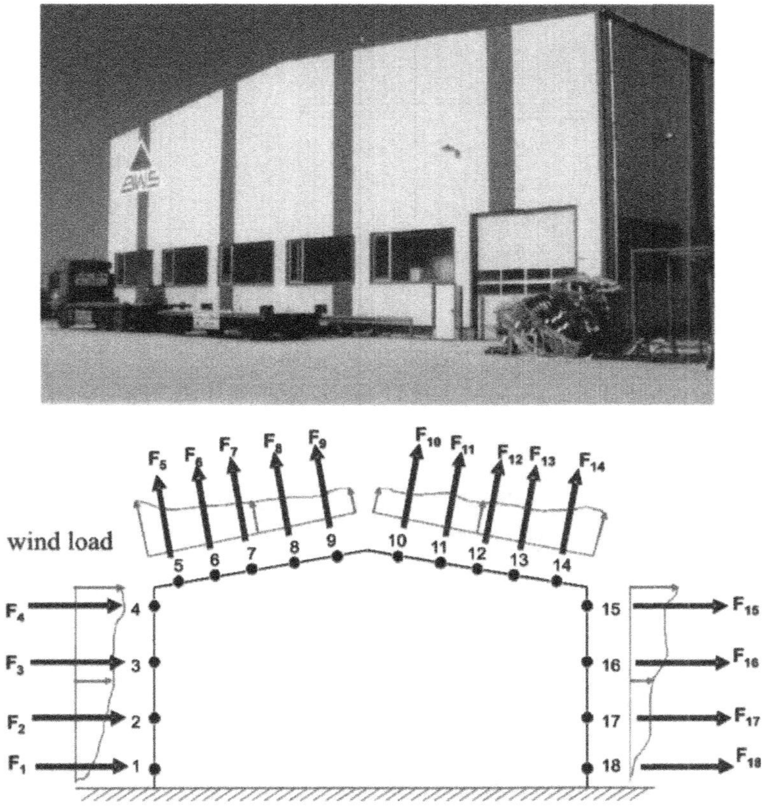

Fig. 3.48 Industry hall subjected to wind loading

Fig. 3.49 Steel arched bridges

the engineering design, the aforementioned industry halls are contemplated again, (Fig. 3.48). In this case, a multi-level approach is applied to appropriately capture the real world behavior of the interactions between loading, structural response and deteriorations. By that, the time-variant computation of the deteriorations and the

Fig. 3.50 Hanger connection plate

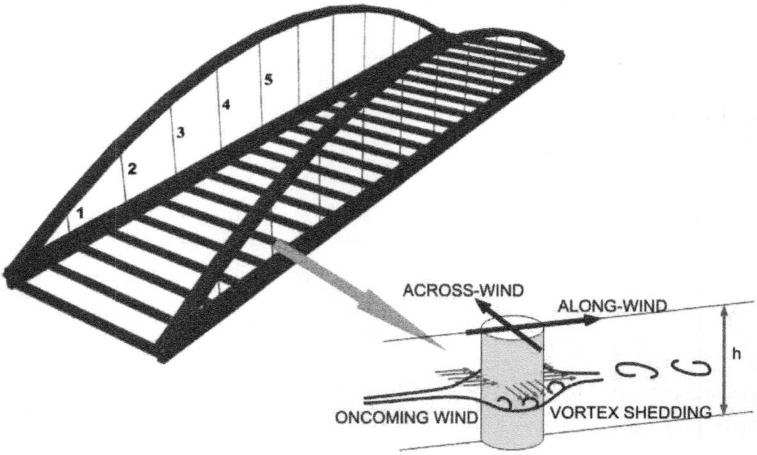

Fig. 3.51 Vortex-induced vibrations (lock-in effects)

damage is accomplished by using two consecutive analysis models: (i) a first-model (Fig. 3.48) captures the stresses due to extreme turbulent wind velocities in a micro-time scale (T $\approx$ 10 minutes); (ii) a second model is taken to estimate the stresses due to wind loading in the macro-time scale (Fig. 3.52).

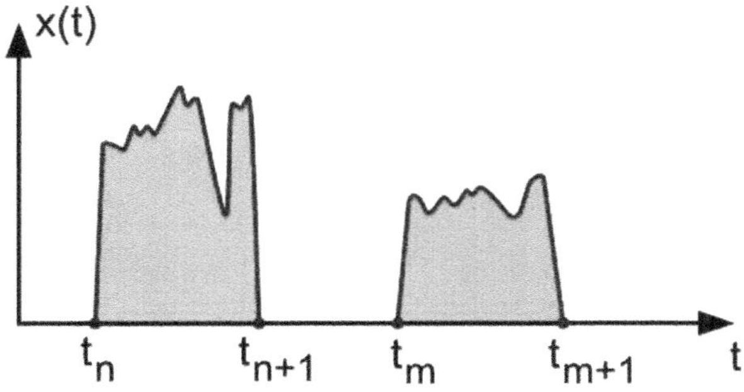

Fig. 3.52 Intermittent continuous stochastic process

Based upon this micro-macro separation, the fluctuations in the micro-model can be assumed as piecewise stationary and Gauss-distributed while the actions over the long term are represented as a stochastic intermittent continuous pulse process (Markov renewal process). The deteriorations induced in the micro-time scale are then accumulated according to the pulse process and using a stochastically modified linear Palmgren/Miner (S/M)-rule. By means of the above described damage accumulation, representing a first passage problem, the failure probability P_f due to fatigue can be computed by a specific Monte Carlo Simulation (MCS). The value for P_f is not allowed to exceed a given admissible limit P_{adm}. This requirement, then, forms a stochastically non-linear and time-dependent constraint for the structural optimization problem, besides side constraints for the selected optimization/design variables.

For the steel frames considered here, according to (Fig. 3.53) only two sizing variables are defined, the height X_1 and the width X_2 of the I-shaped cross-section.

The objective function is the cross-sectional area $A(X_1, X_2)$. Thus we have the optimization problem described in equation 3.10.

$$\min \left\{ A(X_1, X_2) \left| \begin{array}{c} 0.1 \leq X_1 \leq 1.0 \\ 0.1 \leq X_2 \leq 0.3 \\ g\left(\mathbf{X}, \mathbf{Y}, \mathbf{Z}(t)\right) = P_f\left(\mathbf{X}, \mathbf{Y}, \mathbf{Z}(t)\right) - P_{adm} \leq 0 \end{array} \right. \right\} \tag{3.10}$$

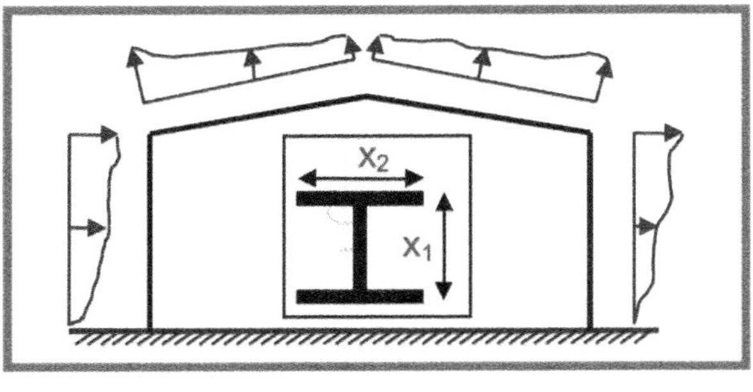

Fig. 3.53 Optimization model for framed system (industry hall)

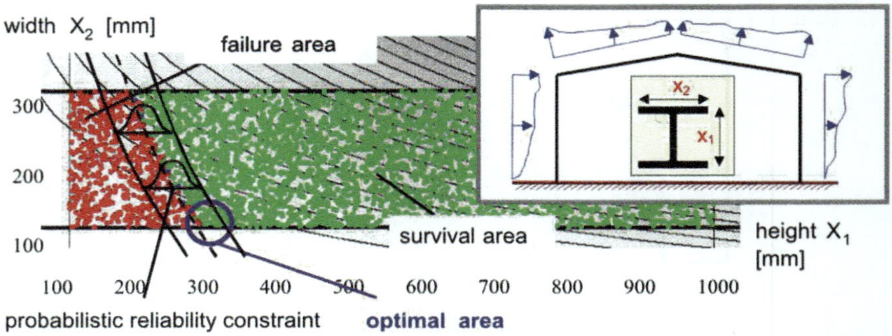

Fig. 3.54 Solution domain for stochastic optimization

The solution of the exemplary design problem can be learned from Fig. 3.54 where the optimization domain is depicted in the X_1/X_2-space.

As illustrated, the stochastic constraint for the failure probability scatters around expectation values (see bell-shaped curves in Fig. 3.54) and creates two segregated sub-domains: (i) a first domain containing green points (no failures), and (ii) a second domain of red points representing failures. The optimum is a set of points (circle) the size of which depends on the selected critical fractile.

3.8.3 Computer-Based Demolition of Complex Structural Systems

Buildings and constructions may reach the expiration of their service life because of insufficient structural quality, changed requirements regarding utilization or simply due to an inacceptable layout. As a consequence, such buildings (departments store, administration buildings, multi-storey buildings etc.), frequently located in densely populated areas, have to be demolished within short time by controlled collapse. This implies that the collapse, triggered through the explosive charges, proceeds according to schedule and does not cause collateral damages, neither to human beings nor to adjacent infrastructure. The following figure exemplifies characteristic blast scenarios that took place in innercity areas such as the explosions for the demolition of (i) a library in Dortmund, (ii) a high-rise residential building in Hamburg and (iii) an administration building in Wuppertal.

All of these demolitions have been attended and analysed for research purposes with the objective of creating a computer-based simulation system by which the real world collapse due to the explosion can be predicted.

Based on this simulation system, an optimal demolition strategy is to be sought in such a way that the discrepancies between the provoked debris pile and a prescribed collision domain are "minimal". Everybody who has watched real world blast demolitions of complex buildings can imagine that a plentitude of imponderabilities can prevent the desired collapse result. Such imponderabilities can be caused by both data uncertainties (structural data, explosion parameters) as well as model uncertainties (range of structural behaviors, contact and impact phenomena

Fig. 3.55 Examples for real world blasts of buildings

during the collapse, wind actions, etc.). If all relevant aspects are to be modeled realistically, then the total aforementioned subcategories for uncertainty modeling have to be materialized, i.e.

- randomness
- fuzzyness and
- fuzzy randomness.

The application of randomness is possible if (i) stochastic variables (basic variables) can be identified, (ii) a sufficiently large universe is available and (iii) if the mathematical principles of randomness are valid. Material properties, such as the modulus of elasticity, Young's module etc. are good examples for a proper randomization. By contrast, fuzziness is more suitable if qualitative issues have to be captured based upon subjectively defined membership functions for uncertain quantities themselves. Herewith, a new type of interval arithmetic can be introduced instead of using conventional numerical concepts. Finally, if items or parts of the stochastic quantities (basic variables or random processes) also contain incomplete information or violate the principles of pure randomness, then fuzzy randomness fills the gap between objective randomness and subjective fuzzyness, within the same common body of representation.

Besides the representation of uncertainty, a further fundamental solution concept is mandatory. Due to the complexity of the simulation-guided optimization problem, the original problem has to be decomposed and distributed into individual subproblems, representing different problem levels with respect to space and time. In the project considered four distinct interacting levels are recognized to map the real world behavior of the collapse cascade appropriately. The subsequent figure (Fig. 3.56) illustrates the spatio-temporal multi-level approach in detail.

As can be seen, on the micro-level the local effects of the explosive charges are modeled using a smooth particle hydrodynamics code and a mesh-free Galerkin method (hereby time scale ranges from micro- to milliseconds while space dimensions [L] are between mm and cm). In a micro-/meso-level sub-model, the shock waves and vibrations are computed based on a finite element method associated with explicit time integration and element erosion (Δt in milliseconds, [L] between cm and m). On the meso-/macro-level of the structural system, an inventory control

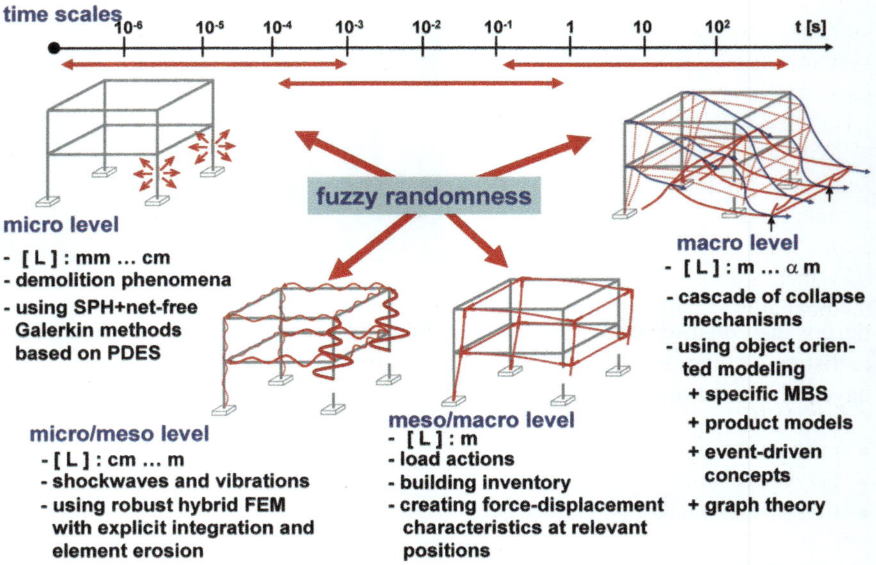

Fig. 3.56 Spatiotemporal multi-level concept used in blast simulation

Fig. 3.57 Collapse simulation using fuzziness

determines the random character of the structural data (concrete parameters, reinforcement assembly, etc.). Also, three-dimensional strain-stress characteristics at specified critical sections are established which subsequently are employed in the collapse cascade of the total system level. The domain of influence is in meters. On the total systems level the collapse cascade is simulated, which spans from a few hundredth to about hundred seconds, and encompasses the total structure, and its fragments after the explosion. The complexity of the collapse cascade requires a flexible approximation model where a multi-body approach fits best, taking into account the essence of the three preceding sub-models. According to the multi-body model, bodies are regarded as objects interacting with other objects: Thus, an object-oriented implementation is accomplished that also allows for an appropriate mapping of the event sequences during collapse. From the graphical diagram in Fig. 3.56 it can be seen, that on all four sub-levels fuzzy randomness is integrated for embedding of the relevant uncertainties already described above. As a prototypical system, a high-rise concrete skeletal building which covers the most significant collapse mechanisms, Fig. 3.57 shows the collapse of the building after the explosion. Also, the arising debris hill is portrayed.

Within the test simulation two quantities, the rigidity of the structure and the friction coefficient in the structure, have been fuzzificated in terms of two triangular membership functions. This approach yields an uncertain range for the debris radius (Fig. 3.58) where the α-value represents the level of membership. The value α=1 means full membership while α=0 (left and right of α=1) marks vanishing membership.

3.8.4 Imperfection-Sensitive Structural Optimization

Imperfections in a structural system, induced during construction and erection, are natural phenomena that never ever can be totally excluded. In general, structural as well as geometric imperfections may occur. Here, the focus is solely on geometric imperfections which lead to more or less aberrations from the planned geometric shape of a structure. Geometric imperfections can become particularly dangerous

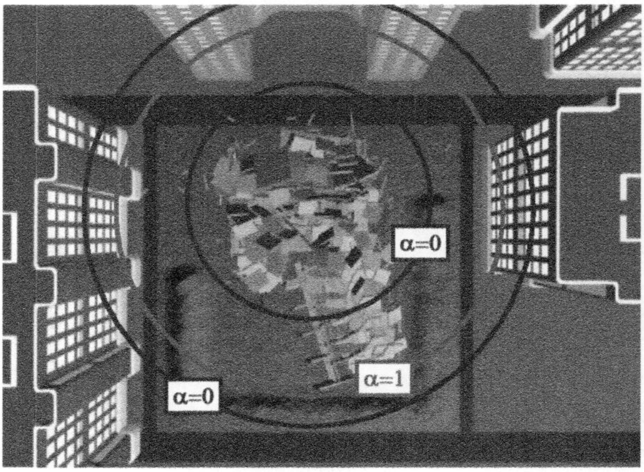

Fig. 3.58 Uncertain debris radius

in structures subjected mainly to compressive stresses because sudden collapse can happen. Typical compression-subjected structures are shown in Fig. 3.59.

If it is attempted to provide an optimum design for the aforementioned structures using a structural optimization model may be inadequate. The reason for this is this the "optimal" shape obtained by numerical optimization may show an extremely sensitive behavior and may fail early to the smallest possible geometric imperfections. To explicitly exclude such catastrophic failures with a reasonable degree of

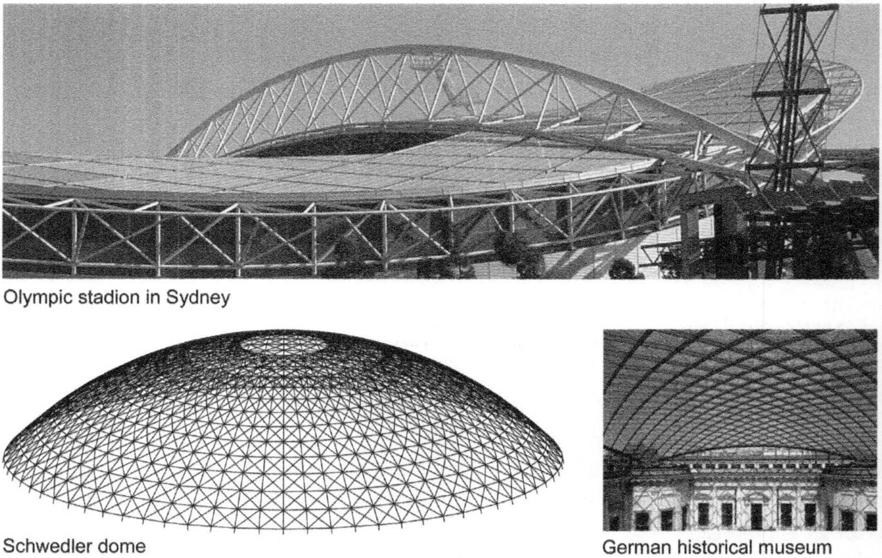

Fig. 3.59 Examples for compression-subjected structures

probability, the uncertainties due to geometric imperfections have to be scrutinized and incorporated into the shape optimization model. Such a model has been created [BAIT03] where again a multi-level approach is pursued. For the reference structure displayed in Fig. 3.59 (left), the relevant uncertainties due to geometric imperfections have been successfully represented by means of stochastic fields, in association with a properly chosen probability distribution and correlation functions. Starting with the prescribed perfect geometric shape of the structure the introduction of random variables (imperfection variables) for the stochastic fields allows the assessment of the worst possible imperfect geometric and, accordingly, the corresponding structural response. To find out the unknown worst imperfect shape it is assumed that the amplitudes against the perfect geometry cannot be arbitrarily large. Accordingly the q-dimensional envelope (ball) in the space of the uncorrelated imperfection variables is determined. Hereby, the characteristic length (radius) of the envelope represents a prescribed fractile being a measure for the gradual assessment of uncertainty. Based on this concept, on a first level of the total optimization model the worst possible imperfect shape is computed by a maximum optimization (see Fig. 3.60), which is designated as anti-optimization.

Having determined the worst imperfect geometric, a conventional structural optimization is carried out on the second level (see Fig. 3.60). As illustrated above, the uncertainty model yields the imperfection model for which a geometrically nonlinear finite element analysis is carried out. This analysis forms the kernel for the structural optimization, in particular, for the constraints of the optimization iteration. The design variables are node coordinates of the arched girder. As objective criterion the strain energy is taken by which the robustness of the structure against sudden failure can be ensured. In Fig. 3.61 the optimization history is displayed.

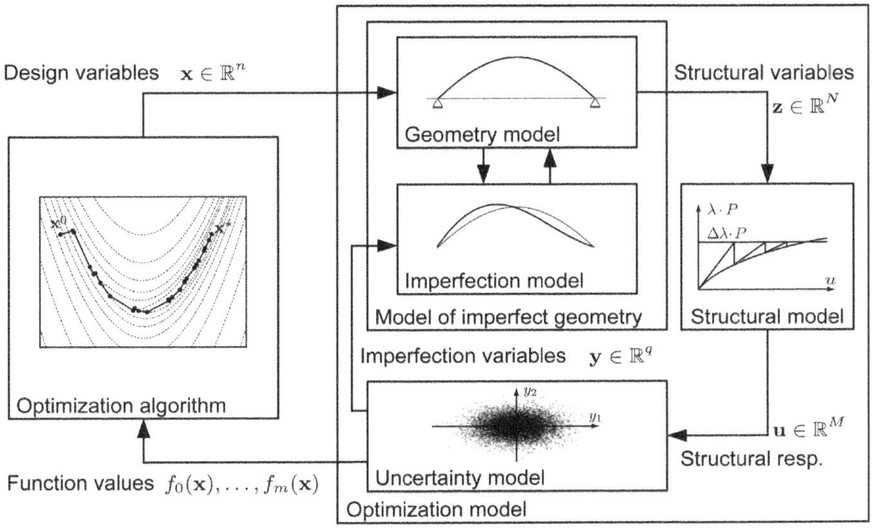

Fig. 3.60 Two level optimization concept

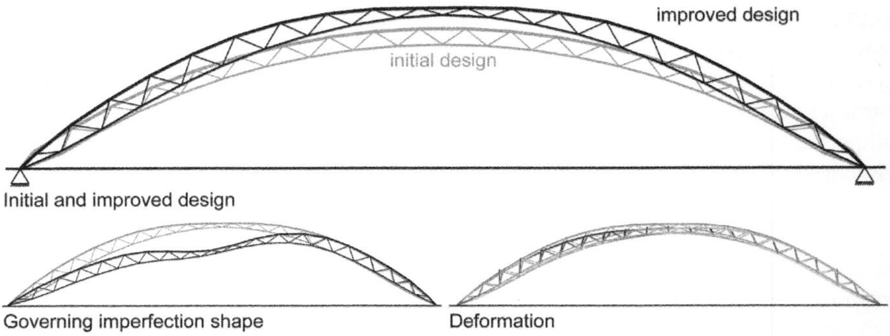

Fig. 3.61 Optimization results

The picture at the top of Fig. 3.61 demonstrates the cascade-shaped reduction of the initial solution where a toggling between the anti-optimization and the structural optimization takes place. The two bottom charts in Fig. 3.61 shows the initial and optimized structure, the corresponding displacements as well as the governing imperfect shape.

3.8.5 Conclusions

The actual endeavors in scientific structural engineering to understand highly complex systems and processes necessitates the introduction of various categories of uncertainty as well the integration of subjective assessments which have to be described in a mathematical fashion. By that, conventional numerical methods as deployed in Computational Mechanics already for a long time can be linked to methods originating from the Probability Theory and Computational Intelligence. The consequence is that such an enhancement of Computational Engineering opens new directions to cope with what has been denoted in Cognitive Science as "restricted rationality".

In the above chapters it has been demonstrated that there are effective and sufficiently accurate ways and means to capture the sophisticated non-deterministic behavior of selected structural systems and processes. Three applications have been scrutinized in which a wide bandwidth of computer models for the acquisition of uncertainty phenomena could be pointed out: (i) the lifespan-oriented design of structures, (ii) computer-based demolition of complex buildings using controlled explosives and (iii) the imperfection-sensitive structural optimization.

The essential benefit in all three cases is the sustainable improvement of the safety of the systems used and the processes activated. Hence, the vulnerability of the systems and processes and the endangerment of human lives can be drastically reduced. In this context, it is significant that the improvements or reductions can not only be evaluated qualitatively but also quantitatively. That is to say, uncertainty can be better gauged and better detected than before, far in advance to hazardous events.

3.9 Structures in Operational Transport: Complexity and Optimization

Herbert Kopfer, Jörn Schönberger

3.9.1 *Introduction*

Operational (or short-term) planning of the fulfillment of transport requests requires both a decision about the used transport resources and a decision about the deployment of the selected resources. Therefore, requests are assigned to resources for fulfillment and the operations of each resource must be determined. Operational transport planning involves assignment as well as sequencing decisions. Even in its pure form, the identification of an optimal sequencing requires checking and comparing a huge number of permutations of the items to be ordered. For n items, there are $n! = n \cdot (n-1) \cdot (n-1) \ldots 2 \cdot 1$ different possible solutions.

To assess the corresponding decisions, each alternative is rated with a meaningful numerical value (costs, travel time, service quality) that allows the comparison of different proposals in the fashion like "... the best solution of...". Therefore, the transport operations planning turns out to be an optimization task which connects short-term planning with the field of optimization in a natural way. Thus, operational transport planning tasks yield an important field for difficult optimization problems and for research on optimizing algorithms.

In the next section we introduce the Vehicle Routing Problem as a basic scenario for operational transport planning. Although in its pure form its complexity is quite high, it does not reflect requirements from practice. In a further section we discuss important aspects of operational transport planning that occurs in many practical situations. The resulting optimization problems are much more complex than the problems of the basic scenarios. Finally, we present some conclusions.

3.9.2 *Basic Scenarios*

3.9.2.1 Traveling Salesman Problem

The fundamental problem in the area of transportation optimization is the famous travelling salesman problem (TSP) comprehensively studied in [LLKS85]. The TSP and methods for its solution in the context of online optimization are again addressed in Sect. 3.11 of this book. It is the most thoroughly investigated problem in the research area of combinatorial optimization. The reason for its popularity is probably that the TSP is easy to understand but hard to solve. The TSP can be formulated as follows:

> *A travelling salesman wants to visit a number of customers at different sites. When he has visited all customers, he has to return back to his starting point. Which way should he choose (i.e. in which sequence should he visit the customers) in order to keep the complete distance of his travel as small as possible?*

The TSP is an attractive example to represent typical sequencing problems in a visual way. There are many applications and extensions of the TSP in different areas of application. The importance of it is also due to the fact that it is to be solved as a sub-problem in many other important problems. Moreover the TSP has frequently been used as a benchmark-problem in order to evaluate new algorithms and is still very important for the performance evaluation of algorithms.

3.9.2.2 Vehicle Routing Problem

The salesman of the TSP merely must visit his customers and does not have to transport anything on his round trip to all customers. So in the area of transportation optimization, the elementary constitutive extension of the TSP is that the customers have a demand for goods that must be delivered to them. This extension leads to the classical Vehicle Routing problem (VRP) which has first been formulated and investigated by [DR59]: The paper of Dantzig and Ramser is concerned with the optimum routing of a fleet of gasoline delivery trucks between a bulk terminal and a large number of service stations supplied by the terminal. The shortest routes between any two points in the system are given and a demand for one or several products is specified for a number of stations within the distribution system. It is desired to find a way to assign stations to trucks in such a matter that station demands are satisfied and total mileage covered by the fleet is a minimum.

For the classical VRP the following preconditions are postulated:

- All goods required by the customers are available at one depot
- All customer requests are known
- The transportation capacity of the vehicles is limited
- The capacities of all vehicles are equal
- The entire capacity of all vehicles is big enough to serve all customer requests
- Each vehicle can deliver each customer request and each customer is served by exactly one vehicle
- There is only one planning horizon, during which all customer requests must be dispatched
- The distances between all customers as well as between the customers and the depot are known
- All vehicles start and stop their tour at the depot

The solution of the TSP is a single round trip, on which all customers are served. Solving the VRP we usually get several round trips since only a part of the customers can be served by a single vehicle due to the limitation of the capacity of the vehicles, i.e. we have to employ several vehicles. As a consequence the VRP is not merely a sequencing problem but a combined assignment and sequencing problem: each customer has to be assigned to exactly one cluster (tour) which is performed by one vehicle serving its customers on its tour. For each tour there must be defined a sequence which builds a round trip (route) followed by the vehicle on its tour. The two sub-problems of assignment and sequencing are not independent. That is why exact solutions of the combined problem can only be reached by solution processes which solve both problems simultaneously.

Optimization Model For a formal representation of the VRP in a mathematical model we introduce the following parameters:

- n number of sites (including the central depot)
- $i = 1$ index of the depot
- $i = 2, \ldots, n$ index of customers, $j = 2, \ldots, n$ index of customers
- $k = 1, \ldots, m$ index of vehicles
- Q capacity of a vehicle k
- $q_1 = 0$ the request at the depot is zero
- $q_i \leq Q$ there is no customer request which is greater than the capacity of the vehicles Q, $\forall i = 2, \ldots, n$, $\forall k = 1, \ldots, m$
- $d_{ij} \geq 0$ the distances between all customers as well as customers and the depot are non-negative, $\forall i \neq j \in \{1, \ldots, n\}$

To store the required assignment and sequencing decisions, we deploy the following two families of binary decision variables

- $s_{ijk} \in \{0, 1\}$ is one if and only if vehicle k drives directly from customer i to customer j
- $y_{ik} \in \{0, 1\}$ is one if and only if customer i is assigned to vehicle k.

Then the classical VRP can be described by the following optimization model:

$$f((x_{ij}), (s_{ijk})) = \sum_{i=1}^{n} \sum_{j=1}^{n} \sum_{k=1}^{m} d_{ij} s_{ijk} \tag{3.11}$$

$$\sum_{k=1}^{m} y_{ik} = \begin{cases} 1, & i = 2, \ldots, n \\ m, & i = 1 \end{cases} \tag{3.12}$$

$$\sum_{i=1}^{n} q_i \cdot y_{ik} \leq Q, \ \forall k = 1, \ldots, m \tag{3.13}$$

$$\sum_{j=1}^{n} s_{ijk} = \sum_{j=1}^{n} s_{jik} = y_{ik} \ \forall i = 1, \ldots, n, k = 1, \ldots, m \tag{3.14}$$

$$\sum_{i,j \in S} s_{ijk} \leq |S| - 1 \ \forall S \subseteq \{2, \ldots, n\}, k = 1, \ldots, m \tag{3.15}$$

The above model belongs to the class of models for Integer Programming. Solving this model, we are searching for values for the binary integer variables s_{ijk} and y_{ik}. The objective function (3.11) consists of the sum of distances between customer sites. A distance between two customer sites i and j is included in the sum, if and only if any vehicle drives directly from customer i to customer j. Equation (3.12) enforces that each customer is assigned to exactly one vehicle and that the depot is assigned to each vehicle. Equation (3.13) ensures that the sum of the demand of all customers served by a vehicle k does not exceed the capacity limit Q of the vehicles. In Eq. (3.14) it is assured that each customer assigned to a vehicle is approached by that vehicle exactly once and that it is left by the same vehicle one time. Additionally, Eq. (3.14) guarantees that a customer is only served by that vehicle that he is assigned to. Finally, Eq. (3.15) prohibits for each tour the existence of short cycles (without visiting the depot) and forces to build Hamiltonian cycles for each vehicle.

Example A typical VRP-scenario is shown in Fig. 3.62. The five customers $i = 1, \ldots, 5$ with the demand $q_i = 2, 5, 3, 4, 6$ units require a replenishment from the central depot 0. Two vehicles are available for serving these customers, each one having a limited capacity of $Q = 10$ units. The travel distances between pairs of involved sites (customers as well as the depot) are given in the distance matrix (c_{ij}):

$$(c_{ij}) := \begin{pmatrix} 0 & 15.9 & 7.6 & 5.4 & 6.5 & 5.2 \\ 15.9 & 0 & 20.8 & 21.2 & 21.3 & 17.7 \\ 7.6 & 20.8 & 0 & 6.2 & 10.8 & 12.0 \\ 5.4 & 21.2 & 6.2 & 0 & 4.7 & 7.2 \\ 6.5 & 21.3 & 10.8 & 4.7 & 0 & 4.1 \\ 5.2 & 17.7 & 12.0 & 7.2 & 4.1 & 0 \end{pmatrix} \tag{3.16}$$

Since the provided transport capacity is tight not all assignments of requests to vehicles are feasible. Figure 3.62 shows one optimal solution in which one vehicle serves customer sites 1,2 and 5 (resulting in the tour $\{1, 2, 5\}$ and the remaining customer sites are visited by the second vehicle (tour $\{3, 4\}$). In both cases, the vehicle's capacity is completely exhausted.

The minimal length route of the first vehicle is $0 \to 5 \to 1 \to 2 \to 0$ with travel distance $5.2 + 17.7 + 20.8 + 7.6 = 51.3$. The second vehicle follows the route $0 \to 3 \to 4 \to 0$, leading to $5.4 + 4.7 + 6.5 = 16.6$ traveled distance units. Overall, the sum of traveled distances equals $51.4 + 16.6 = 68.0$. If we assume that the costs of travelling one distance unit are normalized to one monetary unit then the service of all customer sites causes at least 68.0 monetary units.

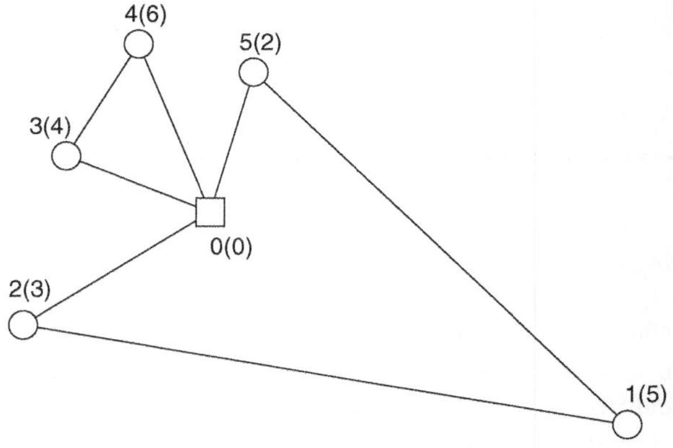

Fig. 3.62 VRP-example with customer sites $i = 1, \ldots, 5$ (their corresponding demand q_i given in brackets) and two vehicles (each having capacity $Q = 10$)

The advantageous combination of visiting the geographically grouped four customer sites 2–5 in one tour is prohibited by the limited capacity of the available vehicles.

Algorithms The application of exact mathematical programming based methods is investigated as well as the qualification of heuristic search paradigms like problem-oriented tree-search-approaches, decomposition techniques or methods inspired by nature.

Suitable methods for the exact solution of vehicle routing and scheduling problems are Branch-and-Bound algorithms [TV02a], Branch-and-Cut approaches [NR02] and algorithms for column generation [BSL97]. With these algorithms problems with up to hundred customers can be solved, if the structure of the problem is good-natured and if there are no complicated restrictions. In all other cases the problems can only be solved sub-optimally by heuristic approaches.

The oldest and easiest heuristic algorithms for the VRP are the Sweep algorithm [GM74] and the Savings algorithm [CW64]. Their performance is very poor, but especially the Savings algorithm is the source of many modern ideas of efficient search algorithms.

The most successful methods for the heuristic solution of complex vehicle routing and scheduling problems are tabu-search algorithms [GL93] and genetic algorithms [Gol89a]. Tabu search algorithms for vehicle routing use special moves for the definition of the neighborhoods of actual solutions [GHL94]. These moves change the sequence of routes and swap requests between tours.

Many Genetic Algorithms use genetic reordering operators for the solution of the sequencing problem and combine the sequencing problem dynamically with the assignment problem. Other types of Genetic Algorithms operate as a meta-heuristic navigating the search of a specialized algorithm for vehicle routing or for a subproblem of vehicle routing. The most successful evolutionary algorithms for vehicle routing are memetic algorithms which are created by combining the Genetic Algorithms with other powerful approaches (e.g. local search) to a hybrid solution method as for instance local search [Sch05].

3.9.2.3 Problem Variants

Table 3.4 shows a classification of important operational transportation optimization problems which can be derived as extensions of the TSP and VRP. There are problems for which the load capacity is irrelevant (unlimited capacity) and problems with limited capacity of the vehicles. For some problems the feasibility of the tours is independent of the sequence of customers; for all other problems the sequence in which the customers are served is decisive for the feasibility of the plans since the actual load of a vehicle increases at each loading location and must not exceed the vehicle capacity Q.

A time window specifies a time interval during which a customer has to be served by a vehicle. If there are time windows postulated for some customers of a routing problem, the corresponding problem type is characterized by an appendix

Table 3.4 Classification of models for vehicle routing and scheduling

		unlimited capacity		limited capacity	
		1 vehicle	>1 vehicles	1 vehicle	>1 vehicles
no sequence dependent use of capacity	without time windows	TSP	M-TSP		VRP
	with time windows	TSPTW	M-TSPTW		VRPTW
sequence dependent use of capacity	without time windows	not relevant		not relevant	VRPB PDP DARP
	with time windows				DARPTW PDPTW

TW, i.e. for instance that the TSPTW is a TSP with time windows for some or all of the customers occurring in this problem [FLM02]. For the Multiple Travelling salesman problem (MTSP) the maximal allowed length of a round trip is limited to an upper length [BH74]. In this case it may be necessary to have more than one single salesman for the service of all customers during the considered planning horizon. Extending the VRP by allowing the simultaneous delivery and collection of goods in a single tour is called the Vehicle routing Problem with Backhauls (VRPB) and is investigated e.g. in [TV02c]. A Pick-up-and-delivery problem (PDP) does not contain a depot where all the transportation goods are located. Rather a PDP consists of several transportation requests and each transportation request requires the transportation of goods from a pick-up-location to a corresponding delivery-location [NB00]. The Dial-a-ride problem (DARP) has the same structure as the PDP with the only difference that for the DARP instead of goods passengers must be transported in hailed shared taxis [Cor06]. The DARP is also discussed in the chapter on online optimization in this book (see Sect. 3.11). A more detailed classification of vehicle routing and scheduling problems can be found in [BGAB83].

In practical applications there are many additional restrictions for vehicle routing problems and some of them are difficult to involve, although they are crucial for the application at hand. Table 3.5 presents a sample of possible attributes and feasible values for vehicle routing problems in practice. The attributes of Table 3.5 usually lead to additional constraints in the related models for vehicle routing. Many of these restrictions yield problems that are far beyond the capabilities of current solution methods. So, new and specific methods are needed to solve complex problems of practical relevance. Extended problems which consider additional aspects of practical relevance are called rich vehicle routing problems [HHJ06]. Many important features of rich vehicle routing problems are hard to consider in models and

Table 3.5 Morphology of vehicle routing and scheduling problems

attribute	value
size of the vehicle pool	one or several vehicles
starting points of the tour and stopping points	one or several depots, no depot
composition of vehicle pool	homogenous or heterogeneous
types of vehicles	dry bulk transporter, trailer, container-truck
belonging of the vehicle	own trucks or employment of a carrier
vehicle capacity	max load, max volume, load space, load length, load width
personal team	one driver, two drivers, variable team
shift-type	one shift or several shifts
EC-social regulation	max driving-time, max working-time, duration of breaks
type of logistic services	delivery and/or pick-up, consolidation, less than truck load, full truck load, transshipment, dangerous goods, additional services
customers requests	completely known, not completely known, deterministic or stochastic
time usage	traveling times, times for loading and unloading
time windows	without time windows, time windows for customers, time windows for drivers

some of them have not been considered at all in scientific publications about vehicle routing.

3.9.3 *Practical Aspects of Real World Problems*

Not only is the solution space of the underlying models affected by practical aspects of planning and scheduling a fleet of vehicles but also the evaluation of solutions. The objective function of basic models consists of the sum of the length of all routes in a plan. But in reality the aspired goal is mostly determined by the minimization of the costs thereby incurred. The length of a route is only a substitute for its costs and the assumption that the costs increase proportionally to the driven Kilometers is too simple. Some models use a combination of route length and required time for route fulfillment as an approximation for cost evaluation. Sophisticated models for the quantification of the relevant costs for the usage of own vehicles are proposed [Erk98] and more elaborate models are needed. For instance, these models have to take into account whether a route of a vehicle is performed by a crew of two drivers or a single driver, since long routes can only be accomplished by two drivers in a daily planning horizon. As a consequence the costs for a route increase by a fixed amount, if the route length exceeds an upper limit of manageable routes for single drivers. Independently of the length of a route, the roadway chosen for the route may also influence the costs because some roads cause toll payments and others do not. While the objective function gets closer to reality and more detailed it becomes more and more difficult for the solution algorithms to solve the emerging problem.

3.9.3.1 EC Regulations No 561/2006

One of the mentioned attributes of vehicle routing problems in Table 3.5 is the comprehension of the EC-social regulation in the process of resource planning. The EC regulations are very important since they are not a matter of efficiency but their compliance is dictated by law. Since April 2007 the new EC Regulation (EC) No 561/2006 concerning drivers' driving and working hours is effective [EU06]. This regulation affects the planning of vehicle tours and routes by restricting the maximum driving times [Ran07]. Although compulsory for all member countries of the EC and therefore of high practical importance this regulation has found little interest in models for vehicle routing so far. Especially the restrictions for driving times during several days and weeks and the optional extensions of driving times are widely neglected.

The restrictions of Regulation (EC) No 561/2006 concern three different time horizons: single driving periods, daily, and weekly driving times. For single driving periods there are restrictions forcing drivers to take a rest period of at least 45 minutes after driving periods of no longer than 4.5 hours. The daily driving time is restricted to nine hours. However, twice a week the daily driving time can be extended to ten hours. Weekly driving times are restricted to a maximum of 56 hours. Still it has to be ensured that the total driving time of two consecutive weeks does not exceed 90 hours. Between two weekly driving times a weekly rest period of at least 45 hours has to be made which can be reduced to 24 hours under certain circumstances.

Neither in practice nor in literature there exist algorithms which are able to reflect such regulations in plans for vehicle routing and scheduling. The main challenge for developing suitable heuristic algorithms for that problem consists in the optimal planning of breaks modeled as flexible time windows which can be shifted but must obey to a set of complicated restrictions. The second severe difficulty consists in the simultaneous treatment of different time horizons. Planning algorithms for the tours of one day must keep in mind the workload of the previous day, the previous week, the actual week, and the planned workload of the next week.

3.9.3.2 Extension by Subcontraction

Although the managing of incoming new requests is the topic of many recent research activities [Lac04, Pan02], the decision whether to accept or to reject a new incoming request has not thoroughly been investigated till now. However, the decision about the acceptance of new requests is very crucial to the quality of the execution plans since a request which does not fit to the portfolio of the already accepted requests causes over-proportionately high costs opposed to relatively low incomes.

Freight forwarding companies have to face the fluctuating demand on the transportation market. Each day a variable number of requests is received from customers at a very short notice. On the other hand, the fixed costs of the own vehicle fleet, consisting e.g. in the wages for the drivers, taxes, and amortization costs, force a maximal utilization of the fleet. Thus, the number of own vehicles is reduced and only a part of requests is fulfilled by the own fleet. All the other orders are outsourced. Using own vehicles for the execution of tasks is called self-fulfillment, while subcontracting means involving an external carrier.

However, the decision for each request is not reduced to "either-or" in the sense of an isolated "make-or-buy" decision supported by adequate comparison methods. Instead, the complex "make-or-buy" decisions evolve into the reference-analysis among the items involved [WK99]. A major impact of such an analysis is noticed within the level and the structure of costs in the outsourcing enterprise [ZÖ0]. The process of constructing a fulfillment plan with the highest possible quality corresponds to solving the integrated operational transportation and freight forwarding problem [Pan02, KK06, KK07].

Caused by the tendency to outsource a part of the transportation tasks to external freight carriers, the need for covering the integrated operational planning problem in practice is soaring. In theory there exist only a few approaches that handle this problem.

The integrated operational transportation problem concerns almost all forwarders with an own fleet of vehicles. For all of the requests to be executed during the next planning period they have to plan the mode of the fulfillment, i.e. they must decide which of the requests should be fulfilled by one of their own vehicles and which should be forwarded to external carriers. In order to minimize their prime costs, they have to solve a usual vehicle routing and scheduling problem for those requests that are dedicated for self-fulfillment. The fulfillment costs for the execution of the set of all subcontracted requests can also be influenced by a skilful operational planning of the engagement of subcontractors. The corresponding planning process is called freight consolidation. The goal of freight consolidation is the minimization of the external freight costs and its degree of freedom consists in bundling the requests, assigning the bundles to carriers, and the arrangement of each single bundle.

The solution of the integrated operational transportation planning problem is a complex decision process which takes place on several levels (Fig. 3.63). The sub-problems of mode-planning, vehicle routing and scheduling, and freight consolidation are strongly dependent upon each other. That is why an exact solution of the entire problem can only be guaranteed by simultaneous solution approaches and good sup-optimal solutions can only be generated by heuristics which take the dependencies between the sub-problems into account.

In practice, the complexity of the integrated operational transportation planning problem is even higher due to complex methods of freight calculation. Most freight forwarders apply several forms of paying for outsourced services. So, there are different types of sub-contraction that should be combined at a single blow by using different subcontractors. Till now, no theoretical approach proposes an integrated solution where several types of sub-contraction are combined. All theoretical approaches in the literature involve only one single type of sub-contraction.

For the freight calculation of bundles with less-than-truckload requests the freight flow consolidation approach (without time window constraints) remains of highest practical relevance. Applying this approach fixed tariffs are used under non-linear

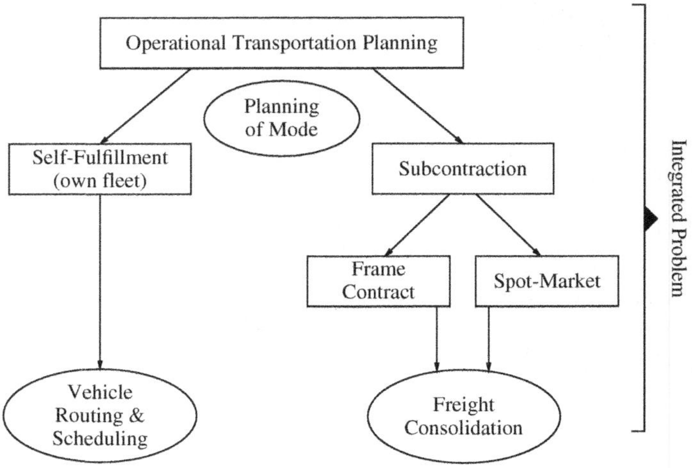

Fig. 3.63 Sub-problems of the integrated transportation planning

consideration of distance, weight of a bundle, and the type of goods that should be transported. The price for subcontracting is quoted on the basis of such tariffs, although it can be modified dependent on the driven direction.

In case of bundles with full-truckloads there are two methods for payment of freight charges which come into consideration. Applying the first method, complete tours are shifted to subcontractors on a fixed tariff basis which is dependent on the distance of the round route to be driven. There are no fixed costs connected with such usage of foreign vehicles, but the tariff rate for the variable costs of each distance unit is higher in comparison to usage of the own vehicle fleet as it covers a part of the maintenance costs (cf. left plot in Fig. 3.64). The second method consists in paying the subcontractors on a daily basis. In this case an external freight forwarder gets a daily flat-rate and has to fulfill all the received requests up to an agreed distance and time limit (cf. middle plot in Fig. 3.64).

Analysis of existing operational transport optimization software systems for freight forwarders shows that the problem is underestimated [KK07]. There is no

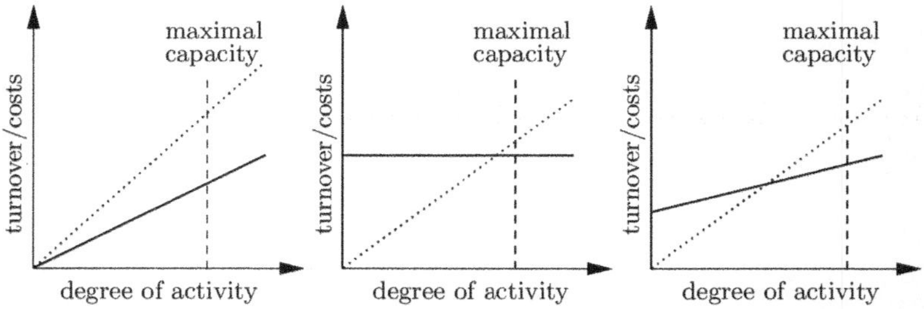

Fig. 3.64 Comparison of costs (*continuous lines*) and turnover (*dotted lines*) for the different types of sub-contraction and self-fulfillment: subcontractors on the tour basis (*left*), subcontractors on the daily basis (*middle*) and self-fulfillment (*right*).

suitable system for freight consolidation on the software market, and a system for the integrated solving of the planning problems for self-fulfillment and subcontracting is not available anyway. Due to the lack of software, the problem of splitting the request portfolio is solved manually, and there is only an appropriate support for the sub-problem in the self-fulfillment cluster. But finding good solutions for the global superordinate problem may be even more important than generating high quality solutions for one sub-problem. In practice, planning of the integrated problem is made hierarchically [JKK06]. In the first place the requests with the highest contribution margin are planned into the self-fulfillment cluster. Here, schedulers can be supported by software that optimizes the sub-problem of building round routes. Then the other types of sub-contraction are considered, also in a hierarchical order. The advantages of simultaneous planning are lost.

Revisiting the VRP-Example We assume now that sub-contraction on a tour basis is possible in the application introduced above. It has to be decided now which subset M_{SF} of customer sites should be covered by the routes of the own vehicles. Only requests contained in M_{SF} are distributed among the vehicles of the own fleet in the tour generation. All remaining requests are collected in the set M_{SC} and are forwarded to a service partner.

Since the decision about sub-contraction is based upon a comparison of the costs of self-fulfillment and sub-contraction, the relative level F of the sub-contraction costs controls the extent of the usage of the latter mode. More concretely, let F denote the charge to be paid to the incorporated service partner for each bridged distance unit.

The results obtained for different tariff levels F are summarized in Fig. 3.65 and Table 3.6. In Fig. 3.65 the subcontracted requests are indicated by a dashed line from the depot to the corresponding customer site. It can be seen that the number of subcontracted requests decreases if the costs for the incorporation of a service partner increase. Table 3.6 shows that with increasing freight tariffs the overall request fulfillment costs rise until the costs for the self-fulfillment of all requests are reached. It can also be seen that the decision to incorporate a freight carrier requires the update of the routing decisions of requests not considered for sub-contraction. This is mainly based on the fact that the consolidation of requests might become

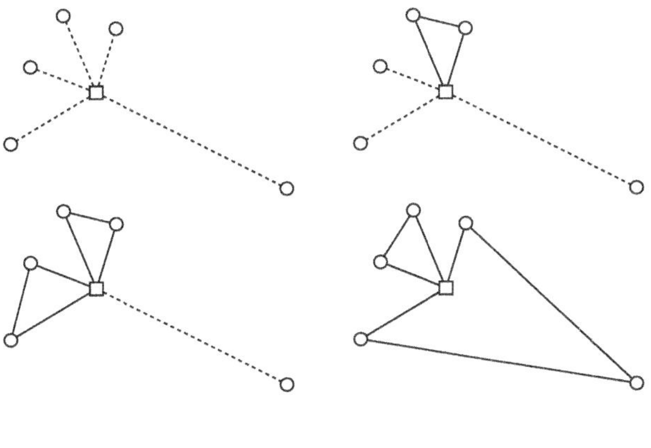

Fig. 3.65 Impacts of different sub-contraction charges: $F = 1.3$ (*upper left plot*), $F = 1.4$ (*upper right plot*), $F = 1.5$ (*lower left plot*) and $F = 2.1$ (*lower right plot*)

Table 3.6 Optimized solutions for different freight charge tariffs F

F	M_{SC}	route veh. 1	route veh. 2	costs
1.3	$\{1, \ldots, 5\}$	–	–	40.6
1.4	$\{1, 2, 3\}$	$0 \to 4 \to 5 \to 0$	–	56.3
1.5	$\{1\}$	$0 \to 3 \to 2 \to 0$	$0 \to 4 \to 5 \to 0$	58.9
2.1	$\emptyset$	$0 \to 4 \to 3 \to 0$	$0 \to 2 \to 1 \to 5 \to 0$	67.9

void if at least one request out of a request bundle cannot be used anymore in the route construction.

3.9.3.3 Dynamics

Recent research activities concerning practical aspects of vehicle routing problems concentrate on uncertainty of the available problem data. Real-time dispatching systems are studied in [GGLM03, FGS04, GNV04, SPG+97]. Surveys on vehicle routing and scheduling problems with incomplete planning data are given in [GP98, Psa95]. Robust planning is defined as the generation of plans that maintain their high or even optimal quality after subsequent modifications [Jen01]. Flexible planning refers to the generation of plans whose quality does not significantly decrease after the execution of algorithmic re-scheduling and alterations of the so far used plans [Jen01]. Robust transport scheduling approaches exploiting explicit probability distributions for future demand are investigated in [BBC+05, Jai88]. Flexible planning approaches typically solve online decision problems. A survey of online vehicle routing and scheduling problems is provided by [Kru01]. Real-world applications are tackled in [FGS04, GNV04, SS98]. Typical unexpected events which necessitate a re-optimization of the current execution plan are situations of congestions, accidents, defective vehicles, and the refusal of acceptance by one of the customers. Especially the last point is often underestimated or even neglected. But the refusal of acceptance has a great impact on the continuation of a tour because the refused goods must stay on the vehicle until the end of the tour and the capacity left for the concerned vehicle may not be enough to complete the tour in the originally planed way. Models and algorithms for online optimization tackle and try to overcome the difficulties caused by incomplete and uncertain knowledge. Only little research effort is dedicated to the breakdown or disturbance of the actually performed transportation processes while most investigations concentrate on incoming new requests and the re-optimization of the current plan on the basis of the actually executed routes and the new (extended) set of orders to be performed. For online optimization, especially for corresponding TSP variants, we refer to the Sect. 3.11 of this book.

3.9.4 Conclusions

Most optimization problems in the area of operational transportation planning are hard to solve. They are subject of intensive research since many years. Including important practical aspects to these problems makes the solving even more challenging.

Mathematical programming approaches (e.g. integer linear programming) fail to produce reasonable solutions for most real-world scenarios due to the complexity of practical vehicle routing problems. This contributes to the success of heuristic approaches. Meta-heuristics which support local search approaches are proposed and prove their applicability. These meta-heuristics suffer from problems related to pre-emptive sub-optimal convergence and are not flexible enough for managing complicated practical situations of real world-problems. To work against this trouble the application of interactive problem solvers is proposed [KS02]. They join the advantages of local search based automatic problem solvers with human inspiration and pattern recognition in order to improve the quality of the problem solutions and to enlarge the set of solvable problems.

To sum up, there is still a lot of research to be done for the optimization of operational transportation planning. On the one hand, the performance of exact and heuristic algorithms for well-known and intensively investigated problems has to be improved. On the other hand the structure and solving of important extensions due to practical aspects will be one major subject of future research.

3.10 Innovation Management: Optimizing the Structures of Innovation Processes

Martin Möhrle, Ewa Dönitz

3.10.1 Introduction

Today, the planning of innovation processes is supported by a large number of instruments, many of which are specifically designed for innovation management purposes, such as innovation evaluation, cost/benefit analysis, and lead user analysis (see [BCW04] [SG02] and [vH05] for overviews). Many other instruments come from classical project management, e.g. project structure planning, deterministic network planning, stakeholder analysis, and project controlling [Ker05]. These project management instruments are regularly incorporated in software tools such as Microsoft Project and others.

Although the aforementioned instruments have indeed proved to be helpful for quite a few innovations, some innovations bear specific characteristics, which lead to a need for more advanced instruments and modeling. The characteristics in question are (i) the multiple risks and uncertainties, by which these particular innovations are influenced and which may be traced back to different origins, such as market demand, technological progress or administrative regulations. Also, there are (ii) multiple goals situations underlying those innovations, so that conflicts between goals are likely to arise. Apart from that, both characteristics may change in the course of time.

First of all, in this chapter the characteristics of innovations, innovation processes, and innovation projects will be discussed in detail. Then, in order to deal with these particular characteristics, an approach will be taken which involves two planning instruments: On the one hand, a variant of stochastic network planning will be suggested, while on the other hand, this will be combined with multi-goal planning. This section is aimed at presenting an instrument that supports the innovation planner interactively in handling challenging characteristics. In the discussion, it will be shown how aspects of uncertainty in connection with bounded rationality occur in innovation management and what impact genetic algorithms may have.

3.10.2 Characteristics of Innovations, Innovation Processes, and Innovation Projects

To begin with, some characteristics of innovations, innovation processes, and innovation projects will be explained to provide a setting for the instruments discussed later on.

Innovations: In the parlance of business administration, an innovation is defined as a new product, a new service, a new production process, or a new combination thereof, which is introduced by a company for the sake of gaining a competitive advantage [MS05]. Accordingly, innovations comprise more than R&D activities, as marketing, production, purchasing, and other – especially managerial – activities have to be combined to create an innovation. As Kleinschmidt, Geschka and Cooper [KGC96] have shown by means of empirical analysis, the rate of success of innovation projects exceeds the average, if these activities are applied in parallel (with different intensity) throughout the entire innovation project. Two further characteristics are typical of innovations:

i. The innovator is faced with risks and uncertainties of different kinds and origins. Three major types of risks and uncertainties must be taken into account: technical risks, market risks, and resource risks [Spi06]. For example: Will the technology under consideration be ready and stable enough for the innovation? Will it be outpaced by another technology? Will the market demand still be there, when the development is complete? Or will it grow to new, unexpected dimensions? Will the company be able to exploit a possible first mover advantage? Will competition between different investments within a company result in a resource stop for the innovation? – These risks and uncertainties have to be identified and considered during the planning process.
ii. The innovator has to deal with a multiplicity of goals, of which a few might complement each other, while many tend to be contradictory (for systems of goals see [SBA02, MD05]). For instance, in the context of product innovation, lead time to market and the total cost of the innovation project are often negatively correlated. So are product quality and price per product unit. Some

authors, like Brockhoff and Urban [BU88], try to subsume this multiple goal situation under the category of a net present value optimization, but for planning in accordance with flexibility issues it seems helpful to argue on the basis of disaggregated goals.

Innovation processes: An innovation process is defined as the generic structure of work packages and work flows within a company and beyond [DP95]. Such innovation processes can be defined on various levels, ranging from single business units to whole industries. For instance, due to legal requirements, there is a typical innovation process for companies in the pharmaceutical industry, as they are obliged to test substances in different phases in order to gain permission to market a new drug, cf. [Ger04, pp. 24–25].

Innovation projects: Whereas an innovation process represents a generic structure, as defined above, an innovation project represents the particular structure of work packages and work flows for one specific innovation. Often, the innovation project is planned by means of adapting a generic structure, however, there still are many specific aspects to be taken into consideration. For instance, some innovation projects necessitate the involvement of suppliers, in which case an expansion of the generic structure may ensue.

3.10.3 Stochastic Networks for the Planning of Risks and Uncertainties

In the planning of innovation processes and projects, the first challenge is to survey and consider risks and uncertainties. An adequate instrument for this purpose is the technique of stochastic network planning, which will subsequently be presented in the form of its variant GERT (Graphical Evaluation and Review Technique), originated by Pritsker (1966), and the related software-tool GERT-Net [Sch02]. In GERT-aided process and project planning there are two distinct ways of approach to the examination of risks and uncertainties:

- One can either establish a stochastic branching with the logic of decision nodes
- or apply stochastic activity durations to the arcs.

3.10.3.1 The Logic of Decision Nodes and Arcs

In GERT-network graphs, decision nodes are employed to enable a branching of activities. The course of the process or project is defined by means of these decision nodes, each of which possesses an entrance logic and an exit logic. Three different elements are permissible as entrances, two as exits [Hen91, pp. 55–56], [Vö71, p. 26], and [Fix78, p. 16]. Consequently, there are six different decision node types in total (see Fig. 3.66).

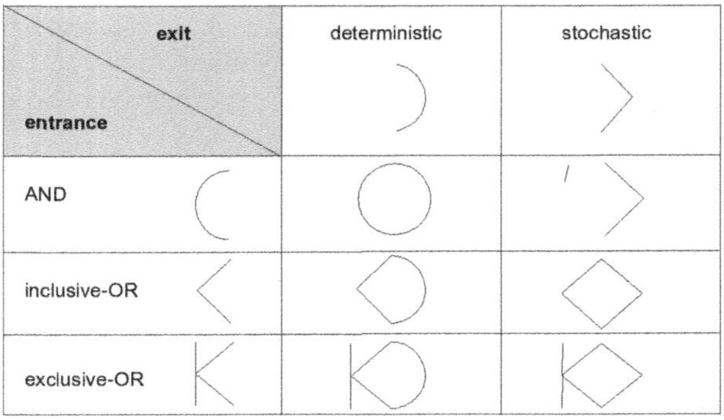

Fig. 3.66 Logic of Nodes. Source: [Hen91]

Three possible elements are at disposal on the entrance side:

- AND-entrance: The node is activated, as soon as *all* incoming activities are completed.
- Inclusive-OR-entrance: The node is activated, as soon as *at least one* incoming activity is completed.
- Exclusive-OR-entrance: The node is activated, as soon as *exactly one* incoming activity is completed.

On the exit side, there is a distinction of two elements:

- Deterministic exit: After the decision node has been activated, all outgoing activities are started, i.e. each activity possesses a probability parameter of 1.
- Stochastic exit: After the decision node has been activated, only one of the outgoing activities is performed. Which activity is to be performed, is decided on the basis of a pre-set discrete probability distribution.

The fact that the duration of activities can be a random variable, is another property of GERT-networks. This random variable is represented with the arcs and can be sustained by different types of probability distribution, e.g. exponential, binominal, normal or geometric distribution [WGW72]. For instance, if an activity undergoes frequent repetition (e.g. in the course of a loop) one thus arrives at different values representing the actual duration of execution.

3.10.3.2 An Example from Automotive Industry

The following example of a complex innovation project from the automotive sector is meant to clarify how GERT modeling works. In the automotive industry great store is set by increasing the probability of a project's success as well as by reducing its duration. There are numerous arguments in favour of the curtailment of project duration [DKS97], e.g. regarding the decrease in the complexity of planning and controlling through the evasion of a temporal overlapping of projects, a reduction of the costs caused by capital employed, the attainment of new time margins and the minimization of difficulties with regard to schedule, and the establishment of oppor-

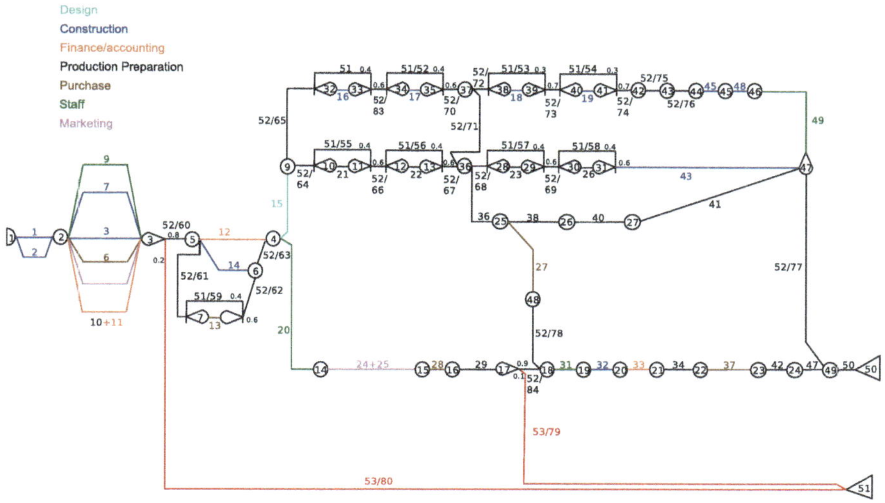

Fig. 3.67 Stochastic network graph for an innovation in the automotive industry. For logic of nodes see Fig. 3.66. Source: [Sch99a].

tunities and available capacities to facilitate a swifter execution of further projects in the future.

This particular aspect was treated by Schreyer [Sch99a]: Starting from a GERT network graph, of the kind on which US-companies base their implementation of projects, Schreyer devised several variants, which differ from the basic network graph either due to the assumption of certain pre-conditions (reducing the duration of some activities by means of learning effects, innovation capacity and the involvement of repetitive elements) or to a consistent reformation of the process structure (reengineering). One of these variants (see Fig. 3.67) is characterized by a marked synchronization of processes [Sch99a].

The different variants were simulated with the software GERT-Net, which is used for the input, arrangement and simulation of GERT-network graphs [Sch02]. The calculation of the GERT-network graph for Schreyer's reengineered variant produced a mean of 253 time units (4.8 years) for the entire project duration, with a probability of success amounting to 71 per cent (see Fig. 3.68). The simulation of GERT-network variants with different parameterization and a comparison of their results enable a detection of coherences and weaknesses within innovation processes projects.

3.10.3.3 Background Information on GERT

GERT-network planning represents a further development of well-known network planning techniques like CPM (Critical Path Method), PERT (Program Evaluation and Review Technique), MPM (Metra Potential Method) and GAN (Generalized Activity Networks), see Fig. 3.69. The conventional, deterministic methods of network planning facilitate a delineation of projects, whose course is clearly

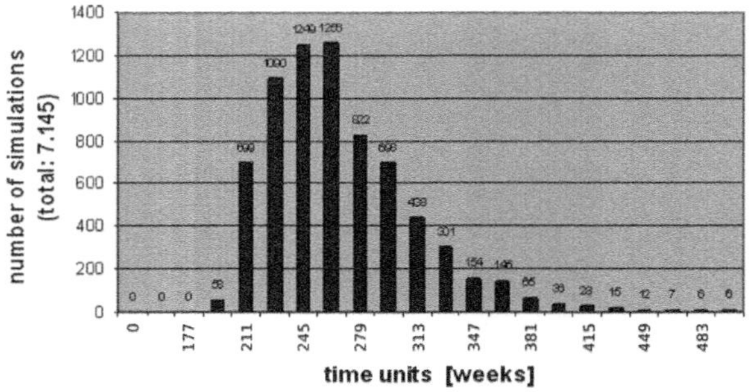

Number of simulations	7145
Success probability	71,45%
Average	253,26
Standard deviation	30,94
Minimum	203
Maximum	482

Fig. 3.68 Frequency distribution of time required for the final node in case of project success. Source: [Sch99a, p. 157].

Duration of activities \ Sequence of activities	deterministic	stochastic
deterministic	CPM, MPM	GAN
stochastic	PERT	GERT

Fig. 3.69 Overview of deterministic and stochastic network planning techniques. Source: [Sch99a, p. 51]

determined as, during their realization, all activities are definitely completed [Fix78]. Stochastic network graphs, however, are also suitable for surveying projects, whose course is not determined completely from the outset.

3.10.4 Planning with Multiple Goals

In addition to the consideration of different risks and uncertainties, the pursuance and coordination of multiple goals represent another challenge in the planning of innovation processes and projects. Subsequently, different systems of goals will be introduced, followed by a discussion of Pareto-optimization within such systems.

3.10.4.1 Systems of Goals for Innovation Projects

In business management, goals are synonymous with desired future conditions in or of an organization. As the definition thereof often relates to various components, and as decision-makers tend to differ in their understanding of the assumed goals, these future conditions are described as systems of goals or multiple goal decisions [BS02a]. In innovation projects, the goal system is often defined with reference to the project management system of goals, which contains the so-called "magic triangle" of quality, cost and schedule [Har04]. This may be useful, as it involves the application of proven systematics. However, it also stands for a limitation of creative

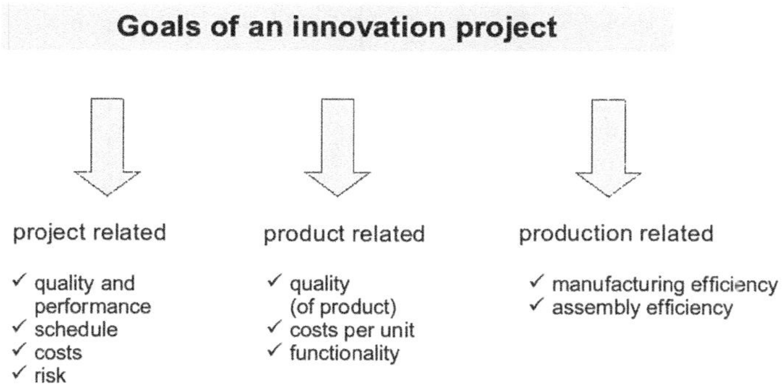

Fig. 3.70 Concrete goal system of an innovation project. Source: [MD05]

scope within an innovation project, as different goals either have to be converted into restrictions or be included in the quality target by way of a benefit calculation.

Therefore, it seems appropriate to plead for the use of a more comprehensive system of goals, such as the one devised by Specht, Beckmann and Amelingmeyer [SBA02]. Here, each goal included in the system contributes to the attainment of economic and non-economic targets within the company. The goals that are relevant for an innovation project can, for example, be sorted into three groups: project-related, product-related, and production-related goals (see Fig. 3.70).

The process of defining goals is a vital component of innovation planning. Based on the company's aims, innovation planning has to include the determination and formulation of specific goals in accordance with the given situation. Consequently, planners are required to monitor their system of goals constantly, in order to maintain its adaptability throughout the course of the project [SBA02].

The outlined system of goals for innovation projects does not result in all goals being complementary (for a similar product life cycle study see [Dow01]). In fact, there are several contradictions. Some examples:

- In most cases the costs of an innovation project, which ends with the market launch of a new product, and the costs per product unit, will evolve in opposite directions: Extensive development and testing are likely to generate lower costs per product unit, but higher project costs.
- The schedule of an innovation project and the functionality of its new product will also be in opposition: The sooner an innovation project has to be completed, the fewer functions its product is likely to possess.
- The same relationship can be found between the schedule of an innovation product and the manufacturing efficiency of the related product.

3.10.4.2 Pareto Optimization

Planning with multiple goals means seeking solutions to problems in which several competing goals are required to be optimized, see [Gö95], and the articles of Küfer and Roosen in this book. Generally, this allows for no such thing as a "perfect" solution, which – in the case of simple problems – is determined by means of a combination of alternatives, and in which all goal functions assume their extreme values [Vie82]. Instead, one tries to find a solution set with regard to different functions of goals, i.e. various solutions that are, in a sense, just as valuable as others. Accordingly, there is no "best" solution to speak of – only one variant over which others have no predominance.

This type of solution is generated with the aid of Pareto optimization [Gö95]. The concept of Pareto dominance was devised by Vilfredo Pareto, who laid the foundations of research into multi-criteria optimization. A target vector with certain values dominates another, if it reaches a higher value for at least one criterium while assuming no lower value for any of the remaining criteria. Consequently, a Pareto optimal set consists of solutions which are not dominated by others. The algorithms used in Pareto-based techniques are aimed at calculating and offering a set of excellent and, preferably, diverse solutions. Then, users are able to select a solution from this set according to their individual preferences.

For instance, the management of a company who has commissioned an innovation project can discuss options of handling the conflicting goals of project costs and costs per product unit with the project manager (see Fig. 3.71). This produces a Pareto set, which does not contain a single "best" point of realization, but requires the management to make a decision in favour of one predominant point on the basis of additional considerations.

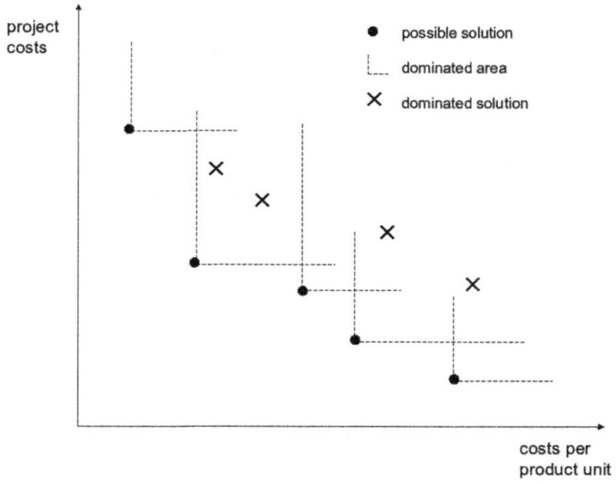

Fig. 3.71 Pareto-set for the comparison of project costs and costs per product unit

3.10.5 *A Concept of Integration: The Stochastical, Multi-Goal-Oriented, Interactive Planning of Innovation Projects*

With stochastic network planning on the one hand, and multi-goal-planning on the other, there are two instruments at disposal, which each permit a separate handling of the aforementioned challenges to innovation planning. What is still missing, is an integration of these instruments in connection with the consideration of changes in the schedule. To deal with this, we suggest a stochastic, multi-goal-oriented and interactive planning process. Here, three theoretical foundations have to be unified. (i) Stochastic network graphs with expanded parameterization are required for the modelling of innovation projects. (ii) Genetic algorithms help to establish Pareto-sets and thus assist a multi-goal optimization. (iii) Decision-makers must be enabled to influence the course of the process or project interactively, especially concerning the definition of a system of goals.

One possible procedure for the optimization of innovation projects will subsequently be presented (see Fig. 3.72). It may be divided into three phases: (i) a basic simulation, (ii) a Pareto optimizing simulation, and (iii) interactive changes in the settings after changes in the environment.

Basic simulation: For the basic simulation the planning problem is first modelled by means of a GERT-network graph with decision nodes and stochastic arcs. If a generic structure for an innovation process exists, it may be used as foundation. Relationships between activity duration and goals have to be specified (see below), and activity acceleration figures are set to 0 (see also below). Then the simulation model is repeatedly run. With GERT simulations it can be calculated

- how probabilities are distributed for success and abandonment of the project respectively,
- how project duration and other variables, which depend on the number of iterations within loops, are distributed in the cases of success and abandonment respectively,

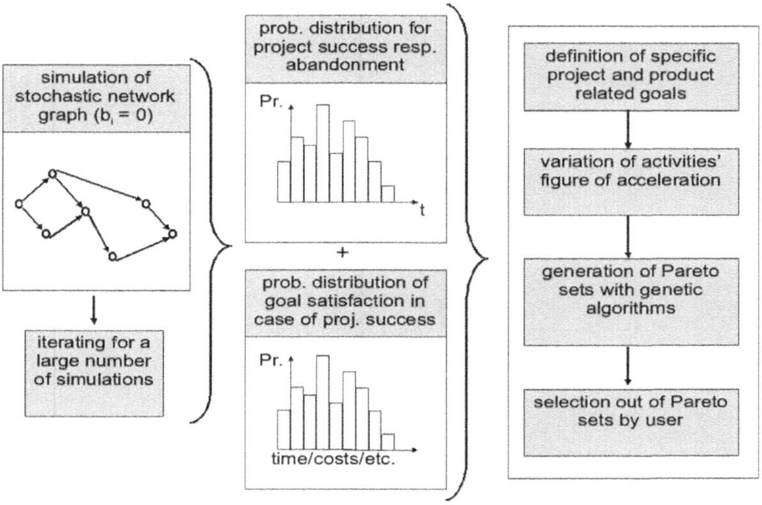

Fig. 3.72 Possible procedures for the optimization of innovations

- and how the remaining variables for the goals are in the basic attempt (in those cases one will find some upper or lower bounds e.g. for costs per product unit).

The basic simulation helps understanding the innovation project and evaluating the parameters used.

Pareto optimizing simulation: Then, the innovation project is to be optimized with regard to its defined goals. This includes a variation of activity acceleration figures and the establishment of relative degrees of goal realization. With the aid of genetic algorithms, Pareto sets of the respective results are drawn up and presented to the user in a comprehensible graphic form. From these Pareto sets the user selects the solution which is best suited to the current situation.

Interactive changes: In the course of an innovation project, goals may change dramatically because of competitor activities, the restructuring of resources or political circumstances. In such cases, the innovation project planning has to be updated by means of appropriate re-optimization.

In the following sections, two idiosyncrasies of the model of procedure illustrated above will be dealt with: First, we will attend to GERT-network planning with expanded parameterization. Second, we will take a closer look at how adequate solutions are generated and selected, and discuss possible ways of representing Pareto solutions.

3.10.5.1 Expanded Parameterization of GERT-Networks with Goal Functions

In accordance with the suggested procedure, a module should be created on the basis of the GERT-Net program, which permits a comprehensive parameterization of activities in an innovation project. In addition to the usual GERT-network activity parameters, i.e. the probability of the completion of an activity p_i and the duration of the activity d_i, two further parameters are to be defined for this purpose: (i) the specific acceleration figure of the activity b_i and (ii) its specific relation with goals f (a_i) (see Fig. 3.73).

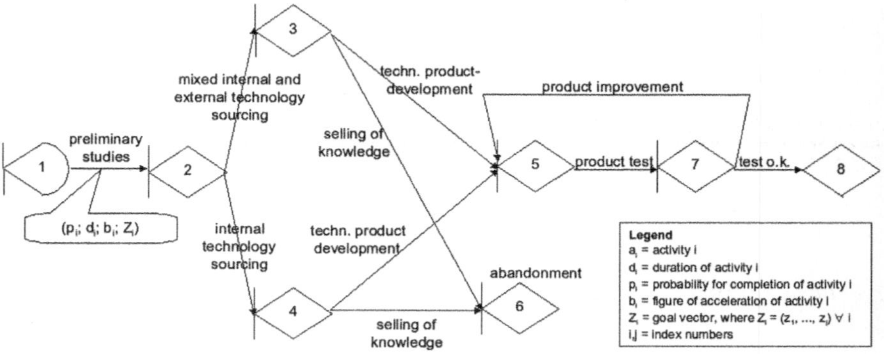

Fig. 3.73 Example of GERT-network graph with expanded parameterization

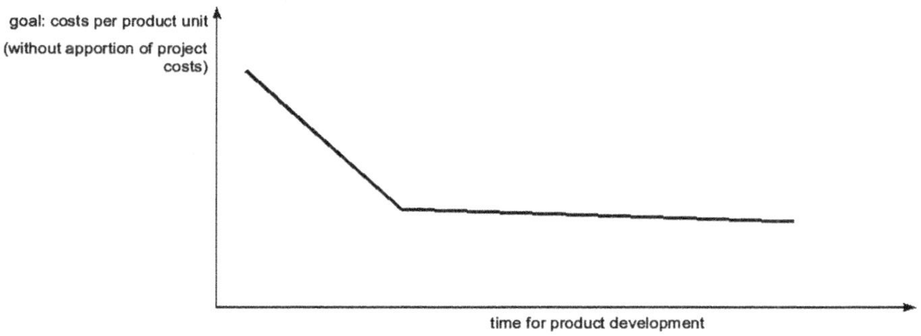

Fig. 3.74 Specific relation between duration of product development and costs per product unit

Acceleration figure: The management is able to influence the duration of individual activities decisively by various measures, which might cause a change in costs. Their scope for action includes the selection of a specific acceleration figure, which shows to what percentage the temporal acceleration potential is being exploited [Sac00]. The duration of an activity can be reduced, e.g. by means of new methods, innovation improvement based on learning effects, the improvement of internal organizational coordination and communication, the utilization of additional resources or outsourcing [Sch99a].

Relations with goals: An activity's specific relation with goals is illustrated by means of functions, and shows to what extent each predetermined goal is influenced by the related activity. To give an example, a possible connection between the duration of product development and the costs per product unit is illustrated below (see Fig. 3.74). This relation is partly marked by reverse trends. In the first section, there is a clear opposition: The more intensive and, consequently, extensive the phase of development is, the lower the costs per unit are likely to be. In a second section, which lies beyond a certain quantity of time for product development, no further effectuation can be detected.

3.10.5.2 Generation and Selection of Pareto Optimal Solutions

Algorithms that employ Pareto based techniques are aimed at calculating and offering a set of solutions which are not only excellent but also as diverse as possible. Those solutions which are not dominated by others, are then presented to the decision-maker. For this purpose, the suggested procedure should be incorporated in a graphic user interface for the figuration of Pareto sets. Such a clear pictorial display is supposed to assist users in making their decisions.

In addition to project related goals, project costs and schedule, for instance, product related goals such as quality, safety and service options are to be taken into account. Some of these goals can be predetermined: In spite of minimal project costs a high quality and a first-rate service are to be attained. All remaining goals are subject to variation. This procedure is illustrated in Fig. 3.75.

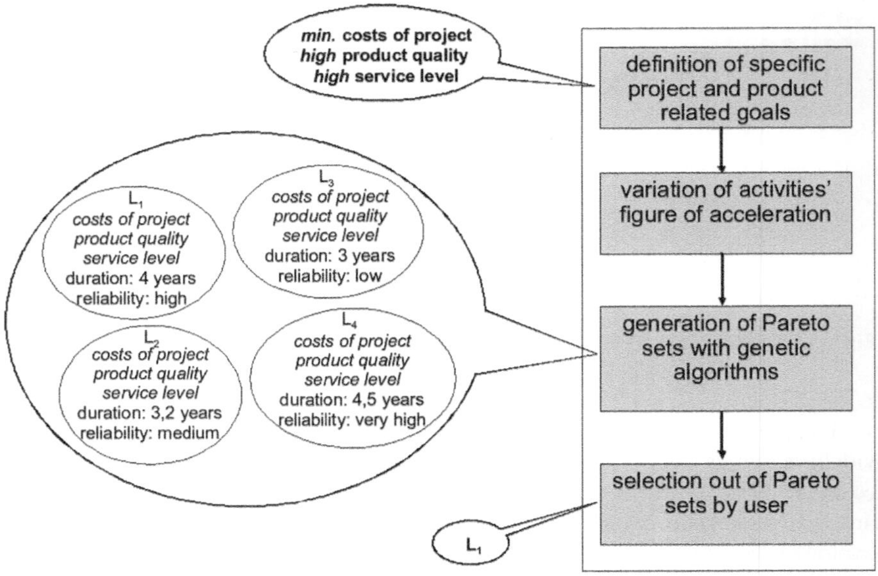

Fig. 3.75 Generation and selection of Pareto sets. The predefined goals within the Pareto sets are printed in italics.

The generated Pareto sets are now to be graphically represented. In this particular case a two-dimensional representation will suffice, as three of five goals are predetermined. If a great number of goals and, consequently, a great number of dimensions are involved, it becomes particularly important to provide a clear graphic display. With more than three goals subject to variation at the same time, an interactive display has to be employed [dFIfTuW03].

3.10.6 Discussion

To facilitate a discussion, two aspects will be singled out: (i) uncertainty in connection with limited rationality, and (ii) the impact of genetic algorithms.

- As the previous chapters have shown, uncertainties and risks are of crucial importance where innovations are concerned. This fact leads to the question, how intensively innovation planners are using stochastic network planning at present. The reply is that they still retain an attitude of reserve towards this method of modeling, not entirely without reason. One cause may be that, so far, no adequate professional software tool is available. Moreover, some sort of bounded rationality may be at play: Possibly, innovation planners tend to avoid stochastic network planning for reasons of communication. To be more precise: If an innovation planner has to settle a date for a specific task with an employee, it appears easier to agree on a certain day than on a period of time. Hence, innovation planners might still prefer the less complex method of deterministic network planning (see Sect. 3.10), although stochastic network planning would provide far better insights into the time and risk frame of the innovation project. In order to overcome this planning dilemma, a combination of both variants appears to be sensible.

- Evolutionary strategies for optimization may be of great help in solving problems with innovation planning. They are especially valuable: (i) if there is a great number of non-linear relations between activities and goals, and (ii) if the goals of an innovation project are shifting – which may be due to various causes, such as competitors' activities or legal changes – a rapid re-optimization of the innovation project has to take place. In such cases, the specific attributes of genetic algorithms [Rec94] can be helpful.

In general, the presented concept is an approach towards a new understanding of the structures of innovation processes and projects. However, it requires further development. Also, prototypes of a respective IT system should be implemented and tested.

3.11 Structuring a Dynamic Environment: Combinatorial Online Optimization of Logistics Processes

Martin Grötschel, Benjamin Hiller, Andreas Tuchscherer

3.11.1 Introduction

In *classical optimization* it is assumed that full information about the problem to be solved is given. This, in particular, includes that all data are at hand. The real world may not be so "nice" to optimizers. Some problem constraints may not be known, the data may be corrupted, or some data may not be available at the moments when decisions have to be made. The last issue is the subject of *online optimization* which will be addressed here. We explain some theory that has been developed to cope with such situations and provide examples from practice where unavailable information is not the result of bad data handling but an inevitable phenomenon.

We begin with some informal definitions concerning online optimization, refraining from giving formal definitions to avoid "technical overkill". Online optimization deals with the following situation: At certain points in time, we have to make immediate decisions based on some data of the past, but information about the future is unavailable. The goal is to make "good" decisions, where the quality of the decision depends on the information available so far, but also on data arriving in the future. In the classical optimization world we would study the problem to be considered, invent a mathematical model supposed to catch the key aspects of the problem, collect all data needed and use mathematical theory and algorithms to solve the problem. We call this approach in this chapter *offline* from now on since the data collection phase is totally separated from the solution phase. In online optimization these phases are intertwined and there are many situations where this is not due to data acquisition difficulties but due to the nature of the problem itself.

A little more formal, in an *online problem*, data arrive in a sequence called *request sequence*, and each time a new request arrives, a decision has to be made. If there are costs or some other objectives involved, we face an *online optimization problem* since we would like to make decisions in such a way that "at the end of the day", i. e., when the last request has arrived and been processed, the total cost is as small as possible.

There are lots of variants of this "basic online framework", and in each case one has to exactly state what the side constraints are, what online precisely means, how costs are calculated, etc. We outline a few of the possibilities that come up in practice.

For someone familiar with *optimal control* the issues indicated here probably sound very familiar, and of course, online optimization is not a topic invented in combinatorial optimization or computer science. Online problems occur, for instance, when steel is cast continuously, chemical reactors are to be controlled, or a space shuttle reenters the atmosphere. There are some differences between these control problems and the *combinatorial online problems* we describe here, however. In the continuous control case, one usually has a mathematical model of the process. This is typically given in the form of ordinary and/or partial differential equations. These describe the behavior of the real system under parameter changes over time. One precomputes a solution (often called optimal control) and the online algorithm has the task to adapt the running system in such a way that it follows the optimal trajectory. For instance, when a space craft reenters the atmosphere it is clear where it has to land. It should follow a predetermined trajectory to the landing place. The measuring instruments determine the deviations from this trajectory, and the algorithms steering the space craft make sure that the craft follows the trajectory using the available control mechanisms on board.

In contrast, the combinatorial online optimization problems we consider have no "trajectory" that one could precompute. Decisions are *discrete* (yes or no), can only be made at certain points in time and not continuously, and in many cases, decisions once made cannot be revoked. In other words, although optimal control and combinatorial online optimization have certain aspects in common, there are significant differences that lead to different problems in practice and require different mathematical treatment.

Let us look at an example. Recall your last move to a new apartment. Moving the furniture into a truck and packing them safely is a nontrivial task. From a mathematical point of view we call this a three-dimensional packing problem. Cutting the pieces of a suit from a cloth roll to minimize waste is a similarly demanding task. An extremely simplified version of these two problems is the *bin packing problem*. We have a (possibly infinite) number of one dimensional bins, all of the same height. Certain items (all with the same "footprint", matching the footprint of the bins, but possibly with different heights) have to be packed in the bins. The goal is to use as few bins as possible. If all items are given in the beginning, we have an offline bin packing problem. If items are generated one by one and we have to pack each item as soon as it arrives and if we are not allowed to repack, we have a "standard" online bin packing problem. It may be that we are allowed to keep only a fixed number of bins, say m bins, open and that we are also allowed to move items between the open bins. But as soon as we need a new bin, we must close one bin and move it away. We call this problem the *m-restricted online bin packing problem with repacking*.

The latter is a typical situation that comes up when you move. A certain number of boxes are open in a room, you pack items into the boxes, are allowed to repack, but due to lack of space, whenever a new box is needed, one currently open box has to be closed and moved away. Again, the goal is to use as few boxes as possible.

There may also be some clock running. Requests appear in sequence and you are not necessarily obliged to serve them immediately (as in the standard problem above), but there may be extra costs incurred with every time unit a request that has appeared is not served. This version of an online optimization problem will be called the *time stamp model*. A natural application is the ADAC Yellow Angels problem to be discussed later.

Given an online optimization problem (e. g., one of the bin packing versions described above), an *online algorithm* is an algorithm that makes decisions respecting the side constraints of the particular online problem that it is designed for. In the standard model of online computation the algorithm has to decide immediately what to do and may not reconsider the decision later (e. g., it packs an item into a bin and cannot repack). In case of the m-restricted online bin packing problem with repacking the online algorithm may repack the m open bins and move items between them. But as soon as a bin is closed the algorithm has no further access to the items in the closed bin. Similarly, in the time stamp model the online algorithm is not always obliged to act immediately at the appearance of a request, but may have to pay a lot if service is delayed.

So, what an online algorithm is allowed or supposed to do depends on the special problem and its side constraints. There is another issue: time restrictions. In the theory of online algorithms, the running time is usually not an issue, i. e., these algorithms may use any amount of computing time or computing power. The only issue of importance here is the lack of information about the future.

If time is precious we speak of *real time problems* and *real time algorithms*. Again, what real time means depends crucially on the time frame of the process considered. Routing decisions in telecommunication (an important real time problem) have to be made almost immediately, while in the Yellow Angels case, ten seconds are an upper bound for the online algorithm to come up with the decision. Control algorithms for elevators may use about a second for a decision, but longer response times would not be tolerated by the passengers.

An area where online problems occur in abundance is logistics which is at the heart of modern production and service facilities. In many real-world applications, logistics systems are automatically supervised and permanently controlled. The online problems arising here usually correspond to the time stamp model (occasionally with a short lookahead) with additional real-time requirements.

As an example, consider an automated warehouse used to store and retrieve goods. Pallets with goods have to be stored and are to be transported from the storage to vehicles for further distribution. The overall goal of a good control is to ensure a constant flow of pallets such that the vehicles do not have to wait too long, to avoid congestion, and to coordinate the traffic. The control decisions are usually discrete, e. g., which pallet to route on which conveyor belt and/or vertical elevator.

This control problem is again online since control decisions have to be made in view of an unknown future and the control has to be updated each time new information becomes known. The time frame for the online algorithms in this area is usually in the range of milliseconds to seconds – depending on the mechanical speed of the stacker cranes or other moving devices. In order to develop good online algorithms it is important to analyze the impact of the control decisions on the current and future performance of the system.

3.11.2 Theoretical Analysis

In this section we describe the approach usually used for evaluating the quality of a given online algorithm. This theoretical concept is called *competitive analysis*. Before introducing competitive analysis in detail, we present some example problems that are revisited later on.

3.11.2.1 Examples for Online Optimization Problems

Online Traveling Salesman Problem. The probably most famous combinatorial offline optimization problem is the Traveling Salesman Problem or TSP for short. An instance consists of a set of locations that are to be visited in such a way that the total distance traveled is minimized, see Sect. 3.9.

To define the online version of the TSP in detail, we need to introduce the notion of a metric space. A metric space is simply an abstract way to "measure" distances. It consists of a set of *points* M and a function $d\colon M \times M \to \mathbb{R}_{\geq 0}$ supposed to give distances between each pair of points. Therefore, d has to satisfy the following properties:

1. $d(u, u) = 0$ for all $u \in M$
2. $d(u, v) = d(v, u)$ for all $u, v \in M$ (symmetry)
3. $d(u, v) + d(v, w) \geq d(u, w)$ for all $u, v, w \in M$. (triangle inequality)

The online version of the TSP is as follows. We are given a metric space (M, d) with a distinguished point o, the *origin*, where the "server" is located at time 0, and a sequence of n requests $r_1, r_2, \ldots, r_n$. Each request is a pair $r_j = (t_j, p_j)$ specifying the time $t_j \geq 0$ at which r_j is released and a point $p_j \in M$. The release times form an ordered sequence in the sense that $t_i \leq t_j$ if $i < j$. The server has to visit all the requested points as they arrive; it may not move faster than unit speed. This problem is called Online Traveling Salesman Problem, for short OnlineTSP.

There are several reasonable objectives for the OnlineTSP, e.g.:

Completion time: This is the total time for serving all requests and returning to the origin o.

Maximum waiting time: The waiting time of a request r_j is the difference between the time when r_j is served, i. e., the server reaches point p_j, and the release time t_j.

Average waiting time: The average of all waiting times.

There are cases where more than one of these objective functions need to be considered. This leads us to multicriteria online optimization, a subject that has received almost no attention in the research literature yet. We will touch upon the issue of "balancing" the objective functions above in the elevator case study in the next section.

Online Bin Coloring. Another simple online optimization problem is Online Bin Coloring, which is as follows. We are given a request sequence consisting of unit size items $r_1, r_2, \ldots$ where each item has a color $c_j \in \mathbb{N}$. The items are to be packed into bins, all of the same size $B \in \mathbb{N}$, as soon as they arrive. Repacking an item later on is not allowed. At each moment there are $m \in \mathbb{N}$ empty or partially filled bins. Whenever a bin is full, it is closed and replaced by a new empty bin. The objective is to pack the items in such a way that the maximum number of different colors in a bin is as small as possible.

3.11.2.2 Competitive Analysis

Competitive analysis provides a framework to measure theoretically the quality of online algorithms. More precisely, it seeks to answer the question what is lost in the worst-case due to lack of information. Competitive analysis was introduced formally in 1985 by Sleator and Tarjan in [ST85]. The term competitive analysis was first used in [KMRS88]. However, Graham already used this method in 1966 for evaluating algorithms for machine scheduling, see [Gra66]. For an overview on competitive analysis, we refer to [BEY98].

In the following, we only consider online minimization problems. Competitive analysis compares the performance of a given online algorithm ALG to that of an algorithm that knows the complete request sequence in advance and can serve the requests at minimum cost. We call this benchmark algorithm *optimal offline algorithm* OPT. For a request sequence σ, we denote the corresponding optimal offline cost and the cost of ALG by $\text{OPT}(\sigma)$ and $\text{ALG}(\sigma)$, respectively. The algorithm ALG is said to be c-competitive for $c \in \mathbb{R}$, $c \geq 1$, if

$$\text{ALG}(\sigma) \leq c \cdot \text{OPT}(\sigma)$$

for all request sequences σ. The *competitive ratio* of ALG is the smallest value c such that ALG is c-competitive. ALG is called *competitive* if ALG is c-competitive for some constant $c \geq 1$ that does not depend on the length of the sequence.

An online algorithm is competitive, if it competes "well" with the optimal offline algorithm on *all* request sequences. That is, competitive analysis is a *worst-case measure*. Hence, in order to show that the competitive ratio of an online algorithm is bad, it suffices to find one sequence of requests on which the online algorithm appears bad compared to the optimal offline algorithm.

We can think of competitive analysis as a game between an *online player* and a *malicious adversary*. The adversary constructs a sequence of requests to be processed by the online player. The adversary has complete knowledge of the online player's strategy that corresponds to an online algorithm, and he intends to construct a sequence such that the cost ratio for the online player and the optimal offline cost is maximized.

Examples. Recall the standard online bin packing problem as described in the introduction. The algorithm FirstFit which packs each item in the first open bin providing sufficient space is known to be 1.7-competitive. Van Vliet [vV96] proved that no online bin packing algorithm can be better than 1.5401-competitive. There are many algorithms known that achieve a better competitive ratio than FirstFit, but none attains the lower bound.

The situation is different for the m-restricted online bin packing with repacking. Lee and Lee [LL85] showed a lower bound of 1.69103 for the competitive ratio of every online algorithm. Galambos and Woeginger [GW93] gave an algorithm that achieves this competitive ratio and is thus optimal. It is interesting to note that the advantage gained by allowing repacking is more than outweighed by the restriction of using at most m bins.

Next we present some surprising competitive analysis results obtained for OnlineTSP and Online Bin Coloring. Consider an algorithm ALG for the OnlineTSP problem on one of the most trivial metric spaces one can think of, the one-dimensional line of real numbers. The server is supposed to start in the origin and we want to minimize the maximum waiting time of a request. We want to be a nasty adversary and to achieve a high ratio between the optimal offline cost and that of ALG and do the following. At time 0 ALG has to decide what the server does. There are three cases: Either the server remains at the origin, it moves to the left or to the right. Suppose the server travels to the right. Then the first request in the sequence arrives at $x = -1$ at time 1. Since the server moved to the right, it will not arrive at the request at time 1 or earlier, so it gets a positive waiting time. The offline algorithm, however, knows where the request will occur. So it can move the server to the position -1 on the line before the request occurs achieving a waiting time of 0. This looks like a "dirty trick", but we have thus shown that no online algorithm can achieve a bounded competitive ratio. This essentially means that all online algorithms are equally bad from the viewpoint of competitive analysis.

Let us now turn to Online Bin Coloring. A "natural" algorithm would put an item with a color already present in one of the bins into the same bin. If the color is currently not present in one of the bins, one would put it in a bin which currently has the least number of distinct colors. Let us call this algorithm GreedyFit. It can be seen [KdPSR01] that GreedyFit achieves a competitive ratio greater or equal to $2m$ (m is the number of bins that can be open simultaneously). The totally stupid algorithm OneBin, which uses only one bin until it is filled completely and puts all items into that bin, achieves a competitive ratio of at most $2m - 1$, making it superior to GreedyFit in terms of the competitive ratio.

Of course, there are many online problems where competitive analysis yields useful results, but for those two problems arising in logistics the results are not very helpful.

Extensions of Competitive Analysis. In the last section we have seen examples where competitive analysis yields surprising results. In various online optimization problems the following phenomena arise:

- There does not exist a competitive online algorithm at all.
- All online algorithms have the same competitive ratio.
- A reasonable online algorithm that performs well in simulation experiments has a worse competitive ratio than an algorithm that obviously performs bad in practice.

The reason for these phenomena is the worst-case nature of competitive analysis. In other words, the malicious adversary is "too strong". Various concepts to reduce the power of the adversary have been proposed in the literature. The most direct way is to restrict the request sequences that the adversary can produce in some reasonable way or to limit its power directly. For instance, the trick used for the OnlineTSP would not work anymore if the adversary is not allowed to travel in a direction where no request is known at present.

Another possibility is to consider *randomized online algorithms*, which are allowed to use random decisions for processing requests. The cost of a randomized algorithm is a random variable, and we are interested in its expectation. Normally, the malicious adversary knows the probability distribution used by the online player but cannot see the actual outcome of the random experiments. As a consequence, he must choose the complete input sequence in advance. This type of adversary is called *oblivious adversary* (see [BEY98] on different types of adversaries). Yet another alternative is to consider random request sequences and doing *average case analysis*, but this requires some probabilistic assumptions on how the inputs look like.

Theoretical analysis can give important insights in possible weaknesses of online algorithms and how to overcome them. However, to really evaluate how an algorithm performs in practice it is often necessary to resort to a simulation of the system. It is certainly no insult to state that the theory developed for online optimization rarely provides good guidelines for the choice of online algorithms to be employed in practice.

3.11.3 Case Study: Elevator Systems

Online problems arise frequently in logistics applications, for instance in high rack warehouses. One particular system was studied in [GKR99], dealing with the distribution center of Herlitz PBS in Falkensee near Berlin. This distribution center stores and retrieves pallets using a complex system of conveyor belts, elevators, and other transportation devices. The overall goal for the control is to ensure a rapid, congestion-free flow of pallets through the system. The pallets to be transported on a production day are not known in advance, which makes this problem an online problem.

In the following, we concentrate on the control of the elevators, which constitute an important subsystem. In the Herlitz distribution center, two groups of elevators work together and it is crucial for the system performance to coordinate each group of elevators well.

Elevator control problems can be seen as a generalization of the OnlineTSP problem, where a request consists of two points instead of one, with the interpretation that the server needs to travel from the first to the second point in order to serve the request. In addition, there is not only one server, but possibly many. Such problems are known as *Dial-a-Ride problems (DARP)*. We call the online version OnlineDARP. The time a request spends in the system is called the *flow time*, consisting of the waiting time of the request and the time needed to serve it.

3.11.3.1 Theoretical Results

Various theoretical results for OnlineDARP have been established. For the completion time objective, there is a 2-competitive OnlineDARP algorithm which was proposed in [AKR00], which means that the solution obtained by this online algorithm may take twice as long as the best possible solution. The authors also show that no algorithm can be better than 2-competitive, in other words, whatever smart algorithmic idea someone may have, there is a request sequence on which the online algorithm takes twice the time of a best solution.

The same chapter also analyzes two general online strategies, REPLAN and IGNORE. The REPLAN strategy computes a new optimal schedule each time a new request becomes known. This is often used in practice. The IGNORE strategy works in phases. At the beginning of each phase, an optimal schedule for the currently known requests is computed and realized. Only when the schedule has been finished a new schedule for the then known requests is computed and a new phase starts. During execution of a schedule all new requests are ignored, which gave the strategy its name. Both REPLAN and IGNORE are $5/2$-competitive [AKR00] and thus not best possible.

We already saw that no online algorithm can be competitive when minimizing the maximum flow time in the case of the OnlineTSP. Since OnlineTSP is a special case of OnlineDARP, this holds for OnlineDARP as well. For the OnlineTSP on the real line there is an online algorithm which is 8-competitive against a *non-abusive* adversary [KLL$^+$02], which is a restricted kind of adversary. For a reasonably restricted class of request sequences, the IGNORE strategy achieves bounded maximum and average flow times [HKR00] which is interesting since this guarantees some kind of stability of the system. In contrast, REPLAN cannot guarantee this stability for this restricted class of sequences.

3.11.3.2 Simulation Results

The elevator control problem has been studied in simulations, too. The particular setting of the Herlitz distribution center was investigated in [GKR99], which compared seven algorithms. It turned out that the algorithms had quite a different performance with respect to average and maximum flow time and that the algorithms used in practice were significantly inferior. Another observation was that algorithms achieving good average flow times give high maximum flow times and vice versa. The REPLAN algorithm achieved very good average flow times, but also very high maximum waiting times. The IGNORE algorithm, on the other hand, gave worse but still acceptable average flow times, but good maximum flow times as well. A variant of the IGNORE algorithm that balances the two objectives was therefore recommended for implementation at the plant. This is the way how we treated the multicriteria aspect of our Herlitz elevator problem. We looked for an algorithm that produces reasonable values for all relevant objectives.

The algorithms considered in [GKR99] were all quite simple from a methodological point of view. More recently, Friese and Rambau [FR06] studied a more involved algorithm based on Mathematical Programming. It is a modified version of the algorithm ZIBDIP which was developed for routing service vehicles (see next section for more on this).

Among other algorithms, they compared the modified version of ZIBDIP called REOPT to the algorithms FIFO and NN (NearestNeighbor), each representing a certain class of typical algorithms. FIFO assigns a new request in round-robin fashion to the elevators, serving each request of an elevator in the order of arrival. FIFO is an example of a simple rule-based algorithm. NN assigns a new request to the elevator giving the lowest waiting time for the new request if it is inserted such that no other request is postponed. The requests of each elevator are served such that the distance from the last to the next request is minimized, hence the name of the algorithm. NN represents a simple greedy heuristic frequently used in practice. REOPT, finally, determines the schedule in an integrated way (the assignment and service order decisions are taken into account simultaneously) such that the average waiting time of all the requests is minimized.

Figure 3.76 displays comparisons of these algorithms. Figure 3.76(a) shows the number of elevators needed to get an average flow time of at most 40 seconds for the given set of requests and the average flow time actually achieved with this number of elevators. It turns out that REOPT achieves this performance with five elevators only, whereas FIFO and NN require eight and six elevators, respectively. The average flow times achieved by the algorithms when they use five elevators only are shown in Fig. 3.76(b). Note that the vertical axis has a logarithmic scale! FIFO, in fact, turns out to be an unstable elevator scheduling system overflowing the buffer space enormously. FIFO is much worse than NN and REOPT. Although NN is able to handle the traffic without overflowing the system, REOPT outperforms NN significantly.

The conclusion of [FR06] is that multi-elevator systems with unit elevator capacity can be efficiently controlled by sophisticated online algorithms.

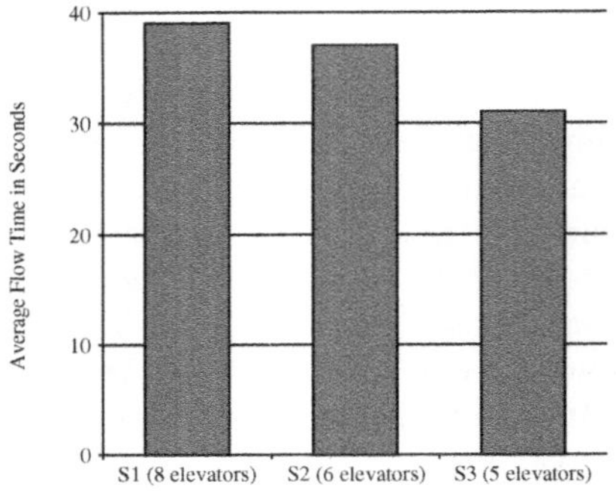

(a) Number of elevators needed to serve a 16 floor system with average flow time at most 40 seconds for FIFO (S1), NN (S2), and REOPT (S3)

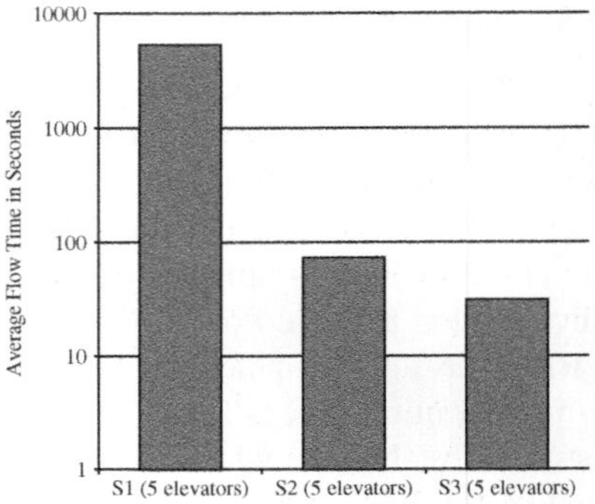

(b) Average flow time achieved by FIFO (S1), NN (S2), and REOPT (S3) on a 5 elevator 16 floor system. Note that the y-axis is logarithmic.

Fig. 3.76 Results of comparison of elevator control algorithms obtained in [FR06].

It will be interesting to see whether similar results can be obtained for the control of passenger elevators. Passenger elevators present other kinds of challenges. First of all, the information available to the elevator control is much more limited, since passengers specify only their travel direction (up, down) when they forward their request. In contrast, for industrial cargo elevator systems the destination is known, which allows much better control. In recent years some elevator companies started to implement destination call control for passenger elevators, too. In such a system, the passenger is required to enter the destination floor instead of the direction only and it is claimed that this additional information leads to an increased performance. We are currently investigating such elevator systems.

Even in destination call systems there are other features distinguishing passenger elevators from cargo elevator systems. The cabin of a passenger elevator usually has a relatively large capacity. This in conjunction with additional requirements, such as that no passenger travels temporarily in the wrong direction, makes the schedules more complex. But even here Mathematical Programming methods can provide powerful elevator control algorithms [TUA05a, TUA05b].

3.11.4 Case Study: ADAC Yellow Angels

The ADAC is the German Automobile Association. It operates a fleet of service vehicles, known as "Yellow Angels", providing help to motorists who had a breakdown. Help requests are registered in one of five help centers, which coordinate the service vehicles that are on duty. Each of the vehicles is equipped with a GPS system that reports to the help centers precise information on the current location of the vehicles. The dispatchers may employ, in addition to ADAC's Yellow Angels, service vehicles from contractors. The overall goal of the planning is to provide a high quality service at low operational cost, two objectives that obviously cannot be optimized simultaneously. This implies that "rules of compromise" have to be found. Good quality of service in our case means short waiting times for the motorists, where the meaning of "short" depends on the system load.

The whole fleet of ADAC service vehicles consists of more than 1600 heterogeneous vehicles. Moreover, ADAC's contractors have some 5000 vehicles that may be employed by the ADAC help centers. In high load situations, there are more than 1000 help requests per hour. These numbers give a rough indication about the large-scale nature of the problem.

This vehicle scheduling problem is a prototypical online problem, since in order to provide a continuous operation the schedule needs to be updated every time a new help request enters the system. ADAC wanted to establish an automatic planning system that provides good vehicle schedules in about 10 seconds computation time to guarantee real-time operation.

From a theoretical point of view, this problem is very similar to the OnlineTSP problem. The main differences are that there are many servers (up to 200 service vehicles per help center) and that the server needs to spend some time at every help request (equivalent to a city visit). For the practical design of online algorithms the theory known so far does not help much, since there are no competitive algorithms known for higher dimensional metric spaces than the real line.

ADAC's requirements suggested to build an online algorithm based on the *reoptimization approach*: Every time some new information becomes available, a *snapshot problem* is constructed that reflects the current state of the system. This snapshot problem is then solved, giving a new schedule to follow until the schedule is updated again.

More precisely, the snapshot problem consists of the following data

- location, time of occurrence and estimate of service time of each help request (henceforth called *requests*),
- current position, home position and service status of each vehicle (henceforth called *unit*),
- set of available contractors,
- information about the operational cost of units and contractors,
- information about the capabilities of each unit, i. e., what kind of help request can be handled by that unit.

The goal is to compute a good schedule servicing all the requests. Here a schedule is a set of tours for each vehicle that is on duty. But what is a "good" schedule? The goals are high quality of service and low operational cost. Hence this problem is in fact a multi-criteria optimization problem. Clearly, these two goals are partially contradicting. To cope with both criteria they are combined by defining the overall cost as a weighted sum of penalty cost for bad service quality and the operational cost. This is called "method of superimposing target functions" in Sect. 2.5.3. Of course, suitable weights reflecting the relative importance need to be determined for this approach. Alternatively, and more fundamentally, one could try to compute further solutions from the Pareto set. This, however, is very difficult to accomplish in the short period of time (10 seconds) allowed in this application. Moreover, there would still be need for an automated mechanism to select one of those solutions since snapshot problems are solved quite frequently.

The issues mentioned here are surveyed in Sect. 2.4.3 of Chap. 2 "The Economy of Modeling". It is simply impossible to take all aspects of this dispatching problem into account. Yellow Angels customers may ask for special treatment, e. g., a seriously sick person is in the car, an important manager needs to get to an airport or a pregnant woman to a hospital, etc. We simply do not know how to rank such requests in general. Therefore, only the major components of the Yellow Angels dispatching problem, namely time and cost, are modeled mathematically and considered in the algorithm, while "special requirements" are left to the human decision maker to be treated in "post processing".

This ADAC snapshot problem is a typical *Vehicle Routing Problem* (VRP) and the solution approach sketched in the following is often used to solve VRPs, see Sect. 3.9 for a survey on transport optimization. To keep things a bit simpler, we ignore the contractors from now on; they can be integrated quite easily. The following method was proposed in [KRT02] and has been implemented in ADAC's service scheduling system.

A *tour* t_u for a unit u is a sequence of requests, each of which can be handled by the unit. The tour t_u gives the order in which the requests are to be served, which in turn allows to compute the estimated waiting times for each customer. This way, a cost value $c(t_u)$ can be associated to the tour, which consists of the "service quality cost" and the operational cost incurred by the tour. For the computation of the cost it is assumed that the unit travels to its home position after finishing the last request of the tour.

For each unit u, let $\mathcal{T}_u$ be the set of all tours for that unit and $\mathcal{T}$ be the union of all the sets $\mathcal{T}_u$. A schedule is now a selection of one tour for each unit such that all the requests are contained in exactly one tour. This selection can be modeled by introducing a binary variable x_t for $t \in \mathcal{T}$; $x_t = 1$ means that the tour is selected. An optimal schedule can now be computed by solving the following integer program (IP).

$$\min \sum_{t \in \mathcal{T}} c(t) x_t$$

$$\sum_{t \in \mathcal{T}:\, r \in t} x_t = 1 \qquad \forall r \in R \tag{3.17}$$

$$\sum_{t \in \mathcal{T}_u} x_t = 1 \qquad \forall u \in U \tag{3.18}$$

$$x_t \in \{0, 1\} \quad \forall t \in \mathcal{T} \tag{3.19}$$

The constraints (3.17) ensure that each request is contained in exactly one tour, whereas due to constraints (3.18) exactly one tour for each unit is selected. Figure 3.77 illustrates the idea of this IP model. In more abstract terms, we can write the IP as

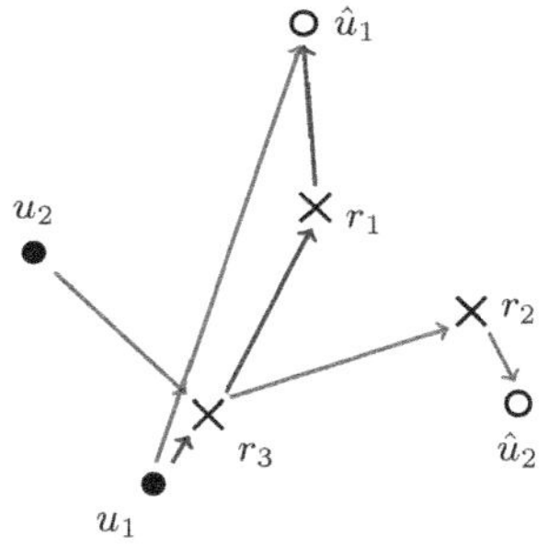

(a) Some tours for two units and three requests r_1, r_2, r_3. u_1 and u_2 indicate the current positions of both units and $\hat{u}_1$ and $\hat{u}_2$ their home positions.

	t_1	t_2	t_3	...	t_N	
	1	0	0	...		r_1
	0	1	0	...		r_2
A	1	1	0	...		r_3
	1	0	1	...		u_1
	0	1	0	...		u_2
c^T	c_{t_1}	c_{t_2}	c_{t_3}	...	c_{t_N}	
x^T	x_{t_1}	x_{t_2}	x_{t_3}	...	x_{t_N}	

(b) Representation of the tours in Fig. 3.77(a) asentries of a matrix A and a corresponding cost vector c.

Fig. 3.77 Illustration of the tour-based model (3.17)–(3.19).

$$\begin{aligned} \min\ & c^T x \\ & Ax = 1 \\ & x \in \{0, 1\}^n \end{aligned}$$

where the binary matrix A and the cost vector c is constructed as depicted in Fig. 3.77(b). An Integer Program of this type is known as a *set partitioning IP*.

One problem with this set partitioning IP is that is has an enormous number of columns as there are extremely many possible tours. However, a technique known as *column generation* [DDS05] in Mathematical Programming is applicable in this case due to the special structure of the constraint matrix. Column generation allows the implicit treatment of all the columns of the matrix without explicitly constructing all of them. This is done by searching for improving tours via the solution of an auxiliary problem, which in this case turns out to be a *resource constraint shortest path problem (RCSP)*.

Both the set partitioning problem and the RCSP are known to be NP-hard combinatorial optimization problems. This means that solving them to optimality is usually quite time-consuming, even if advanced methods are used. In the context of our application for the ADAC, however, the schedule needs to be computed very fast since only 10 seconds computation time are allowed.

A careful adaptation of the solution algorithm to the problem structure facilitates finding good and even verifiably optimal solutions in the allowed computation time. One particular feature of the problem is that the cost corresponding to the waiting time of the requests make long tours expensive, so the tours in the final schedule are usually short, i. e., at most 5 or 6 requests in high load situations. This allows to solve the RCSP by a controlled implicit enumeration of the tours. The set partitioning problem is usually easier to solve if the tours are short. Moreover, since requests that cannot conveniently be covered by a unit can be outsourced to a contractor it is always easy to complete a partial covering by the units to a complete covering

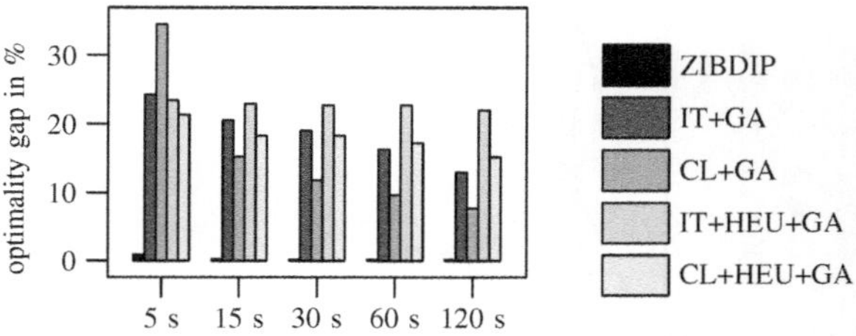

Fig. 3.78 Comparison of ZIBDIP with several variants of genetic algorithms. Depicted is the deviation from the optimal cost in percent after a given computation time. The data is taken from [KRT02].

of the requests. Therefore, good integer solutions, i. e., schedules, are obtained very early in the solution process. These considerations led to the development of the algorithm called ZIBDIP.

To give an impression on what is achieved by ZIBDIP, we cite some results. In a first study, ZIBDIP was compared to some variants of a genetic algorithm suggested by a software vendor [KRT02]. The genetic algorithm GA could be initialized using two different starting heuristics IT and CL as well as a best-insertion heuristic HEU. To assess the solution quality of the snapshot problem, we use the *optimality gap*, which is the ratio of the solution value obtained by an algorithm to the value of an optimal solution. The averaged results for three high load snapshot problems are given in Fig. 3.78. All the genetic algorithm variants finish with solutions with costs that are more than 15% above the optimal value after 15 seconds computation time and are still more than 5% away even after 120 seconds. ZIBDIP, in contrast, is within 1% of the optimal value already after 5 seconds, showing the fast convergence of this method that is required for real-time application.

Figure 3.79 shows the quality of the snapshot solutions obtained by ZIBDIP for one day. Each point specifies the *optimality gap* of the snapshot problem at that point in time. In addition, the figure shows the *load ratio*, which is the ratio of requests and units in the snapshot problem. The optimality gap gets worse for higher load ratios, but does not exceed 7% on the whole day.

Finally, Fig. 3.80 compares the online performance of ZIBDIP with those of two relatively simple heuristics. The first one, BestInsert, builds a new schedule by inserting new requests in the existing schedule such that the cost increases as little

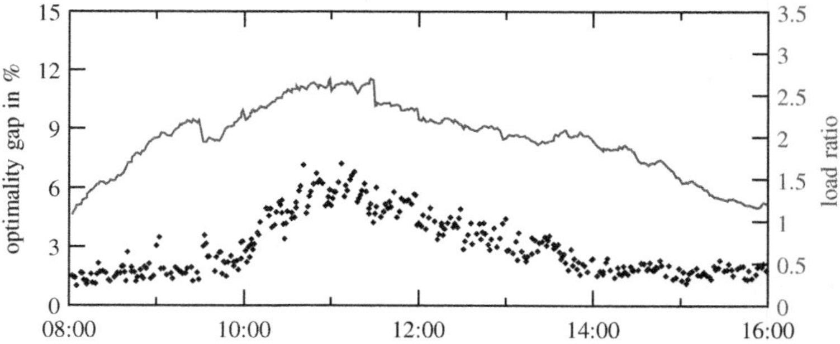

Fig. 3.79 Solution quality in the course of a high load day [HKR06].

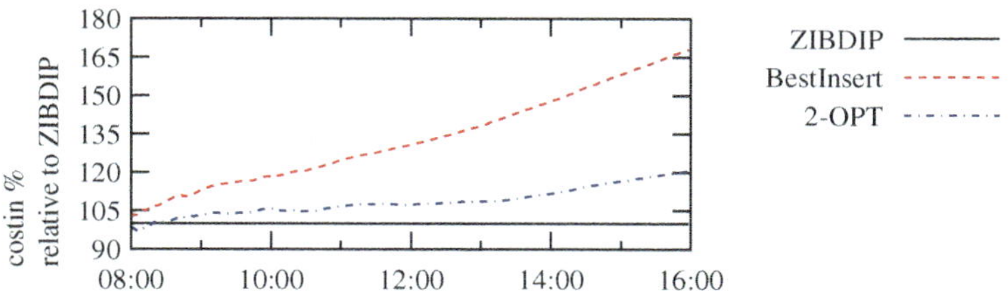

(a) Accumulated cost in the course of 2002-12-13.

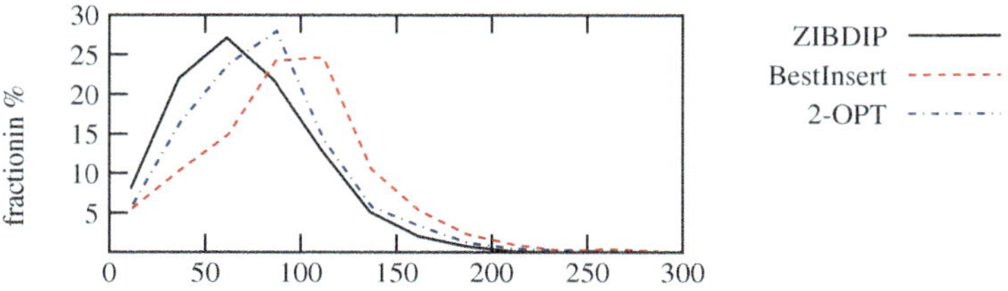

(b) Distribution of waiting time in minutes achieved for 2002-12-13.

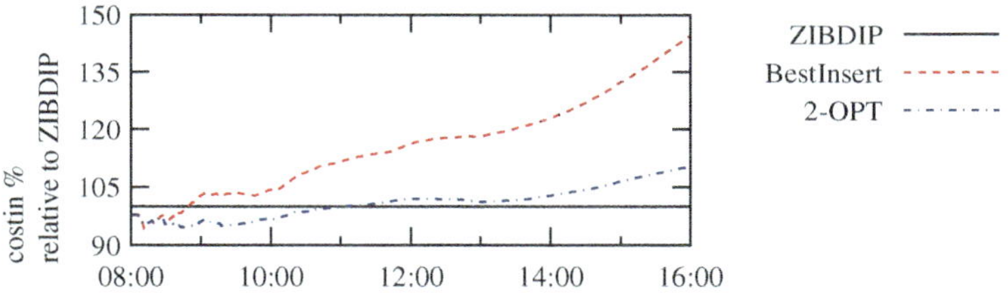

(c) Accumulated cost in the course of 2002-12-14.

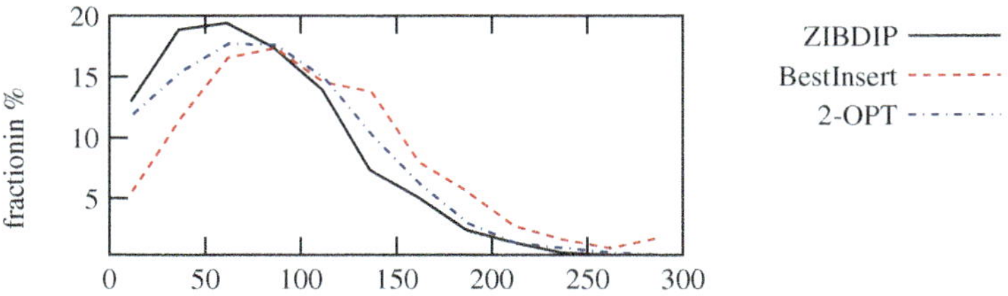

(d) Distribution of waiting time in minutes achieved for 2002-12-14

Fig. 3.80 Comparison of ZIBDIP to two heuristics on two high load days [HKR06].

as possible. The second heuristic, 2-OPT, computes an initial solution via BestInsert and then iteratively tries to exchange requests in one tour or between two tours in order to decrease the cost until no improvement is possible. In the long run, ZIBDIP produces overall schedules that are 40–60% "cheaper" than that of BestInsert and 8–20% "cheaper" than that of 2-OPT. The improvement in the waiting times is significant, too: ZIBDIP serves a larger share of requests with lower waiting time than the two heuristics.

All in all, ZIBDIP is a powerful algorithm and another example that Mathematical Programming methods can lead to real-time compliant online algorithms and codes.

3.11.5 Conclusion

Combinatorial online optimization problems appear in a wide range of practical applications. In most "industrial environments" we have had access to these problems are treated in a "very heuristic" manner, i. e., ad hoc algorithms are employed. Moreover, these algorithms are often programmed by persons who have no experience concerning all the possible pitfalls such heuristics may run into.

The state of the theory of online optimization is not in the best possible shape either. There are various concepts of analysis, such as competitive analysis, but it is questionable whether they provide good guidelines for practice.

At present, it seems that individual case analysis consisting of both, theory and simulation, is necessary. In the examples in this section stimuli for theoretical studies have been gained from simulation experiments, and theoretical investigations guided the design of the algorithms that were finally employed in practice. Overall, serious mathematical analysis and careful algorithm design, based on the many techniques offered by Mathematical Programming, pay off. The examples showed that these efforts lead to much more robust online systems with significantly better performance.

It would be desirable to have a better understanding of the methodology to cope with online problems. The success stories reported here are based on extensive research efforts, all very problem specific and not really guided by "general principles" that would result from a good theoretical underpinning. "Online" is not well understood. Moreover, online problems often have more than one objective – as indicated before. Multicriteria online optimization, however, does currently not exist as a research field. It is, at best, in its infancy.

3.12 Decision Structures on the Basis of Bounded Rationality

Konstantinos V. Katsikopoulos and Gerd Gigerenzer

To the great relief of passengers all around the world airplanes take off, fly, and land every day. Mathematicians, physicists, mechanical and aeronautical engineers have created a large store of know-how on the technical aspects of aviation. The management of the aviation industry, however, remains challenging. Consider, for example, making decisions under uncertainty. We do not really know how to route passengers to jointly optimize time, safety, and fuel efficiency. And, even if we focused on just one objective, say time, the routing problem may be computationally intractable.

Like in aviation, many decision problems in large-scale technical systems resist formulation in optimization terms for various reasons. Typical difficulties are multiple criteria, information that is not easily available and uncertainties of various kinds. Even when they can be formulated as optimization problems, computing an optimal solution often turns out to be intractable. What to do? The approach typically taken in the natural sciences and engineering is to optimize a simplification of the original problem.

This often works well but it can be difficult to know if the simplifications hold and what is the loss when they are violated. More broadly, the optimization approach – with some exceptions, e.g., Taguchi methods – is not tailored to handle issues like uncertainty, robustness, and flexibility that are increasingly recognized as fundamental in the management of engineering systems [Pap04]. Finally, the interface between optimization methods and practitioners often lacks transparency, usability, and acceptance [KOCZ93].

Overall, engineers are trained in the rigorous theory of optimization but all too often, when they graduate, they seem to find themselves using heuristics that worked in the past. At the Max Planck Institute for Human Development in Berlin, our team of life, natural, and social scientists, as well as historians, philosophers, mathematicians, and engineers, has, for more than ten years, been studying heuristics for decision making under uncertainty. Our research program can be viewed as the study of bounded rationality, a popular concept in the social sciences [Sim55]. One of our interests is in modeling the heuristic cognitive processes laypeople and some practitioners – medical doctors and mock jurors – use. This chapter samples some answers to the normative question of how well the heuristics perform. The research has used decision-making tasks that do not directly relate to technical concerns but we also speculate how it can be applied to the management of engineering systems. We start with some general comments on the heuristic view of bounded rationality.

3.12.1 Bounded Rationality: Fast and Frugal Heuristics

Bounded rationality is often interpreted as optimization under constraints, where the constraints are due to impoverished cognitive ability or incomplete information [Con96]. This interpretation is contrary to Simon's [Sim55] who emphasized satisficing, such as picking any outcome that exceeds a pre-determined aspiration level (as opposed to picking only an optimal outcome). Furthermore, this interpretation does not allow much progress as it reverts the study of decision making back to the study of logic, probability, and calculus, while excluding psychology.

Our approach to bounded rationality takes an ecological rather than logical view. It does not study optimal, internally consistent decisions but decisions that surpass aspiration levels with regard to external criteria like speed, accuracy, robustness, and transparency. This fits well with engineering where the focus is not so much on internal consistency but on external performance. Gigerenzer, Todd, and colleagues [GTtARG99, TGtArgip19] model the decisions with a breed of simple rules of thumb, called *fast and frugal heuristics*, which use a minimum of time, information, and computational resources.

The heuristics can be understood from a Darwinian perspective. First, because evolution does not follow a grand plan, there is a patchwork of heuristics, tailored to particular problems. This gives flexibility to the bounded rationality approach. Second, just as evolution produces adaptations that are bound to a particular ecological niche, the heuristics are not rational or irrational, per se, but only relative to an environment. Note that the study of the interaction between the decision-maker and the environment – emphasized in the ecological approach to psychology [Bru55, Sim56] – is missing in the optimization-under-constraints approach.

Finally, and importantly, heuristics exploit core psychological capacities, such as the ability to track objects. This is exactly what allows the heuristics to be simple, yet successful. For example, consider a pilot who spots another plane approaching, and fears a collision. How can she avoid it? A simple heuristic that works is to look at a scratch in her windshield and observe whether the other plane moves relative to that scratch. If it does not, dive away quickly.

In short, in our view, bounded rationality deals with simple and transparent heuristics that require minimum input, do not strive to find a best and general solution, but nevertheless are accurate and robust. What do these heuristics look like?

3.12.2 The Recognition Heuristic

Imagine you are a contestant in a TV game show and face the $1,000,000 question: Which city has more inhabitants: Detroit or Milwaukee?

What is your answer? If you are American, then your chances of finding the right answer, Detroit, are not bad. Some two thirds of undergraduates at the University of Chicago did [GG02]. If, however, you are German, your prospects look dismal because most Germans know little about Detroit, and many have not even heard of Milwaukee. How many correct inferences did the less knowledgeable German group that we tested achieve? Despite a considerable lack of knowledge, nearly all of the Germans answered the question correctly. How can people who know less about a subject nevertheless make more correct inferences? The answer seems to be that the Germans used a heuristic: If you recognize the name of one city but not the other, then infer that the recognized city has the larger population. The Americans could not use the heuristic, because they had heard of both cities. They knew too much.

The recognition heuristic is useful when there is a strong correlation – in either direction – between recognition and criterion. For simplicity, we assume that the correlation is positive. For paired-comparison tasks, where the goal is to infer which one of two objects (e.g., cities) has the higher value on a numerical criterion (e.g., population), the following fast and frugal heuristic can be used.

Recognition heuristic: If one of two objects is recognized and the other is not, then infer that the recognized object has the higher value on the criterion.

The recognition heuristic builds on the core capacity of recognition of faces, voices, and, as here, of names. No computer program yet exists that can perform face recognition as well as a human child does. Note that the capacity for recognition is different from that for recall. For instance, one may recognize a face but not recall anything about that person as a key determinant of its use, Journal of Experimental Psychology: Learning, Memory, and Cognition [CM87].

Intuitively, the recognition heuristic is successful when ignorance is systematic rather than random, that is, when recognition is correlated with the criterion. The direction of the correlation between recognition and the criterion can be learned from experience, or it can be genetically coded. Substantial correlations exist in competitive situations, such as between name recognition and the excellence of colleges, the value of the products of companies, and the quality of sports teams.

Consider forecasting the outcomes of the 32 English F.A. Cup third-round soccer matches, such as Manchester United versus Shrewsberry Town. Ayton and Önkal [AO97] tested 50 Turkish students and 54 British students. The Turkish participants had very little knowledge about (or interest in) English soccer teams, while the British participants knew quite a bit. Nevertheless, the Turkish forecasters were nearly as accurate as the English ones (63% versus 66% correct). Their predictions were consistent with the recognition heuristic in 627 out of 662 cases (95%). More generally, a number of experimental studies have found that if the accuracy of the recognition heuristic, α, is substantial (i.e., exceeds, say, 0.7), people use the heuristics in about 90% of all cases [Nay01].

The recognition heuristic implies several counterintuitive phenomena of human decision making that cannot be deduced from any other theory we are aware of. For instance, recognition information tends to dominate further knowledge, in rats as well as in people, even if there is conflicting evidence [GG02, PH06]. Here we concentrate on the counterintuitive finding that less information can increase accuracy.

3.12.3 The Less-is-More Effect

Assume that a person recognizes n out of N objects. The probability of being able to use the heuristic equals the probability of recognizing exactly one object in a sample of two, or

$$r(n) = \frac{2n(N-n)}{N(N-1)}. \tag{3.20}$$

Similarly, the probability that both objects are recognized, and thus other knowledge beyond recognition must be used, equals

$$k(n) = \frac{n(n-1)}{N(N-1)}. \tag{3.21}$$

Finally, the probability that neither object is recognized, which leads to the necessity that the person has to guess, equals

$$g(n) = \frac{(N-n)(N-n-1)}{N(N-1)}. \qquad (3.22)$$

Let α be the accuracy of the person when exactly one object is recognized and the recognition heuristic is used. Let β be the accuracy of the person when both objects are recognized and other knowledge is used. We also assume that accuracy equals $1/2$ when none of the objects is recognized (and we also assume $\alpha, \beta > 1/2$). Thus, the overall accuracy of a person who recognizes n objects equals

$$f(n) = r(n)\alpha + k(n)\beta + g(n)\left(\frac{1}{2}\right). \qquad (3.23)$$

Definition 1. *The* less-is-more effect *occurs when there exist* n_1 *and* n_2 *so that* $n_1 < n_2$ *but* $f(n_1) > f(n_2)$ *with* $n_1, n_2 \in 0, 1, \ldots, N$.

Definition 2. *The* prevalence, p, *of the less-is-more effect is the proportion of pairs* (n_1, n_2) *with* $n_1 < n_2$ *for which* $f(n_1) > f(n_2)$ *the less-is-more effect occurs.*

The prevalence p of the less-is-more effect varies between zero – no effect – for increasing $f(n)$, and unity – there is always an effect – for strictly decreasing $f(n)$. The prevalence of the less-is-more effect depends on the person's α and β. For $\alpha = 0.8$ and $\beta = 0.6$, simple enumeration yields $p = 1/3$. More generally, the following holds [RK04].

Result 1 The less-is-more effect occurs (i.e., $p \neq 0$) if and only if $\alpha > \beta$. The effect becomes more prevalent (i.e., p increases) as α increases or β decreases. The assumption is that α and β are independent of n.

At first glance, the less-is-more effect might appear paradoxical. But it is not, because less recognition information may simply enable more accurate cognitive processing (via the use of the recognition heuristic). This is condition formalized by $\alpha > \beta$.

As an example, Goldstein and Gigerenzer discuss three Parisian sisters who have to compare all pairs of cities from the most populous $N = 100$ German cities [GG02]. All sisters have $\alpha = 0.8$ and $\beta = 0.6$, but they vary on the number of recognized objects: The youngest sister has $n = 0$, the middle sister has $n = 50$, and the eldest sister has $n = 100$. When $\alpha > \beta$ the less-is-more effect is predicted: for the middle sister, $f(50) = 0.68$, while for the eldest sister $f(100) = 0.60$. Accuracy for $\alpha = 0.8$ and $\beta = 0.6$, based on Eqs. 3.19, 3.20, 3.21, 3.22, interpolated for all n, is graphed in Fig. 3.81 (solid curve; the dashed curve will be explained below).

A less-is-more effect can emerge in at least three different situations. First, it can occur between domains, that is, when the same group of people achieves higher accuracy in a domain in which they know little than in a domain in which they know a lot. For instance, when American students were tested on the 22 largest American cities (such as New York versus Chicago) and on the 22 most populous German cities (such as Cologne versus Frankfurt), they scored a median 71.0% (mean 71.1%) correct on their own cities but slightly higher on the less familiar German cities, with a median of 73.0% correct (mean 71.4%). This effect was

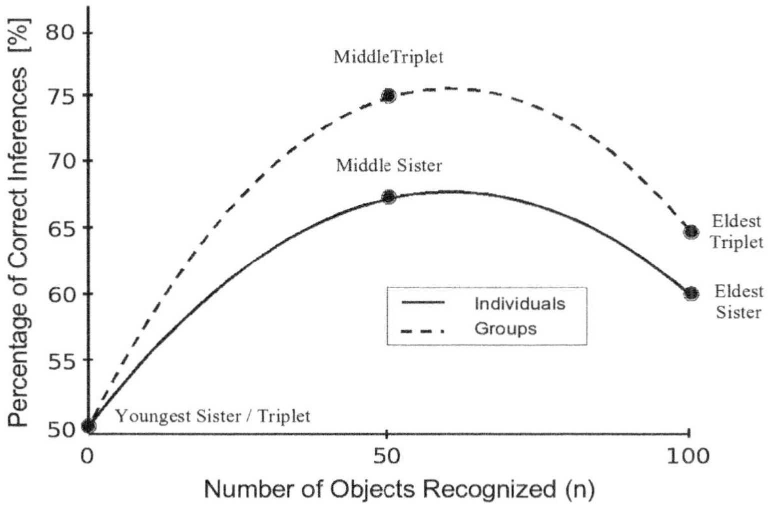

Fig. 3.81 Predicted accuracy as a function of number of recognized objects for $\alpha = 0.8$ and $\beta = 0.6$, for individuals (*solid curve*) and three-member groups that use the majority rule (*dashed curve*).

obtained despite a handicap: Many Americans already knew the three largest U. S. cities in order, and did not have to make any inferences [GG02]. A similar less-is-more effect was demonstrated with Austrian students, whose scores for correct answers were slightly higher for the 75 largest American cities than for the 75 largest German cities [Hof95]. Second, a less-is-more effect can occur during knowledge acquisition, that is, when an individual's performance curve first increases but then decreases again. Finally, the effect can occur between two groups of people, when a more knowledgeable group makes fewer correct inferences than a less knowledgeable group in a given domain. An example is the performance of the American and German students on the question of whether Detroit or Milwaukee is more populous [GG02]. Furthermore, Reimer and Katsikopoulos [RK04] ran a study where groups of people decided together.

In this study, three people sat in front of a computer screen on which questions such as "Which city has more inhabitants: Milan or Modena?" were displayed. The task of the group was to find the correct answer through discussion, and they were free to use whatever means. The correct solution is difficult to prove by an individual group member; thus one might expect that the majority determines the group decision [GH97].

The accuracy, $G(n)$, of a group using the majority rule is calculated as follows. Assume first that the group is *homogeneous* (i.e., all members have equal α, β, and n) and *independent* (i.e., the recognition and inference processes of members are independent given the values of the criterion on the objects). Let $F(i)$ be the probability of exactly i members being accurate and the group, using the majority rule, being correct. Finally let the group have m members, $c(m, i)$ be the number of ways in which i objects can be sampled out of m objects without replacement, and majority $(m) = (m+1)/2$ if m is odd, and $= (m/2)$ if m is even. Then, the following holds, where $i \geq$ majority (m):

$$F(i) = c(m,i)f(n)^i(1 - f(n))^{m-i}(1/2), \quad i = \text{majority } (m) \quad \& \, m \text{ is even}$$
$$= c(m,i)f(n)^i(1 - f(n))^{m-i}, \quad \text{otherwise.} \tag{3.24}$$

$$G(n) = \Sigma_{i=\text{majority}(m),\ldots,m} F(i). \tag{3.25}$$

The application of (3.24) and (3.25) for $\alpha = 0.8$, $\beta = 0.6$, and $m = 3$ is illustrated in Fig. 3.81 (dashed curve). A less-is-more effect is again predicted and p = 1/3. More generally, the following holds [RK04].

Result 2 Assume a homonegenous and independent group, using the majority rule. Then, (i) less-is-more effect is predicted if and only if $\alpha > \beta$ and (ii) the prevalence of the effect equals the prevalence for one member. The assumption is that α and β are independent of n.

Consider now the following conflict. Two group members have heard of both cities and each concluded independently that city A is larger. But the third group member has not heard of A, only of B, and concludes that B is larger (relying on the recognition heuristic). After the three members finished their negotiation, what will their consensus be? Given that two members have at least some knowledge about both cities, one might expect that the consensus is always A, which is also what the majority rule predicts. In fact, in more than half of all cases (59%), the group voted for B [RK04]. This rose to 76% if two members used recognition.

Group members letting their knowledge be dominated by others lack of recognition may seem odd. But in fact this apparently irrational decision increased the overall accuracy of the group. Broadly consistent with Result 2, Reimer and Katsikopoulos [RK04] observed that when two groups had the same average α and β (that were such that $\alpha > \beta$), the group who recognized fewer cities (smaller n) typically had more correct answers. For instance, the members of one group recognized on average only 60% of the cities and those in a second group 80%, but the first group got 83% answers correct in a series of over 100 questions, whereas the second only 75%. Thus, group members seem to intuitively trust the recognition heuristic, which can improve accuracy and lead to the counterintuitive less-is-more effect between groups.

3.12.4 Cue-Based Heuristics

When recognition is not valid, or people recognize all objects, heuristics can involve search for reasons or, in psychological jargon, *cues*. A few years after his voyage on the Beagle, the 29-year-old Charles Darwin divided a scrap of paper (titled, "This is the Question") into two columns with the headings "Marry" and "Not Marry" and listed supporting reasons for each of the two possible courses of action, such as "nice soft wife on a sofa with good fire" opposed to "conversation of clever men at clubs." Darwin concluded that he should marry, writing "Marry – Marry – Marry

Q. E. D" decisively beneath the first column [Dar69, pp. 232–233]. The following year, Darwin married his cousin, Emma Wedgwood, with whom he eventually had 10 children. How did Darwin decide to marry, based on the possible consequences he envisioned – children, loss of time, a constant companion? He did not tell us. But we can use his "Question" as a thought experiment to illustrate various visions of decision making.

Darwin searched in his memory for reasons. There are two visions of search: optimizing search and heuristic search. Following Wald's [Wal50] optimizing models of sequential analysis, several psychological theories postulated versions of sequential search and stopping rules [BT93]. In the case of a binary hypothesis (such as to marry or not marry), the basic idea of most sequential models is the following: A threshold is calculated for accepting one of the two hypotheses, based on the costs of the two possible errors, such as wrongly deciding that to marry is the better option. Each reason or observation is then weighted and the evidence is accumulated until the threshold for one hypothesis is met, at which point search is stopped, and the hypothesis is accepted.

If Darwin had followed this procedure, he would have had to estimate, consciously or unconsciously, how many conversations with clever friends are equivalent to having one child, and how many hours in a smoky abode can be traded against a lifetime of soft moments on the sofa. Weighting and adding is a mathematically convenient assumption, but it assumes that there is a common currency for all beliefs and desires in terms of quantitative probabilities and utilities. These models are often presented as models whose task is to predict the outcome rather than the process of decision making, although it has been suggested that the calculations might be performed unconsciously using the common currency of neural activation.

The second vision of search is that people use heuristics – either social heuristics or cue-based heuristics – that exploit some core capacities. Social heuristics exploit the capacity of humans for social learning and imitation (imitation needs not result in learning), which is unmatched among the animal species. For instance, the following heuristic generates social facilitation [Lal01]:

Do-what-the-majority-does heuristic: If you see the majority of your peers display a behavior, engage in the same behavior.

For the marriage problem, this heuristic makes a man start thinking of marriage at a time when most other men in one's social group do, say, around age 30. It is a most frugal heuristic, for one does not even have to think of pros and cons. Do-what-the-majority-does tends to perform well when (*i*) the observer and the demonstrators of the behavior are exposed to similar environments that (*ii*) are stable rather than changing, and (*iii*) noisy, that is, where it is hard to see what the immediate consequence of one's action is [BR85, GGH^{+}01].

Darwin, however, seems to have based his decision on cues. We will describe two classes of heuristics that search for cues. Unlike optimizing models, they do not weight and add cues. One class of heuristics dispenses with adding, and searches cues in order (a simple form of weighing). These are called *lexicographic heuristics.* Another class dispenses with weighting and simply adds, or *tallies, cues.*

3.12.5 Lexicographic and Tallying Heuristics

We again consider the comparison task in which both objects are recognized. The decision is made on the basis of n binary cues $c_1, c_2, ..., c_n$, $n \geq 2$ (for any object a cue c_i equals 1 or 0). Since the 18th century, a popular decision rule for paired comparisons is the linear rule [KMM02] (2003). In this rule, cue c_i has a weight w_i; weights can be estimated by, say, minimizing the sum of squared differences between the predictions of the rule and the observations. For an object A with cue values c_i(A), the score $\Sigma_i c_i(\text{A}) w_i$ is computed and the object with the higher score is picked. If the scores are equal, an object is picked randomly. *Tallying* is a linear rule where weights are equal to unity, an old idea in psychological measurement [Gul50].

Take The Best is a heuristic in the lexicographic tradition. First, cues are ordered by decreasing validity, where the validity v_i of cue c_i is the conditional probability that the cue points to the larger object ($c_i = 1$ on the larger object and $c_i = 0$ on the other object) given that the cue discriminates between the objects [$c_i(\text{A}) \neq c_i(\text{B})$]. (Without loss of generality it can be assumed that $1 \geq v_i \geq 1/2$). After cues are ordered, the decision maker inspects the first cue. If this cue points to one of the objects then this object is taken to be larger. If the cue does not discriminate between the objects, then the second cue is inspected and so on until a discriminating cue is found; if no such cue exists, an object is picked at random.

One-cue decision making has been observed in high-stake decisions. British magistrates tend to make bail decisions on the basis of one good reason only [Dha03, DA01], and so do British general practitioners when they prescribe lipid-lowering drugs [DH01]. Many parents rely on one cue to decide on which doctor to drive to in the night when their child becomes seriously ill [Sco02].

Both *take the best* and *tallying* are naïve in the sense that they do not take cue dependencies into account. While at first glance they might appear simplistic, simulation studies have shown that naïve heuristics compare remarkably well to statistical benchmarks. Three decades ago, Dawes and Corrigan (1974) convincingly argued that tallying can have greater predictive accuracy than linear regression. Einhorn and Hogarth [EH75] provided statistical reasons for this, including the absence of sampling error in the estimation of weights. Czerlinski, Gigerenzer, and Goldstein [CGG99] replicated this finding in twenty real-world datasets, emphasizing the concepts of *overfitting* and *robustness*. To define these, we distinguish between a learning sample on which a model estimated its parameters and the test sample on which the model is tested. Both samples are randomly drawn from the same population.

Definition 3. *A model M overfits the learning sample if an alternative model M' exists such that M has a smaller error than M' in the learning sample but a larger error in the test sample. In this case, M' is called the more robust model.*

Figure 3.82 shows the accuracy of three heuristics compared to linear regression, averaged across 20 real-world problems [CGG99], e.g., to predict which Chicago public high school has the higher dropout rate based on the socioeconomic and ethnic compositions of the student bodies, the sizes of the classes, and the scores

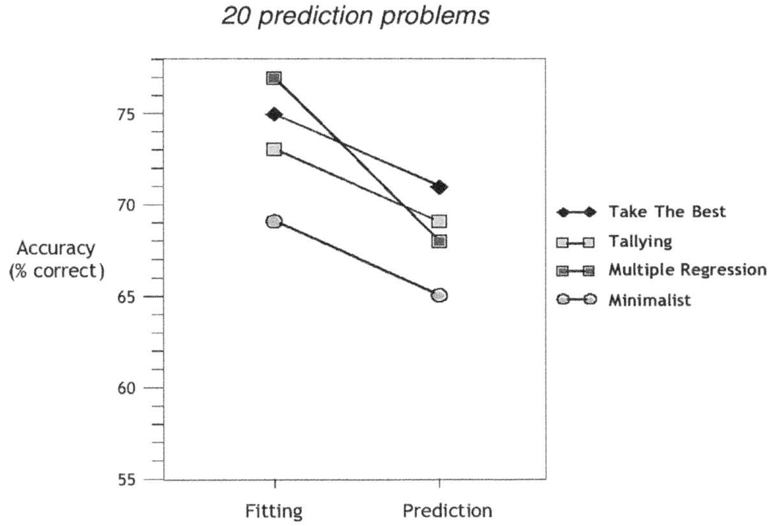

Fig. 3.82 Robustness of three heuristics and linear regression, averaged across 20 real-world problems.

of the students on various standardized tests. (Other problems were to predict people's attractiveness judgments, homelessness rates, adolescents' obesity at age 18, etc.) The three heuristics were *take the best*, *minimalist* (which is a lexicographic heuristic that searches cues in random order), and *tallying*. *take the best* and *minimalist* were most frugal; they looked up, on average, only 2.4 and 2.2 cues before they stopped search. *Tallying* and *multiple regression* looked up all cue information, which amounted to an average of 7.7 cues. How accurate are the heuristics?

Linear regression had the best fit. However, the true test of a method concerns its predictive accuracy, which was tested by cross-validation, that is, the four methods learned their parameters on half of the data (learning sample), and were tested on the other half (test sample). Figure 3.82 shows that *regression* over-fitted the data relative to both *take the best* and *tallying*. An intuitive way to understand overfitting is the following: A set of observations consists of information that generalizes to the other samples, and of information that does not (e.g., noise). By extracting too much information from the data, one will get a better fit but will mistake more noise for predictive information. The result can be a substantial decrease in one's predictive power. Note that both forms of simplifying – dispensing either with adding or with weighting – resulted in greater robustness. *Minimalist*, however, which dispenses with both weighting and adding, extracts too little information from the data.

In general, the predictive accuracy of a model increases with its fit, and decreases with its number of adjustable parameters, and the difference between fit and predictive accuracy grows smaller with larger number of data points [Aka73, FS94]. The general lesson is that in judgments under uncertainty, one has to ignore information in order to make good predictions. The art is to ignore the right kind. Heuristics that promote simplicity, such as using the best cue that allows one to make a decision and ignore the rest of the cues, have a good chance of focusing on the information that generalizes.

These results may appear counterintuitive. More information is always better; more choice is always better – so the story goes. This cultural bias makes contrary findings look like weird oddities [HT04]. Yet experts base their judgments on surprisingly few pieces of information [Sha92], and professional handball players make better decisions when they have less time [JR03]. People can form reliable impressions of strangers from video clips lasting half a minute [AR93], shoppers buy more when there are fewer varieties [IL00], and zero-intelligence traders make as much profit as intelligent people do in experimental markets [GS93]. Last but not least, satisficers are reported to be more optimistic and have higher self-esteem and life satisfaction, whereas maximizers excel in depression, perfectionism, regret, and self-blame [SWM$^+$02]. Less can be more.

Beyond computer simulations, mathematical analyses have also been used to investigate the accuracy of heuristics [MH02, HK05, KM00]. In the case where cues are conditionally independent (i.e., independent given the values of the criterion on the objects), the optimality of *take the best* (that searches cues in the order $c_1, c_2, \ldots, c_n$) and *tallying* can be characterized as follows [KM00].

Result 3 For conditionally independent cues, *Take The Best* is optimal if and only if $o_i > \Pi_{k>i}(o_k)$, where $o_i = v_i/(1 - v_i)$.

Result 4 For conditionally independent cues, *tallying* is optimal if and only if $v_i = v$.

3.12.6 Other Tasks

The paired comparison task is related to other tasks such as deciding whether an object is larger than a certain threshold (classification) or judging how large the object is (estimation). Fast and frugal heuristics have been studied for these tasks as well and have been again found to perform well compared to standard benchmarks [HR08]. For example, Katsikopoulos, Woike, and Brighton [Pro] found that simple, fast and frugal trees can make more robust classifications than discriminant analysis and trees used in artificial intelligence do [BFOS84]. Research on other decision tasks is reviewed by Gigerenzer [Gig04].

3.12.7 Fast and Frugal Heuristics in Technology Development?

According to Moses [Pap04], one of the main goals of the study of technological systems is to deal with changes that occur during the life cycle of these systems. Change can be dealt with actively by building flexibility into the system, that is, allowing the system to perform a number of functions. Change can also be dealt with passively by building a robust system, that is, a system that does not lose much of its performance when conditions vary. Both flexibility and robustness are necessary when there is uncertainty in the system. One might even say that uncertainty represents a chance for improvement in that it motivates flexibility and robustness (perhaps this is what Moses means when he talks about "viewing uncertainty as an opportunity" [Pap04]). The final challenge is to combine these properties with transparency and usability so that the system will be accepted by its users.

Our research program was not designed to study technological systems. Our results about the normative success of heuristics were obtained in decision tasks such as comparing dropout rates in highschools. We did not study how engineering students and practitioners compare, say, two product designs. Of course, at a certain level of abstraction, these are very similar tasks, but we do not want to downplay the potential influence of context. Thus, we see our results as making a methodological suggestion about a new program of research. We believe that a fast-and-frugal-heuristics approach to making decisions in engineering systems may be helpful.

Fast and frugal heuristics tend to be robust. There are more results than those presented here, to this effect. For example, Brighton [Bri06] has pitted the heuristics against powerful machine learning methods (such as Quinlan's ID3 method) using the minimum description length as a criterion of robustness. He found that, in many cases, heuristics compressed the data more than the machine learning methods.

With their focus on external outcomes, heuristics implement more practical intelligence than do mathematical methods that target full internal rigor. Furthermore, heuristics are less ambitious than methods that try to work all the time; heuristics are problem-specific and information-structure-specific. A given heuristic may be applied successfully only to those comparisons, estimations, classifications, or choices with certain statistical properties (i.e., flat or very skewed distribution of cue validities). Taken together, however, heuristics cover a wide spectrum of decision tasks. The set of heuristics has been called the *adaptive toolbox* [GS01].

The adaptive toolbox is a flexible system for decision making: To build the heuristics in it, one combines different building blocks (rules for searching for information, e.g., by validity, with rules for deciding based on the available information, e.g., use only one cue). The building blocks themselves are based on core psychological capacities (e.g., recognition). The toolbox allows the introduction of new heuristics by *(i)* combining existing building blocks in new ways or by *(ii)* creating new building blocks based on newly discovered capacities.

Finally, fast and frugal heuristics are transparent: They are easy to understand and apply (and, hence, are more acceptable). There are two reasons for this. First, heuristics are expressed as clear and simple sequential algorithms (e.g., *Take The Best*). Second, the way they represent the information they use is consistent with people's cognitive representations (e.g., in *Take The Best* validities can be cast in terms of frequencies, not conditional probabilities; see also [HLHG00]). Perhaps for these reasons, some practitioners, such as medical doctors, advocate the substitution of classical decision analysis with fast and frugal heuristics [EEER01, Nay01, KF00]. Just as some successful methods for engineering decision making – Pugh's concept selection [patICoED81] – heuristics can be used to generate, rather than to suggest or impose, new possibilities. We want to encourage academics and practitioners to explore the potential of fast and frugal heuristics in engineering.

3.13 Continuous Structuring in Open Source Software Development

Anika König, Jörg Strübing

3.13.1 Introduction

Recently, research on innovation processes has gained new attention. Within the regime of knowledge societies first of all a lack of scientific knowledge has become obvious: How do innovations proceed? How do we gain new knowledge? How do we structure previous knowledge with respect to current problems? What is the impact of organization on knowledge production?

Within this line of research successful new ways of both organizing knowledge and producing new solutions are of pivotal interest. The driving force here is the expectation that structural features of knowledge organization and problem-solving may come to light, which are not only suitable in their primary domain but transferrable to other fields in which the handling of knowledge and innovation is among the focal activities. With regard to structures, the central issue of this book, investigating innovation practices allows to study structures "in the making". While structures usually do not tell much about their genesis, and thus do not contain sufficient information about their politics and their legitimacy [Sta95], studying processes of structuring "in action" accounts for uncovering the motives, intentions, and perspectives that are the driving forces of structuring in social processes and which are responsible for the actual appearance of the structures under scrutiny.

Research on structures and innovation in the social sciences directs us to a serious shift of perspective as compared to the approaches of engineering, natural sciences or mathematics: Instead of constructing problem solving algorithms to actually solve knowledge problems, in social sciences the focus is on the analysis of ongoing processes of structuring and problem-solving in innovation processes. This has important consequences:

First of all, under scrutiny are not our own preferences on how structuring and innovation processes should proceed. As social scientists we are primarily concerned neither with solving problems of the real world nor with establishing general solutions for solving problems of certain types. Instead, our subject matter is a social field that already is innovative and that already solves its problems – more or less efficient, successful or not, in more conventional or in more innovative ways. Our task here is to find out how the structure generating processes proceed, whether they are successful and, if so, why. Moreover, people acting in that field interpret their and their co-actors actions prior to and independent from our analysis. Thus, our analytic endeavors have to deal with both actions and those interpretations of actions that themselves again motivate new actions and interactions. Alfred Schütz

[Sch53], coined the term "constructs of the second degree" to indicate the special quality of sociological interpretations of what happens in the social world.

For the investigation of structure-building innovation practices the field of new technologies yields especially promising case studies. It is by no means self-evident that processes of technological innovation are performed in innovative manners. Technical or scientific innovations might well be – and are in fact more often than not – developed in fairly conventional and suboptimal organizational settings. However, open source software development has gained special attention within the field of innovation research, because the mode of innovation performed in this field is based on a number of core features that seem to indicate an innovation process that takes place in an unconventional organizational setting.

Our aim in this section is to investigate open source software development as a structure-building process that is not only innovative but that at the same time – and at a certain level of abstraction – shows a number of structural features similar to those of certain mathematical and engineering strategies for solving problems of optimization under conditions of uncertainty, bounded rationality, and multi-criteriality.

Basically, open source software can be defined as software products developed and distributed under certain types of licenses that deviate significantly from common copyright regulations of commercial software. Though there are a number of different copyright models in the field of open source software, the key feature is that the source code of this kind of software is public and may be used and modified by whoever desires to do so due to the so-called "open modification principle" [Lau04]. While not for all open source software, so at least for those published under the "GNU General Public License" (GPL), such as LINUX, there is one precondition: Whoever modifies open software is obliged to publish the modified version under the same license.

While proprietary software is usually developed and maintained by formal organizations, the striking feature of open source software development is (at least as a rule) a minor emphasis on formalized organizational structures or even the complete absence of a formal organizational frame. Instead of a company hiring programmers to develop a software product, the field of open source software development is characterized by individuals or smaller groups starting a non-profit development project and inviting others to volunteer in participation.

The usage of expressions such as "as a rule" or "usually" signifies the large variety of different types of structuring processes within the open source scene. Indeed, we find a broad scale of different organizational settings, due to the fact that in the course of time open source software projects tend to grow both in size and – consequently – in organizational and technical complexity. A project once initiated by a handful of close friends might expand to a worldwide network of participants. This contribution focuses on the Linux Kernel project, one of the largest and oldest open source projects – and of course the most paradigmatic.

A considerable amount of social science research has already been conducted on the phenomenon of open source programming. Research, however, that predominantly deals with the economic and psychological aspects of open source. For economists, the impact of the very special copyright regulations in this domain has been a particular matter of interest [Kol99, Ray97]. Likewise, the question of why people would "invest" a larger amount of work into a product without being its "owner" and without receiving material rewards, such as a salary, has raised questions basically inspired by economic ways of thinking in terms of rational choice and individual profit [BR03, Lut04].

Less research effort, however, has been spent on the process of innovation management and community building within the various open source communities. The crucial issue here is that the process of producing an innovative technological product takes place under the above mentioned very special conditions. Consequently, with the absence of a comprehensive formal organizational frame, such as a company that might organize the innovation process, there is also no formal and reliable hierarchical order within which tasks might be delegated and disobedience to orders might be sanctioned. And, most important, this absence of a formal hierarchy obscures the distribution of decision rights with respect to the acceptance or the rejection of problem definitions, suggestions, and contributions delivered by participants.

Thus, the focus of this contribution is precisely the process of ongoing decision-making in the development process of the Linux kernel and – at the same time – the community-building effects interwoven with this continual process of negotiations concerning the best ways to improve the product under construction.

It is claimed that the Linux kernel community is not to be seen as a pre-established social structure within which the actual development work takes place. Instead, what is seen as a community is a current state of social relations resulting from those interactions programmers continuously undertake in order to jointly develop the Linux kernel software. The self-organization of the Linux kernel community is part and parcel of the working process that at first sight is only dedicated to bringing forth a technical artifact. At the same time, the development process itself has to be organized and structured in a continual process taking into account conflicting requirements, uncertainty about the other participants' proficiency, limited knowledge on alternative solutions, and so forth.

The aim of this contribution is both to sketch the structural conditions of the community that promote the innovation processes taking place in Linux kernel development and to discuss the implications this has for innovation and optimization as linked to structure building. Also, with reference to the framework of this book, we want to contribute to an extended understanding of structures that incorporate the social sciences viewpoint into an interdisciplinary perspective of the subject.

The following section will give a short overview about the basic features of the Linux kernel project, while the third part discusses these features with respect to the question of self-organization and structure-building. This includes the processes of decision-making in technical matters that are closely intertwined with community-building. The final paragraph draws a number of conclusions with regard to the central issues of this book, that is, multi-criteriality, bounded rationality and uncertainty.

3.13.2 The Specifics of Open Source and the Linux-Project

3.13.2.1 Software License Policies

An important aspect of software is the license under which the computer program runs. Particularly nowadays where ownership rights and commodification are increasingly debated – not only in software developers' inner circles but also in society in general – software license policies are gaining increasing attention.

This has not always been the case. Historically, in the early days of computing (i.e. the 1950s), software was a free give-away, delivered by the hardware manufacturers in addition to the purchased hardware [Gra02]. It was a central feature of those – usually only rudimentary – computer programs that their source code was made available to everyone who wanted to read, modify, or enhance it. Hence, before software was demerged from hardware in the late 1960s and its source code often "closed", the users themselves usually adapted the software to their particular needs [Gra02, Web04].

The separation of hard- and software – a process commonly known as unbundling – was a precondition for the emergence of software as a self-contained product and implemented in 1969 [Gra02]. Due to this development, the legal status of unbundled software had to be clarified as well. Consequently, and strongly influenced by official copyright and patent policies, a considerable number of different software licenses was generated by different parties over the following decades.

Software licenses can generally be divided into several groups. One way to classify software is the division between open source and proprietary software. As a commodified good, proprietary software comes without source code, that is, as bare binary machine code, not modifiable by its users: "Its use, redistribution or modification is prohibited, or requires you to ask for permission, or is restricted so much that you effectively can't do it freely" [Fou01]. Open source software, on the other hand, always comes together with its source code which makes it possible to modify it. Usually, open source software licenses allow free copying and distribution of the respective computer program and encourage people to contribute to its advancement.

In addition to the concepts of open source and proprietary software, there are multiple kinds of software licenses, such as for example freeware, shareware, public domain software, semi-free software, etc. [Fou01, Gra02, p. 278]. Consequently, software licensing is subject to continuous debates and negotiations. The same holds for licensing within the context of open source software, since there are also a number of licenses that are all open source but differ with respect to some aspects. These are for example the BSD-licenses (Berkeley Software Distribution) FreeBSD, NetBSD, and OpenBSD, and – most commonly used – the GNU General Public License (GNU GPL) which had been worked out by the GNU project (Gnu's Not Unix) since 1984 [Fou01]. Although in the beginning Linux was released under an own license, since 1993 it also runs under the GPL.

The most crucial features of the GPL – the most prominent and probably also the most widespread license – are that the software may be copied and distributed in any medium, provided that each copy includes a copyright notice. In addition, the program may be modified, although this modification has to be documented thoroughly. The source code, though not necessary for the program's runtime version, always has to be attached to the program, or, alternatively, it has to be indicated where the source code can be easily obtained. Finally, the license is spread to every computer program that is derived from GPL-software which is known as the "viral effect" of the GPL. Nevertheless, licensing under the GPL does not mean that the respective software has to be given away for free – on the contrary, the license explicitly states that "You may charge a fee for the physical act of transferring a copy, and you may at your option offer warranty protection in exchange for a fee" [Fou91].

3.13.2.2 Modes of Organization in Open Source Projects

Open source projects, even if they use the same licenses for their products, internally might be organized in very different ways. According to Kogut and Metiu, the governance structures, as they call it, are fundamental for the development of the software itself, since "organization by which work is delegated influences the product design" [KM01]. There are "democratic" projects as well as hierarchical ones, and both kinds of projects in every single case are organized in a different way. The Apache project for example, one of the more hierarchically organized projects, is headed by a group of important developers. Linux, although there also is a group of developers in charge of the important decisions to be made, in the last resort is still subordinate to Linus Torvalds – the "benevolent dictator" (ibid.). Hence, this mode of organization can be considered as an increase in hierarchy as compared to e.g. the Apache project.

In order to better understand the organization modes of open source projects, Crowston and Howison (2005), based on an empirical investigation, suggest to characterize them by their grade of centralization. They conclude, just like Kogut and Metiu, that the organization modes of the projects under scrutiny are extremely heterogeneous. The projects vary from highly centralized to very decentralized, from "stars" with one central node to so-called "thikets" where every node is connected to every other one [CH05]. Hence, on the basis of the above idea that some projects are hierarchically organized whilst others are more democratic, this would mean that the former are more centralized than the latter.

Regardless of the many differences in internal open source project organization, there are some structural similarities to be found. These similarities basically concern the roles developers can take in the organizational structure of the project. First of all, there are the maintainers. The role of a maintainer is usually awarded due to a developer's accomplishment with regard to the advancement of the software. Maintainers have a high responsibility and many obligations on the one hand, but on the other this role also provides them with reputation and prestige and, if it comes down to it, with the last authority concerning the part of source code they are responsible for. Typically, there are two ways to become a maintainer: first, maintainership is assigned on the basis of meritocracy. This means that who puts much time and effort into the project resulting in a quantitative and qualitative high output, in turn is awarded with a responsible position. Second, whoever creates a fundamental and substantial piece of code, becomes the maintainer for this part or module of the software. Regardless of how they have obtained their position, it is the maintainers who finally decide which pieces of code are included into the program, they are central nodes in the communication and interaction processes, and they are the ones who are responsible for the adherence to the time schedule for the release of new versions of "their" module. However, as soon as the maintainer does not show the necessary commitment any longer, the position is occupied by or assigned to someone else. Hence, the system is very flexible and fluid [Ett04].

In addition, there are the contributing developers. Depending on the project and its size, sometimes this group of contributors is again divided into active, passive, and other kinds of developers. These distinctions are mainly made when the project has many members whose contributions differ profoundly in quality and extent.

Finally, again depending on the particular project, there might be additional roles to be taken. For the Linux project there is, of course, Linus Torvalds – the originator of the first source code version of the Linux kernel. He is surrounded by his close confidents, the "trusted lieutenants", who relieve him from having to deal with everything that is not absolutely central to the development of the Linux kernel (see Fig 3.83). This mode of organization is by and large unique in the open source software world. Very few other projects have a leadership position comparable to that of Linus Torvalds in the Linux project.

However, with respect to the organization modes in open source software development – here in Linux kernel programming – not only the programmers positions in the project organization are of interest, but also what additionally constitutes the structure of the community: a special form of communication that is independent from national borders, languages, and cultures – using the English language as the lingua franca. This kind of communication, usually in the semi-public kernel mailing list (everyone who desires to do so can enlist in this mailing list) or personally

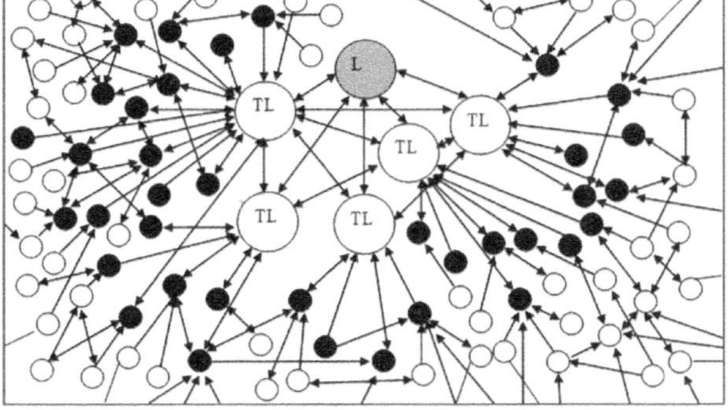

Fig. 3.83 Communication and cooperation between the members of the Linux kernel development community (L = Linus Torvalds; TL = Trusted Lieutenants; • = Credited Developer; ○ = Developer/User; ⟶ ⟷ = Direction of communication, patch exchange, etc.).

via ICQ, email, etc., is the basis for the development of the Linux kernel. Here, negotiations about great and small decisions with respect to all aspects of the program take place.

There is another feature of the projects organization that must be pointed out: the particular importance of the single programmer. While the project would not work without the extremely dense communication (usually there are a few hundred emails posted to the kernel mailing list per day), on another level the programmer is most likely to work alone, namely when programming or testing the software. The results – be it a piece of newly programmed code, the discovery of a bug in someone elses code, a bugfix, etc. – are returned to the community via the list or a maintainer.

Finally, what is of particular importance for the structure of the community and the structure of the program alike is version control and management. Here, Linux differs from most other open source projects since it uses proprietary software for its version control, a program called BitKeeper. Until 2002, Linus Torvalds had organized the project "manually", i.e. without utilizing any computer program intended for this purpose. The excessive demands of this task brought him to finally implement BitKeeper for the management of the many different versions of the Linux software. However, this decision led to deep resentment in the Linux community since many of its members preferred an open source tool to be used for this task. Hence, a heated debate took place, primarily focusing on whether Torvalds decision was politically and morally correct. Nevertheless, it was finally accepted that BitKeeper was used instead of the popular open source version control system CVS (Concurrent Versions System). This incident nicely shows that in the end, even if great parts of the community do not completely agree with Torvalds decisions, he actually does have the final say. It is another question, however, to what extent Torvalds can practically exercise this final decision right. This will be dealt with in a later section.

3.13.2.3 Linux Within the Open Source World

On the one hand, some general features can be observed in all open source projects: maintainers are responsible for different modules or versions of the program, communication is essential for the program development, and although the project cannot be sustained without the community, the work of the single developer is an essential aspect of the entire social and technical structure. Yet, on the other hand, within the open source world, Linux occupies a unique position. This can be ascribed to several different aspects:

First of all, the project is extraordinarily successful – in 2002, Grassmuck estimated the program to be installed on as many as 30 million computers worldwide [Gra02, p. 229]. Today, many major companies and administration departments have "migrated" from proprietary software to Linux; amongst advantages such as the possibility to modify the code, the software is known for its good applicability in networks. In addition, commercial and non-commercial distributions of the program (examples for commercial distributions are SuSE Linux and Red Hat, and for the non-commercial distributions Debian GNU/Linux) also make it possible for the "regular user" to install and use the program without extensive programming skills. Besides, even the commercial Linux distributions are still much cheaper than conventional proprietary software. There might be other open source programs that are as highly qualitative as Linux, but the availability of comprehensive distributions makes Linux particularly popular and, thus, successful and widespread.

Linked to the popularity of Linux is the size of the project. Not only is the program installed on tens of millions of computers all over the globe, but what is at least as remarkable is the size of the community contributing to its development. On the one hand, there are the kernel developers, but they only constitute a part of the community since in addition there are innumerable programmers who in one way or another contribute to the project. This can be in form of a little patch, drivers for new hardware, etc. And finally, a huge part of the community are the passive contributors, the ones who only report bugs but do not write programs themselves.

Although the size of the project itself is already remarkable, what is even more outstanding is that despite and not because of its size the project works efficiently.

3.13.3 Self-Organization Within the Linux Kernel Project

3.13.3.1 Self-Organization and Meritocracy: A Contradiction?

The notion of self-organization in the context of social phenomena has to take into account a serious difference to self-organization with respect to natural or technical phenomena. Different from self-organization in e.g. molecule structures, clusters of microbes, or patterns of algorithms, self-organization of human actors has to deal with intentions, motivations, and interpretations. Human actors might not always – and most often do not at all – oversee the entire range of consequences their actions will bring about. However, when acting, they usually have "something in mind", consciously or not they act towards certain goals and by doing so they employ their own interpretations of their fellow actors intentions, their "expected expectations" [Mea34].

Thus, a study of self-organization within the Linux community cannot treat actors as isolated entities jointly forming certain patterns which result from actor-independent mechanisms of, for example, mutation and selection. Self-organization in society basically means the existence of a certain social order under conditions of the absence of powerful actors equipped with universal knowledge concerting the social process in a desired direction.

Taking a look at the Linux kernel community we cannot refrain from recognizing the central role the originator of Linux, Linus Torvalds, is playing in the development of both the technical artefact and the social aggregation. Also, there obviously is a number of core actors, particularly the kernel maintainers, who possess larger amount of power than other members of the community with respect to the most common and thus most important types of decisions to be taken within the project, for example rejecting or accepting a patch. These facts indicate some type of hierarchy present in the Linux kernel project. However, we would still claim that self-organization is a central feature of the Linux community's social organization and of its innovation process management. Why is this so?

Despite the existence of powerful actors in the center of the development process, there also is a serious amount of power on the periphery: Different from a company or a bureaucracy, within the Linux community members are not members by contract but they are volunteers each and every time they contribute to the project. While it is true that Torvalds or his "trusted lieutenants" have the power to reject contributions as insufficient, ill-structured or faulty, they cannot, at the same time, compel contributions from other actors. The Linux project depends on contributions from a larger community of skilful and motivated members since the project is by far too large to be developed by a narrow crew of programmers. Moreover, the quality of the program as a distinctive feature as compared to the Microsoft Windows operating system is based on a large network of competent user-developers. Also, acknowledging the membership of an actor is not a decision to be taken by central actors but it takes place in the course of communication within the public sphere of the project, for example in the mailing list: A formal boundary between community members and non-members does not exist. The more people are engaged in discussions within the list and contribute to the shared goal, the more they are considered to be members of the kernel programmers' community. E.C. Hughes has coined the term "going concern" for this type of dynamic social aggregations [Hug71]. Instead of being a matter of formal acts, inclusion or exclusion are a matter of developing a mutual understanding of the value an actors contributions have for achieving the however roughly defined common goal. This implies that membership is not a discrete category but a gradual concept: Between a Linux user once sending an email to the list and Torvalds as the center of the whole endeavour there is a continuum of membership intensities, organized by a continuous implicit and explicit evaluation of each members contributions to the communitys core activities.

Since much has already been written about Linux programmers' motivation to contribute to the project without being contracted and paid, we will not discuss this matter in detail. With respect to the question of self-organization, the important point is that the power to overrule decisions in every single case does not imply the same power for all cases to be decided. Contributions from users and independent developers are based on the actors expectation to participate in a joint decision process. Since the Linux kernel community is based on meritocratic principles [Web04], participants tend to accept a larger influence of key actors – as long as their power and status is both justified by a fair amount of merits gained in the project and executed in a restrained manner. The notion of Torvalds being a "benevolent dictator" – as often cited in the Linux literature – signifies this attitude nicely: Since he is the originator of Linux, participants in the Linux kernel development would grant him a last word on every decision to take. However, a look at the Linux kernel mailing list shows that the frequency of Torvalds interference with cases to decide, at least within the list, is remarkably low.

As we have shown, besides personal preferences and habits there also is a functional explanation for this restrained use of directive rights by central actors within the Linux community: Not only are core actors like Torvalds unable to force contributions from other members, they are likewise in need to keep their contributors "in the mode" to participate in the programming endeavour. Although the functional perspective might hold analytically, it is, at the same time, not likely to be the actors basic rationale within the Linux kernel community. Deeply rooted in the tradition of anti-commercial, anti-hierarchical movements the spirit of open source projects like Linux is still egalitarian and – in a grassroots manner – democratic.

With respect to self-organization, the key problem of open source projects lies in establishing a productive balance between central and peripheral clusters of power on the one hand (Torvalds, his "trusted lieutenants", and other maintainers) and the participation rights of "ordinary" members on the other. Essentially, this balance has to be produced and maintained continually throughout the project. In this respect, technical media (like BitKeeper or the Linux kernel mailing list), or informal agreed upon rules of conduct can be recognized as stabilizing features of the community, translating the needs of either group into the needs of the others, and, at the same time, representing a current state of the community's structure.

3.13.3.2 Distributedness and the Need for Effective Communication

At this point we need to take a closer look at how the development work is actually done in the Linux kernel project. In the light of a world-wide distributed community of participants the organization of communication is pivotal for the progress of the project. In our case, geographically distributed participants at the same time means distributed work: Even tiniest pieces of software are manufactured from developers living on different continents, embedded in different cultures and speaking different languages. All the work performed under conditions so profoundly heterogeneous has to be brought together and integrated into one single software system.

Thus, the role of the Linux kernel mailing list cannot be overestimated. It is the medium within which both the exchange of ideas and patches, and the negotiations over decisions concerning the program design basically takes place. It is the virtual locale shared by and accessible to all actors already engaged in Linux development and also for those potential contributors still pondering about the idea of participating in the project. Besides Linux conferences or personal acquaintance with active developers, the Linux kernel mailing list is the central entrance to the Linux community. But more than this, it is the only community-wide arena where requests and claims can be made and negotiated.

A number of typical actions are likely to be taken by actors using the Linux kernel mailing list: Participants might

- complain about malfunctions of the kernel version
- suggest a new or modified functionality
- deliver a patch that supposedly adds or improves the kernels functionality
- discuss advantages and disadvantages of suggestions or patches presented by other participants

- discuss other matters such as general design rationales, programming styles, or product politics

As the Linux kernel mailing list is the central communication forum for kernel related actions, virtually all topics relevant to the whole community are broadcasted through this channel. Of course, there is some more communication between single actors or smaller groups of actors not publicly observable, such as for example private mails or phone calls. However, all discussions and decisions relevant to both the product and the community usually pass the Linux kernel mailing list and are thus public.

Without a central steering committee, even the current agenda hast to be negotiated through the list, though some of the general design perspectives will also be discussed on Linux kernel conferences. Different from formal organizations, negotiating the agenda in the Linux kernel mailing list is an ongoing process where consent is granted only until someone successfully opens up the discussion again. New topics can be brought up whenever someone feels the need to do so. Though obviously a general consent on some basic design rationales exists among most participants, various topics are repeatedly raised with a certain regularity, new suggestions are urged, and varying perspectives on current issues are presented.

The hitherto discussed aspects should have demonstrated how delicately the task of self-organization of the Linux kernel project as a community is embedded into the practical accomplishment of the collective development endeavour, while at the same time it is highly dependent on a communal communication forum. The important point is here that since the Linux kernel community lacks a central feature of traditional communities, that is, a spatial concentration allowing for regular face-to-face interaction including diverse social activities. Instead, nearly all community-building has to be achieved through distant, work-related communication.

In an empirical analysis of thematic threads communicated through the Linux kernel mailing list, we found that while in the foreground negotiation over technical issues of certain patches or the overall design is processed, at the same time strategies are employed to socially grounding the interaction context. For example, a smoothing "social wrapping" [Hen99], such as jokes, greetings or compliments often accompanies potentially conflict-loaded contributions; likewise, actors jump in on harsher statements from other contributors, trying to moderate upcoming conflicts. True enough, similar moderations take place in other work-related types of communication too. However, in spatially non-distributed interactions social relations within the community can be maintained through other levels and forms of joint activity such as having the proverbial chat at the coffee machine or going out for lunch. Thus, in the case of the Linux community social inclusion – as well as social exclusion – can be expressed almost exclusively by Linux-related communication through the mailing list. Repeated positive and supportive reaction to a contributor's postings strengthens his or her position within the community – it produces reputation. However, this is tightly linked to actual contributions to the Linux kernel. The list is not meant to be used for small talk or for the exchange of compliments: "This list is not for the faint of heart" announces the introductory self-description of the list, demonstrating a strict dedication to the kernel as the only subject matter and to software development experience as the only acceptable habit (http://www.ussg.iu.edu/hypermail/linux/kernel/info; 27.07.05). That is to say, the "proof of the pudding" here is the submission and acceptance of patches improving the kernel software. Consequently, larger source code contributions and those that are valued as especially useful, innovative, and elegant increase its originator's reputation. Within the mailing list, an ongoing communication about the acceptance of a patch gives both participants and passive observers (lurkers) a momentary impression of the contributors prestige. However, this is a fleeting impression, too weak to structure the communitys social order. Consequently, there are two more durable representations for the prestige-spending authorship of a piece of software that made its way into a kernel release: First, every patch is "signed-off" by its author, so the other developers working with the code would eventually come across the authors name and electronic signature. Second, each Linux kernel release – like most other open source software – contains a "credit file" that lists the names of the contributing authors. Being mentioned there means that a developer's achievements are made visible to a broader audience compared to the authoring signature of the patch. This is the reason why software companies often tend to have a look at relevant credit files as a candidates recommendation when planning to hire new staff.

3.13.3.3 Fluidity and Durability in Social Order

A most suitable concept for understanding the structure-building processes within the Linux community is that of continual negotiations over status, reputation, roles, and legitimation. Negotiations not to be understood in terms of decision theory or market economy, that is, not necessarily as a formalized rational and strategic act, but as an often unintended and unnoticed process accompanying everyday interactions in every domain and resulting in a "negotiated order" [Str78]. This concept of social order implies that structures in societies are neither given nor fixed but subject to constantly negotiated changes. And, moreover, even maintaining an existing structure requires actors to confirm this order in their way of interacting. Without developers acting among each other with a mutual understanding of a shared commitment to a common endeavour there would be no Linux kernel community. And without, in their practical acting, ascribing certain developers a higher/lower status than others, there would be no "meritocracy".

Of course, interactive ascription and negotiation is only one, more fluid side of generating and maintaining social organizations as structures. The other, more durable side is the representation of a current state of affairs in the organizational memory of e.g. the layout of a technical infrastructure [Sta96]. In the case of Linux this is the organization of the mailing list that enables users to exchange their ideas, positions and information, and the structure of the source code repository (BitKeeper), supplying the participants with the latest version of the program and keeping them informed about changes. Another reification of structure is the modularization of the source code itself, representing a division of labour negotiated and agreed upon at some point back in the projects history. The definition of interfaces as well as the organization of certain functions in certain modules can be interpreted as representations of the underlying social ordering of developers and users. Though information about structures is incorporated in these artefacts it should be stressed that these representations are not identical with the social structure itself. Since they are different from molecules, animals, or plants social actors have to interpret the artefacts as representations of structure. Structure comes into existence only through these mutual interpretations achieved in practical interactions. A simple example would be the structure of the source code: Each suggestion for implementing a new or altered functionality is both an interpretation of the current source codes structure and a potential challenge to it at the same time. The developer's interpretations, their collective negotiations about the suitability of interpretations, and their succeeding actions might – and often do – bring about a change of both the social and the technical structure. However, relying on artefacts as representations of structure implies that actors are not completely free in their actions: Whatever they interpret and negotiate, it is finally dependent on a proof in the "real world", as long it has to work both technically and socially.

3.13.3.4 Structuring of Technical Artefacts as a Social Process

While we have shed some light on processes of developing and maintaining a social organisation's structure, and its embeddedness in the process of technology development in the case of Linux, we have not yet had a closer look at the structuring of the technical artefact itself during the social process of joint distributed work in a heterogeneous community of developers and users. As shown above the basic process of open source software development takes place on two different levels: One is the semi-public negotiation process in the mailing list, the other is the process of writing and testing patches – a process largely performed by single developers.

The existence of these two different and to a great extent separated levels is of importance for the question of multi-criteriality and optimality. One strong motive for developers to participate in the Linux kernel project is the need to adopt the software to local conditions such as e.g. a certain hardware configuration, special applications, or certain tasks to be performed. In our empirical analysis we encountered the case of a developer negotiating a redesign of two kernel modules in order to assemble a number of functionalities in one module while disposing of some other features. The local background for his demand was his work with Linux-based embedded systems such as the ones used in digital watches or in MP3-players. The limited resources available in the application domains of embedded systems call for a rigid selection of only those functions needed for the special purposes addressed. The global view of maintaining and developing the general structure of the entire kernel had to take other issues into account, e.g. established design traditions and rationales that have led to the modularization of the current release, or functional arguments valid for the full-fleshed Linux in contrast to other more specialized usages.

It is right at hand that a universal overarching rationality suitable to manage the different perspectives maintained by the various local actors and groups is not available. Practical solutions have to be developed not only by taking into account different perspectives but also a broader number of additional criteria: While on the one hand the kernel as a whole needs to remain one consistent piece of software (otherwise the joint development efforts would be split up in different projects, thereby suffering a loss of efficiency), it should, on the other, cover as many different purposes, applications, and technical constellations as possible. Furthermore, since the actual modularization represents not just a settled distribution of distinct program functionalities but at the same time a certain division of labor, including shared responsibilities and specialized knowledge domains, new suggestions have to be questioned as to whether they are in line with the established structure or if they can come up with reasons and arguments strong enough to pay off for the resulting organizational and social costs. Also, solutions will be judged with respect to the ratio of stability and innovativeness: System stability is a precondition for the kernel of an operating system, and by and large the old well-established, thoroughly tested system tends to be more stable and reliable than a system with newly implemented functions. However, an aging system cannot possibly keep track with the changing technological environment and with new application needs. Thus, a trade-off between risky innovativeness and the goal of system stability has to be made. A means taken by the Linux kernel community to handle this dilemmatic structure is to work with two separate kernel versions, one (with uneven version numbers) being experimental and thus less reliable while the other (with even version numbers) is a more settled and stable version.

All these considerations form a – not always consciously addressed – background for the negotiations taking place on the kernel mailing list. Thus, whenever a patch is offered, choices have to be made: Take the patch as it is or suggest improvements? Insist on perfect and elegant problem-solutions or on a quick and dirty way to circumvent momentary ill-functions? Discuss only the patch on offer or take an initiative for a larger restructuring of the related part of the system? In processes of "interactional alignment" [Fuj87] the active participants try to fit a new proposal to the various constraints of the different technical, organizational and social levels. However, these processes are ubiquitous, they are not a privilege of open source development. Nevertheless, in cases like the Linux kernel project decisions have to be made in a semi-public participative process, that is, the other actors have to be convinced, they cannot simply be overruled.

In the case of the Linux kernel project the negotiation process is supported and stabilized by a rationalistic attitude typical for its participants (mostly software engineers and other "technology people"): In the light of a logic-based subject-matter like software they are convinced that this very subject matter itself is helpful in sorting the most suitable solutions out from ill-structured ones. There is a strong sense among developers that disputes can and should be resolved "objectively" and that the best dispute resolution mechanism is simply to "let the code decide" [Web04]. However, a look at the negotiations taking place in the Linux kernel mailing list reveals that this belief in the certainty of logic does not automatically lead to unequivocal decisions. While a larger number of cases can obviously be settled by checking the performance of a patch compared to the former state of the program, a still considerable amount of decisions lack such clear logical grounds. Though there is a limited number of clearly defined performance ratios that might be argued with, negotiations over patches rely on a broader set of – partly even contradictory – criteria, and some of them will be evaluated differently depending on the participants' perspective.

One might ask: Why should Linux kernel developers care at all about other participants' perspectives and requirements? Since all developers are free to pick a kernel version and to produce their own custom tailored versions by implementing locally desired features or structures, why should they bother about lengthy discussions? As shown above, from the point of view of Linus Torvalds and his core team, integrating a heterogeneous and distributed crowd of developers is a necessary precondition to get the work done and the quality of the product thoroughly evaluated. But how about others, like e.g. the afore mentioned developer of embedded systems? He could have implemented the new module structure as required and have his system running. He could even get manufacturers of watches or – for that matter – intelligent refrigerators to use his Linux version. His problem would be, however, that from this split off point on he would have to maintain his version without the established social and technical structure of the Linux kernel community. This invaluable advantage of a supportive knowledge base and work force of thousands of experienced developers is one of the pivotal incentives for staying within the community by negotiating towards compromises instead of producing individual or local versions.

3.13.3.5 Similar Goals, Different Solutions

While most often negotiations within the list have to deal with acceptance, modification, or rejection of one particular patch offered by a single developer or a small team, in principle it is also possible that different patches dealing with the same problem are presented to the mailing list more or less simultaneously, offering different solutions for the same problem or adding the same functionality. Practically, however, this happens rarely. The unlikeliness of a parallel production of different solutions to similar goals may be seen as an indicator for the strong inclusive effect of the Linux kernel mailing list and the BitKeeper source code repository. Contributors to the kernel source code are and need to be members of the mailing list plus they seem to monitor the discussion in the list thoroughly enough for not starting to develop a patch that others are already on the way to accomplish. Likewise, through the BitKeeper system all members have easy to use and real-time access to the most current source code version, thus new developments will not pass a developer's attention unnoticed.

Despite the mailing list and BitKeeper system, in principle it would be possible that two developers nevertheless start writing different codes in order to solve the same problem independently from each other. This might happen because participants do not regularly or necessarily publish their intentions to develop a certain patch or feature. This is especially true for patches that solve problems claimed by the author of the patch. Here most often problem claim and suggested solution come in a package, ready-made for instant application. Though we did not perform a reliable check on this issue, a rough scan of the contents dealt with in the Linux kernel mailing list indicates to a large degree the absence of simultaneously developed competing patches.

Even forking, that is, the deliberate split off of one ore more parties from a former joint development project in order to develop different versions of the software (e.g. in cases of conflict over the design rationale or over the most elegant way to approach a problem) seems to be a rare event in the Linux kernel community. Different from the afore discussed option of producing a custom-tailored version of the kernel and thereby "leaving" the community, forking would not result in an irreversible loss of the community. Rather, for the kernel project it would produce the need to decide whether to integrate one or the other solution, none of them at all, or both. The possibility to accept two or more solutions for the same feature is most often limited by technical constraints. First of all, implementing alternative solutions for the same function results in a product more complex than (technically) required and thereby causing e.g. runtime problems and disproportionate complexity. Secondly, because of temporal or other resource limits especially in the kernel of an operating system a larger number of functions do not allow for dual or multiple structures at all. On the other hand, there are aspects of the program that call for adaptability, and that is, for the users choice between different alternative functions already implemented as options. For Linux this is true for instance for the user interface.

Thus, unlike evolutionary adaptation processes in other domains, in Linux kernel development competition usually does not take place on the level of different developers solutions (competing patches) but instead on the level of competing arguments and evaluations judged by peers via the mailing list. Through the respective maintainer the community chooses not among various solutions but among the most convincing statements pro or contra the application of a certain solution offered by a contributor.

3.13.4 Outlook

Though so far we have discussed social and technical innovation processes in the Linux kernel project, the aim of this contribution is more general, i.e. to use this project as an example to illustrate a number of issues related to social structure-building. A number of key concepts, with regard to structure generating processes in other disciplines have been mentioned in Chap. 2 of this book, such as multi-criteriality, optimality, bounded rationality, uncertainty. How then are these key concepts relevant and applicable if the domain of structure-building is society?

Analytically, we can recognize different types of structures in social phenomena on different levels of scale (micro, meso, macro; or – for that matter – interaction, organization, society), some of which are known to the actors and some not. Although social structures are obviously part of the adaptation of society and its parts to prevailing conditions, they are far from being optimal – as can be noticed in discussions over e.g. social inequality, urbanization, education systems, or globalization issues.

The case of the Linux kernel community shows aspects of structure on all these levels. It is part of a social movement and it represents a special form of globalized production based on a new type of organization combined with certain rules of interaction. In all these dimensions the structure of the Linux kernel community deals with structural conditions given in its ecology: Near-monopolistic market situations in the domain of operating systems; skill levels, traditions and fashions among programmers; the logical structure of programming languages; aspiration levels of different classes of users; availability and form of technical infrastructure, and so forth. Although the Linux kernel community can be understood as a response to these environmental conditions, its form and existence may not, however, be explained by these conditions. Quite obviously the social structure of this community is not determined by its ecology but there are elements of both coincidence and active choice besides those given conditions: Given its ecology, nobody could seriously have predicted the emergence of open source communities and their way of producing software. To generalize this issue: Human problem-solving does not result in completely determined and therefore predictable solutions – neither with regard to the subject-matter nor to the social organization of the issue of the problem-solving process employed. This inevitably results in forms of uncertainty to be taken into account in both the participants actions and in the social scientists' analysis of these processes.

What is remarkable about social structures is the way how they are developed and maintained: Though structuring happens through active interactions of humans, in most aspects this is achieved unconsciously. Actors perform various activities following goals or modes or routines, and in doing so they produce social structure as a by-product. As in the Linux case: In their deliberate efforts to produce an operating system in a way that is remarkably different from the way Microsoft produces, a growing number of developers and users by and by create and maintain a unique social structure which combines social community with aspects of effective work organization on a largely informal base.

Every effort to optimize the organizational structure of the community has to deal with a heterogeneous set of conditions resulting from considerable differences in the various actors' perspectives: Contributing maintainers for instance might want to have immediate writing permission for the current source code version in order to by-pass lengthy decisions processes in the mailing list. For other participants this would be a threatening prospect since they would loose their joint control over the kernel source code. Likewise, Torvalds and his core team of trusted lieutenants could easily be thought of as preferring direct and exclusive decision rights over design rationales and patch application in order to both speed things up and to make the kernel a unified whole. Following either one of these perspectives would not lead to optimality. It is right at hand that the question of optimality cannot be judged from a universal rational point of view.

This is where multi-criteriality also comes into play. Whilst for some software developers the criterion of immediate individual writing permission is the foremost goal with respect to their contribution to the entire software system, for others joint control over the kernel source code is the most important criterion. In the Linux case, in order to meliorate and optimize the version control system, the means of negotiation is employed as the best suitable means to reach the momentary stage of optimality. In this example this would be "as many rights as possible and as much community control as necessary". It has to be kept in mind, however, that due to the characteristics of social systems, this stage of optimality at another instance might be something completely different.

The perspectives taken by the different types of participants in the Linux kernel community each represent a specific mix of requirements that would locally be judged as the optimal way of organizing the community and developing the source code, though the different local claims might conflict with each other. Thus, participants necessarily act based on a specifically bounded rationality. The final means that leads to the emergence of both the community as it is and the kernel in its current state is the continual stream of negotiations. No single actor or group of actors would have been able to establish the actual solution to the problem of organizing the kernel development. The knowledge structure emerging from these negotiations can be seen as an "intersection of independent lies" as the biologist Richard Levins puts it [Lev66]. The American sociologist Leigh Star explains the sociological essence of this notion: "... Each local truth is partial and flawed; no a priori specification can encompass any global truth, but when (...) actors join local truths they create a robust emergent negotiated order" [Sta96]. The kind of cooperation discussed for the Linux case consists of such a joining of local truths – and this exactly is a pivotal dimension of inevitable uncertainty significant in our case: One can never know with certainty what the others know, want and will do.

In discussing matters of structure-building in society we have chosen the case of the Linux kernel project as an example. It might be asked, however, as to whether this example is representative or at least typical for all structure-building processes in societies? Of course it is not. There are various forms of how humans develop social structures: Some are more overt and deliberate while others have the character of only happening occasionally and without the actors' active knowing. The same holds for social structures themselves: Some are very formal, strict, and forcing, while others are more informal and weak. However, the focal point of our argument is that structure-building in the social domain basically takes place in a continual negotiation process and that any social order constantly has to be maintained by interactions in order to exist – these points not only hold for the Linux kernel community but for society in general with respect to all its facets.

3.14 Structure Generation Under Subjectivity: Selected Examples from Acoustics

Peter Költzsch, Volker Bormann

The field of acoustics is strongly prone to take subjective criteria during structure generation into consideration as most, if not almost all acoustic systems strongly relate to the human auditive perception. As soon as a perceptual factor plays an important part in the assessment of a system, subjectivity will indispensably have to be dealt with in any optimizing procedure. We will elucidate how this subjectivity may be taken into account, partially by relating it to objectively measurable quantities, partially by just admitting a certain range of undecidedness or fuzziness.

3.14.1 Room Acoustics

Visitors of concerts or theaters expect a technically good presentation of the performance at every location of the respective auditorium. They expect "good acoustics" wherever they are seated. There is no precise definition of good acoustics in a room, though. This concept is rather described in fuzzy expressions, such as audibility, acoustic properties of a room, acoustical intimacy, liveliness, reverberance, fullness of tone, sound volume, timbre, acoustic "warmth" of the room, spatial impression, auditory spaciousness, and acoustic envelopment of a listener (a listener should be bathed in sound from all directions), etc.

Room acoustic criteria have both subjective and objective aspects as is discussed in the literature, e.g. [AS93, And85, And98, ASNS97, Ber92, Ber94, Ber96, Ber03, FKS84, FSW87, FV98, Kut00, Rei79, Sch99b]. The subjective aspects relate to the perceived acoustic quality of a room, such as the perception of acoustic spaciousness, or the clarity of the sound of the orchestra at music performances. The objective aspects are governed by the physical description of sound field parameters corresponding to the subjective quality assessment. Both kinds of criteria are also dependent on type and usage of a room and on the type of an acoustic performance. Criteria for different kinds of music and for speech are to be distinguished.

The acoustic design of a room or a hall for the purpose of musical performances is usually part of an overall design. Thereby objective methods are used including objective criteria as well as subjective elements which accompany them. "Good acoustics" of a concert hall results from the interaction of subjectively assessed acoustic impressions in general and of subjective sound perceptions in detail as well as objectively measurable sound field parameters at those points where listeners and musicians are placed.

An evaluation function representing the goal "good acoustics" results from individual and from collective experiences with acoustically good or bad rooms. Furthermore such a goal function can be derived from hearing tests in a room and from tests in a synthetic sound field in an anechoic room by mainly investigating separated sound effects.

As mentioned before, at present no standard is at hand for an overall acoustic quality assessment of a room or for "good acoustics". Especially the mapping of objective factors that can be measured to a possibly multi-criterial subjective assessment is still unsolved. The acoustic literature offers a set of criteria for the assessment of the acoustic quality of rooms regarding musical performances (for example in [FV98, FS93, FSW87]).

3.14.1.1 Acoustic Assessments of Concert Halls

To illustrate the issue, we cite subjective assessments of acoustic quality for two concert halls, which were investigated with help of 50 – 60 test listeners at 5 selected representative seat areas:

	Neues Gewandhaus Leipzig (1981)	Schauspielhaus Berlin (1985)
Hall:	fan-like plan	classical shoebox form
Volume:	21,000 m^3,	15,000 m^3,
	1900 seats, 11 m^3/seat	1430–1670 seats, 10.5–9 m^3/seat

Assessments on the criteria loudness, duration of reverberation, clarity, and spatial impression were requested from the test listeners. These criteria seemed to be most important. For the assessments a scale of estimation was used. It was divided into 5 classes, with the middle class representing neutrality or no complaints. Neighboring classes represent perceptible complaints, then disturbing complaints in both directions. From the results conclusions were drawn on the acoustic quality of the hall at individual seats as well as of the whole hall. In Fig. 3.84 the results of the subjective tests are shown. The results are averaged over the seat areas.

As a further illustration acoustic optimization was carried out for the new Semper Oper in Dresden with subjective goal functions. The optimization was necessary because of the reconstruction of the opera-house after its destruction in World War II. It was reopened at February 13th, 1985. The large hall of this opera-house is of the type of a theatre hall with a semicircular base layout and four balconies. (Volume: 12,500 m^3, 1,300 seats, 9.6 m^3/seat). The hall was to be designed for performances of operas as well as for concerts. (KRAAK [Kra84]). On basis of existing documents a model of the historic hall of the Semper Oper was prepared on an scale 1:20, see Fig. 3.85. In this model the impulse responses of the room were measured as objective criteria. The measurements took place binaurally with an artificial head. It was measured in the empty room and also with a simulated fictitious audience in the opera-house. From the measurements the characteristics of sound quality were determined. These are reverberation quantity, clearness index C50, clarity index C80, spatial impression index, loudness, echo criteria, centre time, and reflection measure.

For subjective assessment hearing tests were carried out before opening the rebuilt Semper Oper. The tests included 86 persons in case of opera tests and 81 persons in case of concert tests. Overall 28 tests were carried out with different

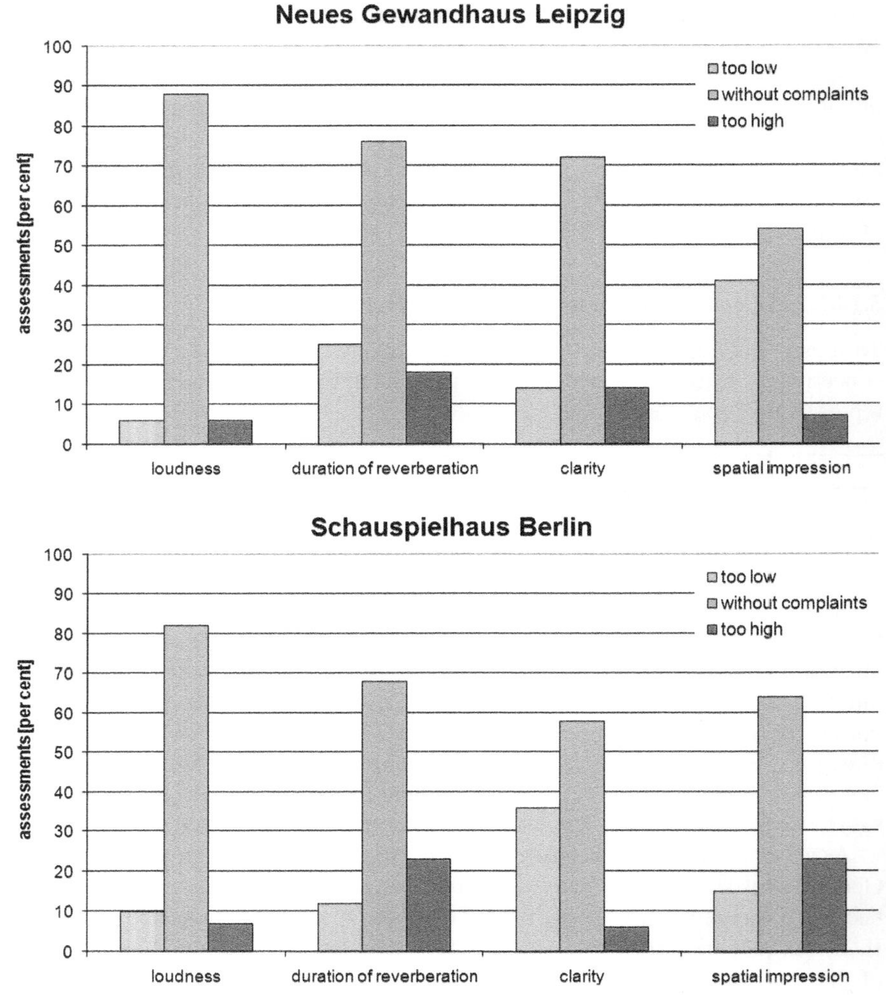

Fig. 3.84 Selected results of subjective assessments on the acoustic quality of two concert halls. In both cases the populated grand hall was evaluated. The diagrams show the percentage distributions of these assessments for different acoustic qualities (FASOLD/STEPHENSON [FS93])

variants (spatial details, musical motives etc.). The test persons were 28% experts of room- and electro-acoustics and sound editors, 22% musicians, and 36% music-interested concert visitors. The criteria of assessments were dynamics (loudness, balance), temporal structure (reverberance, clarity [transparency], blend, echo), spatial structure (spatial perception, spaciousness), frequency composition (timbre, change in timbre), and acoustical overall perception. The quantity of hearing impression related to these characteristics had to be marked on a scale.

The following results were achieved: For concert performances the objective characteristics which were derived from impulse measurements were found on the

Fig. 3.85 View into the auditorium of the model (1:20) of the historic Semper Oper Dresden. The model is equipped with absorbing material to simulate the audience. (photo: R. DIETZEL, TU Dresden)

average to be in an optimal range. The differences between various positions in the audience and the differences between an occupied room and an empty room are relatively small. The subjective assessments yielded the majority of criteria as optimal. So for concert performances it was found as acoustically extremely good. For the opera case the objective tests showed the reverberation time as substantially dependent on the equipment of the stage. The other objective criteria were found partially in the optimal range and partially slightly off the optimal range. The subjective assessments showed in summary the following results:

- Loudness, reverberation duration, mixing and spatial perception were estimated as optimal or as nearly optimal.
- The transparency was assessed lower than for the concert case.
- The speech intelligibility was assessed between good and satisfying.
- The intelligibility of recitatives was assessed as better than good.

The comparison of acoustics of the old with the new Semper Oper (Fig. 3.86) was interesting: The similarity in architecture of both opera houses reinstalled optimal values for spatial perception, mixing, and loudness. Therefore it can be stated the new opera house has nearly attained the former acoustic quality. The auditorium of the new opera house is slightly increased, leading to a slight reduction of the compactness and an increase of the reverberation time. But the slope of the parquet in the background and the view to the stage had improved the acoustic conditions, causing the good subjective assessments.

3.14.1.2 Methods for Acoustic Optimization

Acoustical optimization can be performed with theoretical as well as experimental methods. Computer simulations are based on physical laws of geometrical

Fig. 3.86 View from the stage into the auditorium of the new Semper Oper (Photo: R. Dietzel, TU Dresden)

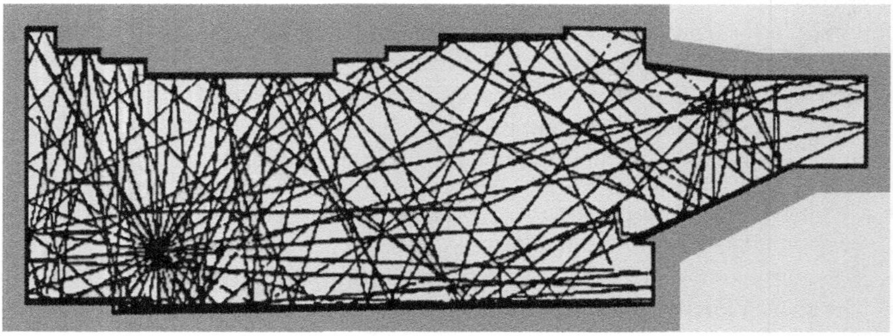

Fig. 3.87 Computer simulation with a sound particle method (STEPHENSON in FASOLD et al. [FV98]). The "sound particles" emerging from a point source on the left side in the sketch are traced to their incidence on the surrounding walls and on internal objects like seats or a populated auditorium, where specific laws of absorption, reflection and scattering are applied, creating new (but weaker) particles as secondary sound sources

room acoustics (ray acoustics). Here wavelengths are considered as short in comparison to the characteristic dimensions of a room. The main types of simulation are the mirror image source method (Fig. 3.86), ray tracing models or sound particle simulation methods (Fig. 3.87) including Monte-Carlo methods (VORLÄNDER/MECHEL [VM02], STEPHENSON [Ste04]) The essential goal of computer simulations is to find out the room's impulse responses, Fig. 3.88. From this the interesting acoustic criteria of a room can be calculated, e.g. clarity index C80 in Fig. 3.89.

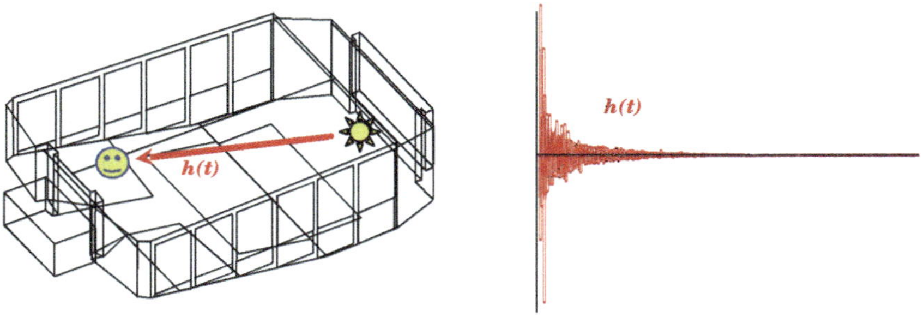

Fig. 3.88 Computer modelling in room acoustics. The left image shows the acoustic system with positions of a sound source and a receiving individual. The right image shows the impulse response h(t) received by the listener upon creation of a very short sound impulse (VORLÄNDER [Vor04])

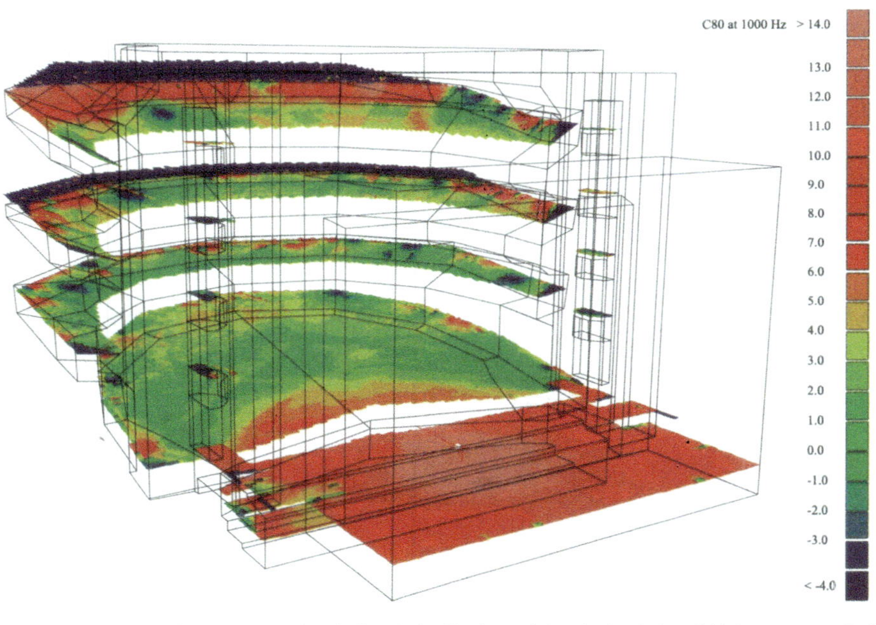

Fig. 3.89 Results of a computer simulation: Distribution of the clarity index C80 in a concert hall with three balconies [Vor04]

The results of a computer simulation can be realized by an auralization, i.e. the conversion of a theoretical waveform into an audible signal. That means the computed sound field at a seat in the auditorium of a virtual room is practically created by respective sets of loudspeakers whose placings are taken into account in the calculational realization. This auralization is carried out with the help of the calculated room impulse response, further with a "dry" (free of echo) recorded sound (e.g. music signals in anechoic rooms) as well as with outer ear transfer functions. Computer simulation with auralization thus enables us to hear into a yet not existing room.

Besides computer simulations there are also experimental model techniques. The procedure is based on the similarity criterion of the Helmholtz number (see [Köl03]). According to the reduction of geometry of a room (usual scale 1:20) the sound wave lengths are reduced in the same relationship. The frequency range is reciprocally transformed to the geometrical scale reduction (see example Semper Oper in Fig. 3.85).

In a such model room the reflection and absorption properties of room surfaces must also be modelled. Sound sources like shock excitation type spark transmitters with selectable sound radiation characteristics are used. The receiver is a model artificial head. It has two sound pressure receivers. Room impulse responses are determined in model measurements equivalent to computer simulations. For the same geometric and acoustic settings the desired room acoustic characteristics are calculated (e.g. reverberation time, clarity index etc.). Additionally, the model offers the possibility of auralization (hearing into the model).

Model measurements are very expensive and time consuming. The advantage, however, consists in clearness of the model. This is a crucial advantage in co-operation of acousticians with visually oriented architects. As of today these diffraction and scattering processes are the trouble items of computer simulation. Even the details of configuration within a room can be varied fast and simply with model measurements. Examples for that are inclinations of walls or of ceiling sections.

Natural sound fields can be reproduced in anechoic rooms (free-field rooms) with electro-acoustic equipment. For this purpose synthetic sound fields are generated with help of loudspeaker devices. Figure 3.90 shows an example.

This research method for optimizing room acoustics has the advantage of delivering systematic variations of characteristics of a single sound field. Furthermore the influence of variations of these components on room acoustics quality can be subjectively judged.

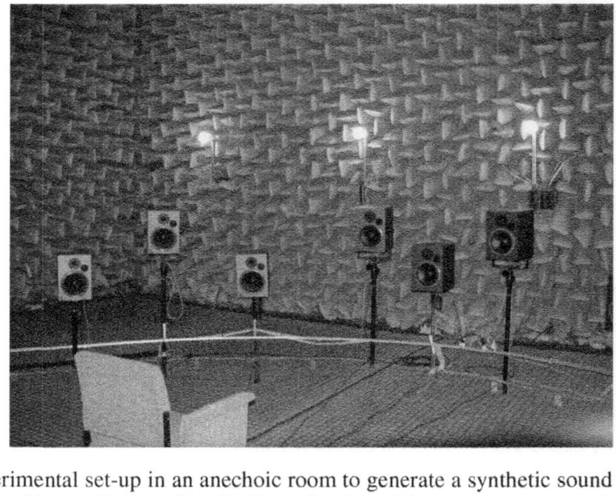

Fig. 3.90 Experimental set-up in an anechoic room to generate a synthetic sound field. The problem addressed in this case is the subjective investigation of the apparent source width of an orchestra (following M. BLAU in [Köl03])

3.14.2 Further Examples

3.14.2.1 Optimal Adjustment of Hearing Aids

Modern hearing aids have a large number of adjustable parameters. Setting them appropriately helps to compensate the deterioration of hearing capability individually in a variety of circumstances. Therefore the problem is to find out an optimal parameter constellation "by hand" in order to maximally use the rest-hearing ability. Such a hearing aid adaptation is successively carried out in a "dialogue" between the adapting acoustician and the hearing-impaired person. For developers and manufacturers of hearing aids the optimization includes objective goal functions, such as miniaturization of a hearing aid, minimization of energy consumption, expansion of the range of adjustable parameters, and adaptation to requirements of users.

In reality, instead of objective criteria, the multi-criterial optimization by a hearing-impaired person is primarily concerned with subjective characteristics of his or her hearing impression. These are loudness, pleasantness (pleasant impression, being agreeable), timbre, distortions, speech intelligibility, noise, background and ambient noise. Furtheron, non-acoustic subjective criteria frequently play a crucial role, e.g. comfort of wearing, low maintenance, being user-friendly, service comfort, low purchase price, visibility in public.

For the adjustment processes often the so-called characteristics spider is applied as a suitable tool. With its help the multi-criterial optimization with subjective goal functions can be demonstrated, see Fig. 3.91.

3.14.2.2 Acoustic Quality of Loudspeakers

Assessments of the acoustic quality of loudspeakers are mostly performed by subjective criteria. In hearing tests the overall subjective impression is rated as a

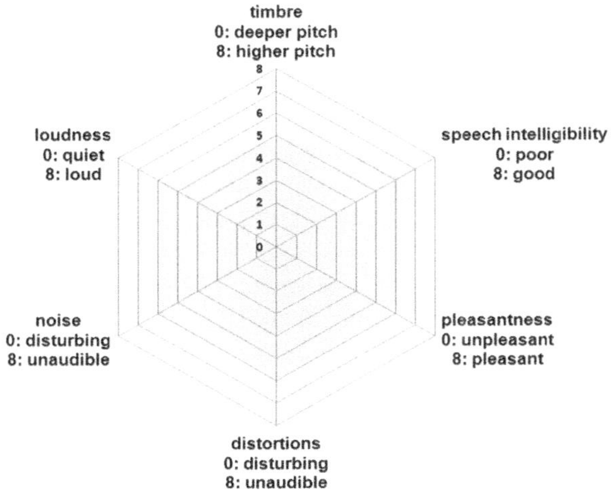

Fig. 3.91 Characteristics spider for judgements of natural sound samples with an evaluation example (following G. FUDER in [Köl03]). The general axis direction (inferior to superior) is outwards, so larger covered areas represent better overall solutions.

subjective goal function of an acoustic optimization [Kli87, Kli88, Bec92, Oli03]. There are objective criteria for processes of loudspeaker optimization, such as sharpness, bass reinforcement, clarity, change in timbre, and distortions. They are mainly used to identify defects in the sense of deviations from an optimum, as there is a certain correlation of subjective inferiority and an objective miss from typical values of each of them. The fulfillment of all partial objective criteria is not identical to a subjectively perceived quality by the way of this optimization, though.

The acoustic qualification of test listeners is of great importance to this problem. Since this is generally relevant for optimizations with subjective goal functions, a more intimate discussion of loudspeaker optimizations may be useful.

Trained experts as listeners are used in hearing tests of manufacturers. Such listeners represent subjective "measuring instruments". The requirements to the listeners are hearing experience (experienced HiFi hearing, regular visits of concerts, musical activities by themselves). These experiences determine critically and discriminately the listener's conceptions of an ideal sound expression and the skill of being able to estimate a sound impression. Additionally, assessment experience of trained listeners is an essential advantage. This includes familiarity with scaling methods and other conditions of hearing tests. In tests for loudspeakers as consumer products untrained persons (laymen) are preferred, as they can be regarded as adequate representatives of the customers at a later date. Since the latter are not evenly distributed throughout the population, demographic and socio-economic factors must be adjusted in these tests accordingly.

Already the training process can influence the listener. Trained listeners are in general more critical than untrained listeners. For example they valuate acoustically bad loudspeakers worse than laymen. So judgements of trained listeners are not representative for unbiased and untrained listeners. Expensive acoustic improvements of loudspeakers which were required by trained testers would often not even be noticed by untrained ones. Trained listeners are best in subjective tests with regard to reliability of assessments, though. During the test sessions attention must be payed to a potential temporal and physical exhaustion of the testers, as this may cause mental and physical fatigue and thus reduced attention and motivation.

In a concrete case, four loudspeakers were tested in an extensive investigation [Oli03]. In order to estimate a first setting of objective criteria for an optimization, sound signals were measured in an anechoic room at conditions of a right angle to the diaphragm areas of the loudspeakers and in a hearing window of horizontally ≈ 30 degrees and vertically ≈ 10 degrees. Additionally measured quantities were sound power, directivity index of sound radiation, and transfer functions.

In hearing tests an assessment scale was used for a perceived quality as a general impression (scale of assessment: 0 to 10, representing absolute rejection to total preference). The results distinctly correlate subjective judgements in preference ordering of the loudspeakers concerning their sound quality with the results of the objective criteria measurements. This holds for both the valuations by untrained listeners and those by trained ones. Nevertheless the test designers conclude that, as of today, hearing tests are the final crucial arbitrator of the acoustic quality of loudspeakers.

To improve the desired correlation between subjective and objective assessments, the subjective valuation of loudspeakers should not only be performed with the target of a single rating value in mind. Instead, individual subjective sound characteristics should also be described and scaled. Verbal terms are used for sound characteristics, like volumes, sharpness, clarity, height or bass reinforcement, spatial feeling, change in timbre, bass clarity, and brightness. Objective assessments, on the other hand, refer to descriptions of loudspeakers properties at concrete listening conditions to be determined by measurements. The goal is to express the impression of listening by acoustic conditions which can be measured at a place of listening.

Both subjective and objective tests results can be represented in a characteristics spider similar to Fig. 3.91, even though they may not map directly to each other. It may be worth trying, though, to test certain combinations of objective properties on coincidences with subjective valuations. Should such a coincidence be found, future results of subjective quality assessments may be forecasted eventually. Even then the subjective loudspeaker testing will further be required as they address predominantly the taste of customers even if the technical quality is secured.

Even though there are noticeable correlations between subjective valuation and objectively measurable sound properties of loudspeakers a lot of work still has to be done to create a somewhat quantitative mapping between these two. The reported results on the coincidence of subjective assessments and physically measured quantities at certain listening places and boundary conditions are at least encouraging for this effort.

3.14.2.3 Acoustic Quality of Products and Acoustic Design

For many products the emitted sound is a perceived quality feature [Jek04, Jek05, BJ98, BJ96], [BH02, BH03]. It can be relevant to customers in a purchase condition. Manufacturers and buyers aim at and expect a sound that is a characteristic feature. In this context the sounds are frequently regarded as acoustic visiting cards, "product sounds", or corporate sounds.

Nevertheless the sound property of a certain product is ambivalently judged. Some persons lay more stress on usability qualities and regard the emitted sound as very minor attribute. Others prefer quiet products. There are still others preferring louder noise and expect a typical sound of a product.

Further factors are important in sound valuation because the emitted sound implies an information content. This content is of concern to safety factors (e.g. in case of deviations from a normal machine noise or noise of an approaching vehicle as a warning signal to pedestrians). Noise can indicate operating conditions and necessities of maintenance or repair.

Unquestionably, emotional and very different meanings are linked to sounds in everyday life. This depends on age, living standards, attitude towards life, hearing habits, and experiences. A consequence is the preference of some types or marks of products by different parts of a population. So, some prefer extravagant and striking sound properties and others prefer particularly discrete sound qualities. Some examples are: Which volume and sound is a mixer in a kitchen allowed to have?

What strength of noise, how loud, and which timbre may a vacuum cleaner have according to suction power or normal operating conditions? Which acoustic signal suggests overload, dust bag emtying or even technical maintenance? What sound should be created for a motor vehicle, ideally specific for a type of car (e.g. the so-called "Porsche sound")? What kind of sound should be heard in a car stemming from the windscreen wipers or the direction indicator?

In an actual case these considerations lead to the question what really matters in an optimization with respect to so many subjective and objective criteria, not forgetting to take into account generally accepted limits of noise (e.g. noise with motor vehicles) even if the potential customer would rather transgress them. These problems of acoustic quality and acoustic design are rather similar to those addressed in the before mentioned cases. This leads to more general questions: Can the chosen functions of subjective goals be definitely formulated? Is it possible to arrange a goal function in a form which can be measured? How large is the fuzziness of goal functions? How to identify adequate test listeners?

Problems of acoustic qualities and acoustic design are characterized by the criteria [Jek04, JB05a] which limits of sound emissions are to be considered with respect to hearing damage and annoyance, which threshold values are to be considered as a safety factor to avoid risks or dangers and to judge operating conditions, which subjective preferences of noise are to be considered in very different situations of everyday life, and which sound quality of a product and at production is to be considered satisfying to requests of manufacturers and users.

To progress on these issues empirical investigations are necessary in scientific research and in developments. Evaluation methods must be established for acoustic product quality and acoustic design by means of field studies and laboratory tests, to derive rules for a targeted design of product sound. The aim is a code of practice for acoustical optimizations using subjective goal functions which are adapted to the products.

3.14.2.4 Subjective Evaluations of Acoustic Qualities of Musical Instruments

For musical instruments the important acoustic criteria are pitch, response, timbre, projection etc. These can be derived from different measurements. As of today, acoustic evaluations of musical instruments are traditionally conducted in a subjective manner. But more and more a mix of subjective and objective criteria is used which complement each other. The trend clearly leads to the application of objective criteria which can be physically measured.

The statement of an expert is a generally accepted opinion: "An optimization is principally possible only if subjective criteria are described objectively.... (But) the condition arrived at is still completely insufficient today." Therefore in this case the correlation between objective criteria and subjective evaluations is of crucial importance [Bal03, Gä95, Mey77, Zie95, Zie96, Zie00, Zie02].

For subjective test procedures it is important to include judgements both by musicians and by listeners. However the test results of listeners and of musicians can deviate considerably from each other. With respect to practical ranking of instruments musicians have a plain opinion: The satisfaction of the professional musician must be reached. Objective criteria can only be used as a support to subjective impressions.

The discrepancies between performer and listener, e.g. with respect to guitar sound valuation, are well-known: they use rather diverse criteria for their assessments. Nevertheless, a wide variety of instruments, large production output, and the aim to arrive at high acoustic quality gives the motivation to try to objectify the optimization procedure, creating an almost ideal test bed for melioration based on measured quantities and correlation with subjective valuations. In order to approximate this aim, the hitherto poorly defined subjective criteria must be defined more closely by (combinations of) physical quantities. Beforehand the subjective criteria should be assessed by trained listeners and professional musicians. Additionally, musical experts, manufacturers, and, if necessary musical historians have to contribute. A consideration of non-acoustic criteria will presumably be required as well.

A final example of optimization in this area is the quality judgement of bells [Fle00, HT87]. According to expert opinions only subjective judgement is important. As an expert puts it: "The subjective judgement of a bell expert is above all results of physical measurements." [Fle] In practice acoustic judgements on the basis of subjective criteria are exclusively carried out by special bell experts, church musicians, historians, but not by physicists or laymen. Both of the last would only be able to assess the sensory pleasantness of the sound of a bell that is just poorly defined. The acoustic goal function of a bell is highly subjective and of a multi-criteria nature. At present there is no objective procedure that would yield the same rating hierarchy as subjective evaluations, even if a set of objective, measurable optimization criteria would be provided.

3.14.3 General Aspects of Structure Optimization with Subjective Goal Functions in Acoustics

3.14.3.1 Subjectivity vs. Objectivity

An important part of acoustics is concerned with a sensory perception: the hearing. In some cases hearing is even connected with seeing as previous examples already have pointed out. In this domain the "human factor" always plays an important part and resulting structures are decisively influenced by individual perceptions. Accordingly, many acoustic problems cannot be investigated without an explicit consideration of individuals. Acoustics, electrical engineering and information technology, mechanical engineering, civil engineering and architecture as well as psycho-acoustics, social sciences and medicine all work together in the pragmatic assessments of technical solutions, leading to the typical difficulties and problems of an "a priori" interdisciplinary science.

In this field of application methods and criteria of the sciences are connected necessarily with methods and criteria of the humanities. This combination, including the human perception, leads to interesting new questions not common the sciences of nature and engineering:

- How do individuals assess results of physical processes leading to a negative subjective perception?
- Which consequences does this evaluation have for use of suitable objective, physical methods and criteria?
- How can perception processes of individuals be modelled? How can these psycho-acoustic models be coupled to physical models of processes? How can this connection of physical and psycho-acoustic models be used for an acoustic optimization with subjective goal functions?
- Which influence do individuals (tester, a group of testers) have on optimization processes with regard to their non-objective evaluations? Can the subjective part of such an optimization be objectified by suitable procedures?
- Is it sensible to objectify processes that include human perceptions? What are objective processes in human sciences in contrast to those in the sciences of nature and engineering? Can a perception by an individual be a kind of a subjective measuring instrument?
- Can the fuzziness of procedures and results be avoided or is it an immanent part of such problems and their solutions? Are subjective optimizations less reliable than objective optimizations?
- Can an immanent fuzziness sufficiently be justified by the ubiquitous variety of human experiences and valuation of optimization results? In acoustical optimizations we principally perceive a notion of uncompleteness because of subjective aspects of goal functions.

As of today the acoustic quality of a room – e.g. a concert hall or an auditorium – can not completely be characterized by objective quantities. Larger assessment groups in concert halls and auditoria, test groups in acoustic laboratories, test individuals of all ages and of different training conditions regarding the qualification for hearing tests are necessary to determine the different acoustic characteristics.

So, e.g. for a concert hall, spatially resolved tests including all seat ranges on parquet and on the galleries have to be carried out for different types of music to investigate all effects. In addition special acoustic investigations in laboratories are frequently applied. Based on their comprehensive knowledge, experts can deliver "definite" subjective estimations of the acoustic quality of such a hall. Finally, conductors, concert masters, first violinists and soloists will judge the acoustic quality of it by their authority, their standing, and their aura.

In order to acoustically optimize a concert hall today the above mentioned methods of modelling (computer simulation, model measuring techniques, and synthetic sound fields) are used. Furthermore, extensive empirical results of many acoustically excellent concert hall designs can be taken into consideration. Besides that, hearing tests are made with methods which include human judgements in fictitious halls. Thus first-class acoustic solutions result by the interaction of objective and subjective methods and goal functions.

At present a limitation or mitigation of the subjective valuations in an optimization process by the increasing objectification of partial criteria is not noticed. On the contrary, subjective goal functions and methods for their determination are constantly further being developed. Perhaps the struggle for "good acoustics" of a concert hall is an example for an optimization process in which both, the goal and the way to it, optimally intermingle. The goal is the creation of good acoustics of the hall. The way to arrive at this goal is an interaction of objective and subjective methods and criteria at the interface of natural and human sciences.

The examples of the acoustic quality of musical instruments and of loudspeakers point into the same direction. In both cases the application of objective criteria and methods make very much progress. Especially many automated measuring systems and test procedures are applied. This development is strongly pushed by commercial interests, high production outputs, common and diverse use, and vast numbers of installations (e.g. loudspeakers are located in every home and in every car). The qualitative standard of this objectivation is based on constantly refined subjective goal functions and psycho-acoustic understanding. Nevertheless at present and in a near to medium future subjective evaluations of goal functions play still a large role for acoustic mass products.

Subjective factors are also of great importance if the optimization goal is acoustic quality and acoustic design of products. At present the group-specific acceptance of acoustic properties of numerous products and hence their design goals is being investigated. The respectively required psycho-acoustic and psycho-social approaches are still in their infancy. This is also true for the development of assigned subjective criteria, methods of judgement, and the start to objectify partial criteria. Due to commercial impact the methodology is most advanced in the field of passenger car sound design. It may be estimated that the systematics in this area may be very useful for the entire field of acoustic quality and acoustic design.

Finally, the adjustment of hearing aids is a completely different case. Here a technical equipment must be adapted optimally to the acoustic perception of a hearing impaired individual. In this case acoustic factors must be considered on the one hand, e.g. loudness, timbre, distortion, speech, intelligibility. On the other hand the influence of non-acoustic factors as pleasantness, wearing comfort, low maintenance, suitability for users, operability, service comfort, low purchase price, visibility etc. must be taken into account. Several of these factors cannot be objectified. That means this optimization "for an individual" is unavoidably and to a large extent determined by subjective influences.

3.14.3.2 Pareto Sets

As already described in detail, room acoustic quality is an extremely complex criterion. It can be described only by a multi-dimensional characteristic vector and is therefore prone to pareto-optimality trade-off. Accordingly efforts are undertaken to identify just a few relevant attributes for the complex quality assessment. These attributes should be physically precalculable. They should be measurable in experiments and should correspond to subjective judgement in hearing valuations as well.

A successful identification of such attributes would vastly reduce the expenditures in the acoustic optimization of concert halls and the like. Furthermore such a reduction to an objectified standard attribute assessment set may replace the subjective averaging evaluation of the total acoustic quality of a hall presently only realized by questioning a larger group of testers.

Promising starts on this way are to be seen in the reports of ANDO et al. [ASNS97], SAKURAI [Sak00], SATO et al. [SOT+02]. These authors had selected a set of four most important physical characteristics:

- sound pressure level (listening level), created by the pressure amplitude of the early reflections and the following reverberation,
- initial time delay gap, i.e. the time between the direct sound and the first reflections at the place of a listener,
- subsequent reverberation time, obtained by the decay rate of the sound energy between -5 dB and -20 dB,
- interaural cross correlation coefficient.

Detailed investigations showed that these four attributes are independent in their influence on subjective evaluations of the acoustic quality of a hall. With the help of the law of comparative judgements (pair comparisons) a linear scale of values is obtained for the four quality factors. Then an overall quality factor for the acoustic quality of a concert hall is derived by adding the scale of values under certain conditions. The four selected quality factors and the overall quality factor are useful for an acoustic characterization of a concert hall as was shown by the authors. On this behalf they compared the computational results (use of the mirror image source method), the results of measurements in the finished and equipped concert hall, and the results of hearing tests in all areas of seats.

Two conclusions of special importance are drawn from these investigations. First there are large individual differences with respect to personal preference of the four selected quality attributes. That refers particularly to the amount of the pressure level of sound and to the individual preference of the reverberance impression, by which the subjective notion of the initial time delay gap and the reverberation time are determined.

Secondly, consistent numerical computation rules for the determination of the four acoustic attributes for all areas of seats, as well as the overall acoustic quality, allow to apply computational optimization methods to such a problem. SATO et al. [SOT+02, SHT+04] have used an evolutionary optimization method with help of genetic algorithms for such calculations. As an example a concert hall of a typical shoe-box type was acoustically optimized step-by-step. The results were successfully compared with the space proportions of the Wiener Musikvereins-Saal which is internationally well-known for its excellent acoustic properties.

The reported procedure can be used as a starting point to the following investigations:

- Are in fact selected acoustic attributes the factors which decisively determine the acoustic quality of a concert hall?

- Can the overall acoustic quality criterion, being derived from objective and subjective investigations, further be improved? The aim is a rational restriction of the complexity of the quality criteria.
- What is the acceptable limit of fuzziness in a single attribute, not leading to a noticeable influence of the acoustic quality criterion of a room?
- How can the inter-individual differences of subjective evaluations be integrated better into the overall criterion of the acoustic quality of a hall, in particular the differences between laymen and experts as well as between listeners and musicians?

For a multi-criteria optimization there are several goal functions: A whole Pareto set of solutions exists in which criteria tradeoff can only performed by subjective stressing (see Sect. 2.5.3). This is frequently the case in acoustics, e.g. with the standard example of the acoustic quality of a room. Numerous objective partial goals, like reverberation time, clarity index, early lateral energy fraction, spatial impression index, loudness etc. can be identified. Each of these objective criteria has an optimal range, with a variation width resulting from subjective experiences. According to experience the acoustic quality of a room should be optimal if the values of all criteria are within their optimal ranges.

The span of the optimal range can be relatively large, however, and the settings of individual criteria within their ranges influence the variation width of others considerably. So the subjective general impressions of the acoustics of a hall are distinctly different in cases where partial criteria settings far off the conventional value sets, but each within its usual optimal range, are chosen. In such a case single acoustic characteristics are investigated by hearing tests, thus determining the levels of and adherences to single subjective goal functions. These are e.g. loudness, reverberance, clarity, spaciousness, acoustically wrapped up, intimacy, timbre, etc. By this procedure they are treated as partial subjective goals of a respective pareto optimization.

The general optimization problem is thus divided into an objective and a crucial subjective part. The objective part contains the design of the objective partial criteria in the optimal ranges, based on computations, measurements, or experiences. The subjective part selects the solution set from the Pareto set of the objective partial criteria, based on subjective tests by inclusion of human perceptions.

Chapter 4
Transdisciplinary Perspective

Indeed, we live in a world of structures. We depend on them, we generate and destroy them, and we are part of the evolution of nature and culture along with the associated permanent processes of destruction and resurrection of structures in our physical and mental environment. What can we learn about structures and structure generating processes from looking at case studies in such different disciplines as engineering, biology, logistics, psychology, sociology and many others as we did in the preceding sections?

First, we realize that in all disciplines there is a fundamental desire to understand the mechanisms by which structures originate or are generated, survive or, eventually, are destroyed. Understanding these mechanisms may allow to control and optimize the associated processes, at least to a limited extent, for the purpose of developing a better world. Second, it becomes clear that there are a number of generally valid features which limit the control and optimization of structure generating processes, such as multicriteriality, uncertainty and subjectivity. These features can be identified in all disciplines, as soon as we turn from abstract model situations to those of the real world. We then usually are well advised to augment our model analysis by heuristic knowledge, established by experience, in whatever field we consider.

So, evidently, there is a strong and powerful transdisciplinary basis for analysis of structures and structure generating processes in all disciplines.

4.1 Methodological Survey

Within the framework of general properties of structures and structure generating processes there are many specific features in the various disciplines. Technical structures lend themselves to optimization, at least in a restricted sense, when a system of values can be associated to them. Accordingly, mathematical analysis, in various levels of sophistication, is the most adequate method in the technical sciences. In fact, it is usually the very task of analysis there to provide strategies to optimize technical structures and structure generating processes with the goal of maximum economic profit. Similar features are frequently observed in the natural sciences

K. Lucas, P. Roosen (eds.), *Emergence, Analysis and Evolution of Structures*, Understanding Complex Systems, DOI 10.1007/978-3-642-00870-2_4,

and economics. This is, of course, quite different in the realms of biology and, even more so, in social sciences. There appears to be no elaborate system of multiple values in biological evolution in general, beyond the maximum of reproductive fitness, although there may exist some hierarchically lower target functions that are persecuted provided that the main target fulfillment remains unchanged within the limits of statistical fluctuations.

In social processes change is not based on abstract, objective structures but rather on the society's or a community's subjective perception of them and their values. Mental reflection in non-formal languages is the most common type of analysis here.

4.1.1 Classification of Structures

Structures are manifestations of interactions. The crucial processes generating, maintaining and, eventually, destroying structures are the internal interactions between the elements of a system and the external interactions with its environment. Only in simple cases are the interactions between these elements understood in detail: Simple structures in fluids such as snow flakes or Bénard cells are dependent on the special and rather well understood interactions of the molecules within a fluid and also on the particular interactions of the fluid as a whole with the environment.

In most practical cases, however, a thorough understanding about the interactions and the evolution of an observed structure is not achievable, although the relevant interactions can be identified in principle. Examples of such more complex interactions were identified in biological applications. So, in Sect. 3.4 a theoretical RNA structure optimization and its limitations on the basis of evolutionary algorithms was discussed where the assumed goal was minimum free energy. Further, in Sect. 3.5 it was shown that molecular structures with desired properties can be generated by in vitro evolution in the laboratory by setting appropriate environmental conditions as discriminating sieves.

Again rather different interactions are responsible for the structure generation of an automobile or other technological structures. Often, structure generating or rather structure destroying processes require even interactions spanning various time and space scales, as demonstrated in a case study on computer-based demolitions of buildings in Sect. 3.8 Finally, entirely different interactions are generating social structures. Such interactions were studied in Sect. 3.13 in an effort to understand the continuous and successful self-structuring processes between the involved human beings in the Linux open source software development. Here both objective facts, such as personally achieved software development goals, and also interpersonal, subjective valuations, such as discussional behaviour, play interlocking roles for the placement of the structural members within the community.

Depending on the types of interactions two different types of structures came to appearance, static and dynamic structures. While for static structures any external interactions must be eliminated, these are essential for dynamic structures because they provide the flow of energy necessary for stabilizing them. Static structures are abstract phenomena, which have only relevance for short times in the real world.

Even such a seemingly static technological structure like a bridge is, on a life-time scale, really depending on interactions with the environment in all phases of its generation, its maintenance and its destruction. It is thus a dynamic structure, i.e. a structure depending on interactions with the environment and changing with them, as are all structures in the real world, be they of a material or an immaterial nature. A severe consequence of recognizing the important structures of our world as being dynamic is that their generation processes, being dependent on a multitude of interactions, are unpredictable and thus also non-reproducible in detail. While this is almost trivial in such disciplines as sociology, psychology and other humanities, it appears odd in the technical and natural sciences. For example, in production technology, we are used to mass generation of most of our everyday products, thus representing prototypes of reproducibility. But the contradiction is only feigned. In detail no single structure of mass production is fully identical to any other one, although the major appearance and functionality may be.

Finally, a common feature of all dynamic structures and thus all structures of the real world is their transient nature. They need a permanent interaction with the environment. If this interaction is modified or interrupted the structure will be changed or even destroyed, such as a bridge without maintenance, a power plant without import and export of energy, a cultivated flower garden without care that may turn into a biotope, or a scientific thought or theory without feed back by the community.

4.1.2 Understanding Structures

In order to understand structures we design models, mostly deliberately simplified. They are the scientist's primordial access to the understanding of the real world's complexity. There is no hope to analyze its structures and structure generating processes in all its details, even though the networking aspects and raised complexity issues in modeling efforts gain more and more interest in recent research activities. There is also no sense in an extensive modeling depth since we are normally not interested in such a detailed knowledge. Instead, in the applied sciences we are investigating particular issues and we want to answer particular questions. The approach generally taken is to design a model of the process or the structure we want to analyze. So, in a strict sense, we do not analyze reality, we rather analyze models of reality, even though we take care that the model represents reality at least in an adequate manner. This holds for all fields of sciences and humanities, in particular also for the fine arts and architecture, where models of the final work represent preliminary steps to the final result. The model that is set up cannot be categorized as right or wrong, but rather as useful or not useful for the analysis to be performed. In most cases models are not even supposed to describe structures or processes of the real world but rather only basic dependencies of real world behavior on the parameters under variation. This is again true for all disciplines, although in the technical and natural sciences models can frequently be designed so close to reality that their predictions of structures and processes can quite well be used in practical construction problems or melioration efforts.

An impressing example for this approach was given in Sect. 3.1 for the field of production engineering where the cooling process of a die by a cold liquid flowing through bores has been considered. The underlying model of representing the three-dimensional heat flow from a die surface to cooling water, streaming in a tube grid, by light radiation from the tube locations on to the die's surface is, from a pure scientific approach, simply wrong: The representing spreading functions are not equal. However, the difference is, in practical dimensions, so small that the systematically obtained results by an evolutionary search strategy, mimicking biology's evolution processes in a quite simplified manner, yield significant practical improvements to prior knowledge.

A much more subjective example was discussed in acoustics: Within a certain range of indifference, combinations of objective target values of physically accessible quantities may be used to model the acoustic impact of a performance hall structure on the sound perception in the auditorium, even though the coupling of physical values and perception is by no means stringent as of today. So a physical "sound design" prior to creating a new construction will certainly help omitting grave errors that otherwise would require heavy redesign in later stages of the building process. Here again, the early practical structure selection work, enabling the investigation of a large number of competing designs, is performed on simplified representations of reality, not reality itself. Even though the existing models are by no means perfect and certainly cannot claim to represent the governing subjective "final" taste (potentially changing with generations of listeners), the representation is certainly helpful and may, on the long run, help to adapt objective representations even more to the subjective notions.

Another very important aspect of structure understanding is the deliberate inclusion of uncertainty issues into respective models and studying their effects on practical predictions by parameter variation. This reflects the fact that in general none of the parameters or features of a model can be specified accurately. In the technical sciences this may be due to a lack of knowledge associated with a production process or with external interactions. Compared to the natural and technical sciences uncertainty is an even bigger issue in the social and economic sciences. Accordingly, this scenario technique is even more habitual to gain an understanding of the structures in non-technical areas.

4.2 Striving for Optimality

Whenever we can define quality measures, such as optimal structures of biomolecules or optimal structures of power plants, and as soon we can set up a formal model on the basis of mathematical equations or at least in the form of better/worse relations, the goal of optimization can be approached by a variety of methods, no matter which structure and which discipline is considered.

Very powerful methods have been developed to systematically optimize just one goal, as the so-called fitness of a suggested realisation is a global property easily to be attributed by a one-to-many comparison, even though it may require significant efforts to actually create a new realisation and perform the classifying comparison. Typical examples for this class of structural optimization are the isolated optimization of certain technical properties of construction elements, as mentioned in the various case study contributions.

A both practically more relevant and technically more demanding optimization issue is the trade-off between conflicting goals, the problem of multicriteriality. Typically, structures to be generated underlie a set of various frequently conflicting measures of value. Automobiles are not only constructed with the goal of maximum speed, but also low fuel consumption. In radiotherapy the intensity of radiation is targeted for optimal destroying capacity of the tumor but also for minimum damage to its environment (Sect. 3.2). A residential heating system should provide a cosy, comfortable environment to live in whilst not requiring a too energy-expensive operation (Sect. 3.6). A concert hall is to be designed in a way that both allows for speech and musical performances (Sect. 3.14).

In such cases a rational procedure is to obtain a so-called pareto set of solutions, unless a hierachical order can be associated to the measures of value. Methods mostly derived from evolutionary optimization, sometimes supplemented by targeted mathematical methods, are available that allow to compute the members of a pareto set, although a quantitative measure of quality is not always easy to define. An example for multicriterial tests from molecular biology was treated in Sect. 3.4 where it was found that in RNA molecules not only the molecular structure is a relevant measure of value, but also various kinds of stability, which may be included in the finding of an optimal representation. Yet another example was addressed in Sect. 3.10 on innovation management, where a pareto set associated with the conflicting goals of project costs and costs per product unit was discussed.

A major aspect in the as yet rather limited public acceptance of pareto methods is their inherent ambiguousness with respect to the decision on which individual solution is to be taken for a problem in question. Even with a fully developed pareto set the final decider is left alone with the arbitration of subjective weighting of partial goals. This was a frequent point of criticism when respective methods have been suggested in economics and production planning. Other fields of application show similar deciders' reactions (cf. Sect. 3.3). It is therefore still usual to superimpose partial goals by an "objective" linear goal weighting, thus reducing the solution space to just one "really optimal" solution. In these cases more efficient mathematically based optimum search strategies may be put to work directly from the start if the linear weighting factors are thought to be known in advance.

As of today, the public perception of decision processes seems not yet ready to appreciate the consideration of inherent multi-criteriality. This goes hand in hand with the non-availability of generally efficient pareto set determination methods, leading to a strong desire in the human optimizers to apply single-criterion-methods. Multi-criteriality always involves a certain amount of subjectivity, which is generally disregarded in our modern "objective" society. Presumedly here an educational approach is necessary to bring the general idea of ubiquitous tradeoff necessities to a broader auditorium.

More fundamentally, although we can think of an optimal technical structure, such as an optimal ballpen or even a power station, after having decided upon one combined goal, we will find it most difficult to imagine an optimal plant or animal, or even more so, an optimal society. At least the evolution of plants and animals so far followed the Darwinian pattern of mutation and selection, with the goal of adaption to an ever-changing environment. The idea that this development proceeds towards a teleological, predetermined goal of optimality is generally disregarded today, at least in the domains of scientific views on the world.

So at first sight, there seems to be a discrepancy of natural and technical/artificial structures. However, the historical perspective reveals that the evolution of technical structures has much in common with the evolution of nature. Although at every moment an optimal structure with respect to economic profit or practical function may be thought of, resembling an optimal adaption to the environment, it will be difficult to predict the future development since the quality measures themselves develop in unpredictable ways. A prominent example is the development of power plant technology. While in the beginning of industrialisation the clean power plant exhaust emissions (water steam and carbon dioxide gas) bore no relevance whatever, they play major roles in the present discussion of energy provision long-term sustainability.

Comparable paradigm changes may be observed for private transportation solutions. While in the last century goals like individualistic automobile design, spacyness, velocity, strength and elegance were regarded as topmost qualities, nowadays low gasoline consumption, handyness in tight traffic situations, accident safety and endurance govern the valuation discussion of four-wheeled automobiles. Interestingly, quite the opposite development took place in the two-wheeler regime, where earlier practicability considerations lost more and more ground against elegance, (mostly unusable) power availability and fashionable designs. The two last examples, being both so closely related in terms of purely practical application and so strongly differing in public valuation, show that rather unforeseen quality measures guide even the technical development of artificial structures into directions not predictable from their historical roots. In a more general view this may be regarded as a kind of adaptation to changing "environments" with discernible parallels to natural evolution.

In addition to these unforseeable changes in public or even objective valuation as time goes by and circumstances change, the transient nature of rather strictly definable, aggregated goals may be a crucial part of a technically definable problem. In the worst case goals may change on the same time scale or even faster than the results of an anteceding optimization process are realized. Such situations occur frequently in logistics, in the scheduling of elevator systems or with the "yellow angels", as discussed in the Sect. 3.11.1 on online optimization. Here effective, mathematically-based treatment methods are generally at hand. Such processes are basically different from those discussed before as the goal change is, with stochastic limits, foreseeable and part of the problem definition. One might interpret the formerly discussed historical paradigm changes as a kind of online problems, though, by widening the viewpoint towards inclusion of further goal aspects beyond any momentary optimization effort. There even is an emerging discipline that frequently deals with this kind of methodology transformation, even though it does usually not regard itself as being an optimization business: technology assessment. But in a more general view, taking into account unforeseen changes due to a wide application of a new technology is nothing else than a systematic consideration of a more general technical and societal goal set.

4.3 An Hommage to Statistics

Limitations in the stringency of models for structures and structure generating processes may be of a practical as well as of a principal nature. While a practical limitation of stringency is mostly based on economic reasons and discussed in more detail below, the principal limitations shall be scrutinized first.

In Sect. 3.8 problems of statistical weather impact on buildings, as well as material variance effects in a targeted destruction of buildings, have been discussed. Here the stochastic nature of the physical phenomena cannot be ruled out by whatever effort is scientifically or technically undertaken. Similar effects, although caused by a deliberate restriction of the underlying model, paying tribute to the insecurities of creative tasks as well as the uncontrollable effects of real markets and competition situations, are treated in innovation management (Sect. 3.10) when trying to create a basis for a sensible subdivision of complex invention and melioration processes assessment.

These cases are treated by a kind of simulation access: A multitude of individual, stochastically chosen scenarios are being set up as a kind of "environment" for the system in question, and it's behaviour is simulated for each of them. For each simulation run a different "practical" result is observed: A building to be destroyed by a blasting may collapse at the right place, or it may fall on other buildings standing aside. An industry hall may collapse due to an extraordinary strong storm load, or it may withstand the simulated load over the whole simulation period as the load may have been extraordinarily soft.

The statistical decision of success or failure in the course of a larger number of innovation steps may lead to the desired success of the whole process after a set time interval, or the statistically occurring negative outcomes of individual steps have led to a negative result of the individually modeled process. Typical kinds of uncertainty are related to the technology under consideration, such as the uncertainty whether it will be ready in time and stable with respect to other competing technologies on the market. Here the instrument of stochastic network planning has proven to be adequate.

In any case there is no sense in interpreting the outcomes of an individual simulation run of the systems in question. Rather, a larger ensemble of simulation runs, each of them subject to stochastical variations of influencing factors of an identical model setup, must be considered. There is no such thing as a secure answer to very practical related questions, such as "Will the buildings next-door be impaired by my blasting?" or "Will the new machine be ready for production in two months?"

Giving simple answers such as "yes" or "no" to such questions in certain situations is, within the predictability of the model, not honest. Even if any practical decision maker would like such answers, as he has no statistical ensemble of processes but just the one incident to decide for, any scientific advisor should refrain from giving them and should not progress beyond the statistical limits of the respective model, even if the answer is not satisfying to the asker.

If decision makers are generally not satisfied with answers that an existing model may yield, they may well be open or induced to invest into the prediction quality by additional model development and/or calculational hardware. A typical example, not discussed as a case study in this book, is the weather forecast that has been improved significantly during the last decades. Significant branches rely on its predictive quality, such as agriculture and tourism. More and more real life effects have been integrated into the models of atmospheric processes that yield the forecast, but at the expense of ever-increasing calculation times. Even on the most recent hardware they require significant amounts of time relative to the actual course of the predicted phenomena.

A more fundamental limitation, however, is the increasing uncertainty introduced into the model in the course of increasing detailedness. Again, in computer models for weather forecast, the interactions of atmospheric processes, when looked at in detail, must be parametrized on the basis of incomplete knowledge both on the model detailedness as well as on the data acquisition level: Depending on the scale of the forecasted region the atmosphere is represented as grid points with distances of several kilometers and respective time steps. Readings of weather data are even sparser. Many practically important weather phenomena such as hailstorms or thunderstorms are induced by smaller scale structures, such as hill tops or atmospheric circulations over cities, seas and rivers, though. So differences, both between model and reality as well as between different simulation runs starting from deliberately stochastically disturbed initiation data, pile up over the course of the prediction runs and eventually lead to diverging prognoses of temperature, rain, and wind distribution at progressing time points in the future.

As more certain information can be built into a forecast model its performance will become more reliable, although there are definitely limitations to this process. In particular, the realism of models is strictly limited with respect to the detailedness of the answers obtained from them. Again, any simple weather forecast model will easily predict that the noon temperature of the 25th of December will be lower than that of the 25th of July, but it will not be able to predict whether there will be sunshine or rain at these days. Any statement transgressing the model divergence level, such as the centennial weather calendar, is not scientifically sound, even though it may be very appealing to the more simple mind.

As a consequence, the creators of any scientific model should very sensitively watch all levels of insecurity or stochasticity, both in experimental data feeds as well as in modeling assumptions leading to any kind of prediction. Practical importance lies not only in the positive prediction of the phenomena in question but – maybe even more – in the statement of the limits of prediction or exactness. Labels like "scientifically proven" should be communicated with very much care and, if possible, always in combination with the boundary conditions under which the respective statement was obtained.

4.4 Subjectivity and Bounded Rationality

All structures of the man made world are finally generated on the basis of unprecise and subjective value perceptions. The success of an automobile on the market is to a large extent depending on parameters quite independent of its functionality, such as perception of elegance, social standing and the like. The same is true for works of fine arts or philosophy.

So, in the execution and analysis of structure generating processes we frequently have to incorporate a certain subjectivity of value assessment. A typical example was treated in Sect. 3.2 on intensity-modulated radiotherapy. The subjective nature of optimality is due to the specific choice of objective functions by the treating oncologist. A subjective choice has to be made from the associated pareto set, in the sense to balance between increasing the radiation dose in the tumor at the expense of also increasing it in the surrounding organs, and vice versa. The particular choice depends on the oncologist's judgement of the clinical situation and is based on his clinical experience. Another example was treated in Sect. 3.3 where an interactive multicriterial decision tool was presented and tested on the problem of selecting a handheld GPS receiver from a large variety of offers on the market. It was shown how the subjective choice of target values eventually reduced the number of offers to a practically manageable value.

A particularly prominent field where subjective value assessments are relevant is acoustics, as discussed in Sect. 3.14. Almost all acoustic structures depend on human auditive perception. Any optimization procedure must therefore take subjectivity into account. Acoustic quality, similar but more difficult than the subjective perception of thermal comfort, can also be measured in terms of some physical quantities, although the final decision about the optimal structure is normally made by human tests.

Coupled to stochasticity, but not to be mixed up with it, is the principal limitation of efforts that we can engage on any given problem of understanding or optimization. Our whole life is full of decisions that must be taken on the basis of incomplete knowledge. In the discussion of bounded rationality systems (Sect. 3.12) respective optimization problems were shown that cannot be dealt with thorough in-depth analyses. It is shown by various examples that incomplete knowledge of a structure or a process invites to use fast and frugal heuristics, which abandon the idea of modeling a process in great detail. There are many systems, in particular those depending on complex, in particular human decisions, in which analysis can be performed more economically by such heuristics than by involved but necessarily restricted model simulations.

A much more technical optimization target, but nonetheless strongly infected with subjective, unsharp, not easy or even impossible to obtain knowledge is the early design of chemical engineering processes (Sect. 3.7), as long as not only the scientific process is concerned but the implementation with consideration of numerous competitive, financial, market-related and other side effects. Here again, relatively simple rule-based decision trees (expert systems) are known to work quite well, even if they work more like rules of thumb than a targeted optimization system.

One of the most important things in bounded rationality and its treatment methods is to know where to stop: As the respective rules of thumb are typically only developed by experience in a certain limited domain the potential of an unrevisited methodology transfer is very limited. Nonetheless custom-developed rule sets frequently do their work and must not be disregarded in favor of a causality-based excessive simulation approach. Accordingly, in repetitive optimization tasks it may help to set up such rule sets without a sense of necessity of causal justification. Even if they do not yield optimal results for a given individual task, they may at least help to identify sensible starting conditions for in-depth searches, or deliver good results with a (relative) minimum of labor.

Many purely numerical optimization problems point to a similar direction of a sensibility of restricted understanding, even if a completely mathematical formulation could be created. An in-depth analysis of mathematical interrelations of a model may deem too costly in many cases, especially so if an existing, but bulkheading, simulation model may be exploited for optimization purposes. The comparison of residential heating designs (Sect. 3.6) may serve as an example for such a case. It was based upon a very complex and detailed model of connected energy sources, couplings and holdups whose internal structure is hidden from the simulating user. A thorough mathematical analysis with the implementation of a rigorous and fast optimization schema would have required a large amount of manpower. Instead, an evolutionary approach has been put to work that uses the simulator as black-box target function provider for any proposed set of boundary conditions. Even with this deliberate disregard of the underlying mathematical structures pareto sets could be determined, although with extended calculational efforts. The costly human effort has been circumvented and replaced by technical procedures that may be interpreted as a kind of bounded rationality.

4.5 Convergence

From all this we recognize basic general properties of structures, structure generating processes, and our involvement with (re-)organizing them in all disciplines. With full recognition of and respect for the disciplinary peculiarities of goals and methods we can identify a broad transdisciplinary basis of analysis. As soon as we take a more fundamental view beyond our usually restricted models, we realize that the methods of analysis and optimization converge, regardless of the different disciplines, to a valuation of the realms of heuristics as coequal to the ubiquitous more and more detailed mathematical modeling.

This is no contradiction to the fruitful efforts of even more mathematics and in-depth computer simulations which can be observed in many fields of sciences and even humanities today. Whenever possible, the more formal simulation and optimization approaches in modeling structures and their evolution, including boundaries and uncertainties, are extremely productive in detail. There are question domains, however, that cannot be approached with this class of methods, even in highly technical or scientific domains. Rather than leaving these questions bashfully out of the scope of disciplinary research activities, alternative approaches may be included and even fruitfully intermingled with rigorous approaches to address them as well in an appropriate manner.

The proper approach of analysis depends to a large extent on the system, the questions posed and the answers expected. And in this respect we should address multi-criteriality a last time. As was shown in many cases, most relevant practical questions deal with more than one target. Even though joint optimizations of several targets at the same time may be feasible by technological innovations, for any point in time and any state of development a competition of aims, to be decided by pure subjectivity, is ubiquitous. On the other hand man might not yet be ripe to accept this fact deliberately, even if deep inside the fact is widely perceived. There are even political implications of this schism: The more an ubiquitous multi-criteriality is accepted as a fact of life the less prone both individuals and groups should be to follow pied pipers who promise any kind of uncompromising optimality.

Contributing Authors

Thomas Bartz-Beielstein

Thomas Bartz-Beielstein is a Professor of Applied Mathematics at Cologne University of Applied Sciences. He received a PhD from the Faculty of Computer Science at the Technical University Dortmund. His PhD thesis "New Experimentalism applied to Evolutionary Computation" was supervised by Prof. Dr.-Ing. Hans-Paul Schwefel. Dr. Bartz-Beielstein is a member of the "Fachausschuss Computational Intelligence der VDI/VDE-Gesellschaft für Mess- und Automatisierungstechnik (GMA)" and a program-committee member of the leading conferences in the field of evolutionary computation. He has published several books on experimental research and more than 50 articles in journals and conference proceedings.

Katharina Bauer

Katharina Bauer studied chemical engineering in Dortmund and finished her degree in December 2005. Since February 2006, she is working at the department of Biochemical and Chemical Engineering at the Technical University of Dortmund on her PhD. In her research she focuses on the development of a systematic procedure for the synthesis of industrial purification processes

K. Lucas, P. Roosen (eds.), *Emergence, Analysis and Evolution of Structures*, Understanding Complex Systems, DOI 10.1007/978-3-642-00870-2,

Volker Bormann

Volker Bormann studied ergonomics at the Technische Universität Dresden from 1976 and received his diploma in 1980. From 1980 up to 1985 he has been a PhD at the Technische Universität Dresden. From 1985 up to 2004 he was staff member in the institute of ergonomics and in the institute of acoustics and speech communication at the Technische Universität Dresden and was involved there in several industry projects. From 2004 Volker Bormann is employee in the Gesellschaft für Akustikforschung Dresden mbH.

Ewa Doenitz

Ewa Doenitz works as a research associate at the Institute of Project Management and Innovation (IPMI) at the University of Bremen, where she wrote her PH.D. thesis on scenario technique and options on how to improve consistency analysis. Her fields of research are innovation process development and techniques for future research, both based on statistical and operations research methods.

Volker A. Erdmann

Since 1980 Prof. Dr. Volker A. Erdmann is professor for biochemistry and molecular biology at the Free University of Berlin and the Founder and Director of The Berlin Network for RNA Technologies. His research specialities are the development of RNA technologies for the fields of biotechnology and medicine (diagnostics and therapeutics). The results have been published in more than 435 publications. Dr. Erdmann received in 1966 his M. Sc. degree in biochemistry from the University of New Hampshire, Durham, NH, USA and in 1966 his Ph. D. degree in Germany from the University of Braunschweig in conjunction with the Max-Planck-Institut für Molekulare Medizin, Göttingen, with a thesis on the chemical modification of RNAs. In 1987 Professor Erdmann was awarded the Leibniz Award for Scientific Excellence from the Deutsche Forschungsgemeinschaft and he is also a member of the Berlin-Brandenburgische Akademie der Wissenschaften (BBAW).

Gerd Gigerenzer

Gerd Gigerenzer is Director at the Max Planck Institute for Human Development in Berlin and former Professor of Psychology at the University of Chicago and John M. Olin Distinguished Visiting Professor, School of Law at the University of Virginia. He is also the director of the Harding Center for Risk Literacy, Berlin, Batten Fellow at the Darden Business School, University of Virginia, and Fellow of the Berlin-Brandenburg Academy of Sciences and the German Academy of Sciences. He won the AAAS Prize for the best article in the behavioral sciences and the Association of American Publishers Prize for the best book in the social and behavioral sciences. His award-winning popular books *Calculated Risks: How To Know When Numbers Deceive You*, and *Gut Feelings: The Intelligence of the Unconscious* were translated into 18 languages. His academic books include *The Empire of Chance, Simple Heuristics That Make Us Smart* (with Peter Todd & the ABC Research Group) and *Bounded Rationality: The Adaptive Toolbox* (with Reinhard Selten, a Nobel Laureate in economics). *Rationality for Mortals*, his most recent book, investigates decisions under limited time and information. He has trained U.S. Federal Judges, German physicians, and top managers in decision-making and understanding risks and uncertainties.

Martin Grötschel

Martin Grötschel currently holds the Chair of Information Technology at the Institute of Mathematics of TU Berlin, is Vice President of the Konrad-Zuse-Zentrum Berlin, and Secretary of the International Mathematical Union. He has chaired the DFG Research Center Matheon "Mathematics for key technologies" from 2002 to 2008. He is a member of the Berlin-Brandenburg Academy, the Leopoldina, acatech, and a foreign associate of the US National Academy of Engineering. He has received various academic awards including the Leibniz-Prize and the Beckurts Prize. His areas of mathematical research include optimization, discrete mathematics, and operations research. He has contributed to application areas such as production planning, logistics, public transportation, and telecommunication.

Dietrich Hartmann

Dietrich Hartmann chairs the Institute of Computational Engineering at the Ruhr-University of Bochum in the Department of Civil and Environmental Engineering to which he was appointed as Full Professor in 1987. From 1982 to 1987 he held an Assistant Professorship for Structural Mechanics and Structural Optimization at the Technical University of Dortmund, intermitted by a one year stay as Visiting Professor at the University of California at Berkeley in 1983/1984 and funded by the Alexander von Humboldt Foundation. He has been founder member of the Center of Expert Systems (ZEDO) in Dortmund and the Research Center for Multidisciplinary Analysis and Applied System Optimization (FOMAAS) in Siegen. Furthermore, he is a member of the Northrhine-Westfalian Academy of Sciences in Düsseldorf as well as the German Academy of Technical Sciences (acatech). His covered research fields include computing in engineering, numerical simulation, structural optimization, computational intelligence in engineering, particularly agent-based systems, lifetime-oriented design concepts for structures, damage assessment, structural health monitoring and computer-based demolition of structures. Having been active in numerous joint research projects, he is author of more than 200 technical publications, including technical papers and text book contributions.

Benjamin Hiller

Benjamin Hiller studied Computer Science at the Technische Universität Ilmenau and received his diploma in 2004. Since then he has been a PhD student at the Zuse Institute Berlin, where he was a member of the DFG research group "Algorithms, Structure, Randomness". His research interests include combinatorial optimization, online optimization, and logistics. In addition to his academic research, he has been involved in several industry projects that employ advanced mathematical techniques to solve real-world problems.

Konstantinos V. Katsikopoulos

Konstantinos V. Katsikopoulos works on decision theory and bounded rationality. He has made contributions to the mathematical modeling of heuristics that people (laymen and experts) should or do use to make decisions with limited resources of time, information, and computation. He is a faculty research scientist at the Center for Adaptive Behavior and Cognition of the Max Planck Institute for Human Development. He holds a Ph.D. in industrial engineering and operations research from the University of Massachusetts. He is a past recipient of a German Science Foundation Fellowship for Young Researchers (on the topic of psychology of technical creativity), which he spent as a visiting assistant professor at the Massachusetts Institute of Technology. He has authored articles in engineering, mathematics, economics, and psychology journals and has been an associate editor of IEEE Transactions on Systems, Man, and Cybernetics.

Peter Költzsch

Peter Költzsch, Professor of Technical Acoustics, Emeritus at Dresden University of Technology, has worked in the field of acoustics and fluid mechanics for more than forty years. He previously held the position of the chair of fluid mechanics at the Freiberg University of Technology. He has experience in research and science of sound, fluid mechanics and the interaction between flow and sound. In the last years he dealt with the history of acoustics and fluid mechanics, beginning in the era of Pythagoras up to Lighthill. He is a member of the Berlin Brandenburg Academy of Science and Humanities as well as of acatech, German National Academy of Engineering. He is the author or coauthor of more than 180 articles in acoustics, noise control, and fluid mechanics as well as lecturer in a series of lectures in acoustics, fluid mechanics, flow acoustics, similarity and dimensional analysis, and currently history of acoustics.

Anika König

Anika König studied Sociology and Social Anthropology at Technical University Berlin, Free University Berlin and the University of Amsterdam. From 2004 until 2007 she was a researcher at the Department of Sociology at Technical University Berlin in a research project on the performance of visually supported oral presentations. In 2004, she conducted a preliminary study on cooperation in open source software projects for the Berlin-Brandenburg Academy of Sciences and Humanities. Since 2007, she is a PhD scholar at the Department of Anthropology at the Australian National University in Canberra.

Herbert Kopfer

Herbert Kopfer is the head of the "Chair of Logistics" at the University of Bremen since 1992. He previously was professor for Business Informatics at the University of Siegen and he is a visiting professor at the University of Rennes I since 2002. He has been the chairman of the Section "Logistics" of the German Academy Association for Business Research (VHB) from 2000 to 2004 and he is a member of the International Federation for Information Processing (IFIP), Working Group 7.6 "System Modeling and Optimization" since 2002. His research interests include the optimization of complex adaptive systems with interrelated autonomous subsystems and their realization in logistic systems and supply chains. Since 2004 he investigates the optimization of multi-echelon autonomous logistic systems in the Collaborative Research Center (CRC) 637 of the DFG (Deutsche Forschungsgemeinschaft/German Research Foundation).

Karl-Heinz Küfer

Karl-Heinz Küfer graduated with a PhD thesis in 1992 on a problem in stochastic geometry influenced by probabilistic theory of polyhedra. This work was continued during his habilitation with the analysis of the probabilistic complexity analysis of multicriteria optimization until end of 1997. In 1998 he joined the department of optimization of the Fraunhofer Insitute for Industrial Mathematics (Fraunhofer ITWM) as a scientific advisor and established a research group on multicriteria optimization of radiotherapy. Since 2004 he is heading the department of optimization where 25 scientists do work on industrial multicriteria optimization problems from different branches.

Klaus Lucas

Klaus Lucas has been the chair of Thermodynamics at RWTH Aachen University since 2000. He previously held positions at the University of Stuttgart, at the University of Duisburg, and as the Scientific Director for the Institute of Energy and Environmental Technology in Duisburg-Rheinhausen. He is vice president of the Berlin Brandenburg Academie of Sciences and Humanities as well as a member of acatech, German National Academy of Engineering. Also, he is as consultant to various industrial companies in the field of energy and chemical engineering. His research interest include the molecular modeling of fluids, energy system analysis and optimization and phase and reaction equilibria in fluids. He is the author or coauthor of more than 150 technical articles in scientific journals and is the author of textbooks on engineering thermodynamics and molecular models.

Jörn Mehnen

In 1995 Jörn Mehnen began his research career at the Department of Machining Technology (ISF) at the University of Dortmund, Germany, having received his MSc in Computer Science (Dipl.-Inform.) from the University of Dortmund. There he also obtained his PhD (Dr.-Ing.) and received the admittance for teaching (venia legendi, Habilitation) in 2005 as a Privatdozent (Priv.-Doz). In 2007 he joined Cranfield University, United Kingdom, as a Senior Lecturer within the Decision Engineering Centre. His scientific background focuses on the application of computer scientific and mathematical methods — especially evolutionary algorithms and statistics — to problems in mechanical engineering.

Thomas Michelitsch

Thomas Michelitsch studied applied computer science at the TU Dortmund University and finished in 2003 with a diploma degree. He started working as a scientific employee at the Institute of Machining Technology (ISF) of the Faculty of Mechanical Engineering in Dortmund. Since 2007 he is head of the team "Optimization" at the ISF. His research interests include computer simulations and optimization algorithms applied to mechanical engineering problems and the evolutionary design of mold temperature control systems in particular.

Martin G. Möhrle

Martin G. Möhrle is Professor and director of the Institute for Project Management and Innovation (IPMI) at the University of Bremen, Germany while at the same time holding the chair of Innovation and Competence Transfer. He previously held a similar chair at the Technical University of Cottbus from 1995 to 2001. His area of work is technology and innovation management, and his preferred topics are industrial patent management, technology roadmapping, future research, multiproject management, and methodical invention (TRIZ). From 2004 to 2007 he was speaker of the German academic TIM committee, which consists of more than 100 university professors working in the field of technology and innovation management (TIM). Recent publications include a book on technology roadmapping (published in 3rd edition by Springer) and an encyclopaedia of technology management (published by Gabler). Prof. Möhrle is member of the editorial advisory boards of the journals R&D Management and Creativity and Innovation Management.

Mike Preuss

Mike Preuss is Research Associate at the Computer Science Department, University of Dortmund, Germany (since 2000), where he also received his Diploma degree in 1998. His research interests focus on the field of evolutionary algorithms for real-valued problems, namely on multimodal and multiobjective niching and the experimental methodology for (non-deterministic) optimization algorithms. He is currently working on the adaptability and applicability of computational intelligence techniques for various engineering domains and computer games.

Peter Roosen

After his diploma Dr. Roosen acquired his PHD in combustion engineering at the Aachen University of Technology. For many years he worked there as a group and project leader and senior engineer in the fields of optical combustion measurement techniques and evolutionary computation for chemical engineering optimization problems.

Since 2001 Dr. Roosen works as a freelance scientist and consultant in different scientific and technical fields, among others the development of pareto optimization algorithms, self-adapting therapy systems for brain insult remediation, algorithm-centered web system development, and thermodynamics-oriented didactical work.

Gerhard Schembecker

Gerhard Schembecker received his diploma as well as his Ph.D. in Chemical Engineering from Technische Universität Dortmund in Germany. He is co-founder of the consultancy firm Process Design Center (PDC) having offices in Dortmund (D), Breda (NL) and San Diego (USA). As President and CEO of PDC he has worked in more than 100 industrial process synthesis projects. He holds several process innovation awards. In September 2005 he was appointed Professor for Plant and Process Design at the Department of Biochemical and Chemical Engineering of Technische Universität Dortmund. His research interests focus on the conceptual design of (bio)chemical processes with special interest in chromatographic and crystallization processes.

Karlheinz Schmitt

Karlheinz Schmitt received the Diploma degree in computer science from the University of Dortmund, in 2000. From 2000 to 2007 he worked as a Research Associate at the Chair of Systems Analysis, University of Dortmund, Germany. His main research interests include single- and multi-objective optimization, co-evolution, parallel computing and bio-inspired computation. Since 2008 he is chief executive of a software company for advertising and communication.

Jörn Schönberger

Jörn Schönberger received a diploma in mathematics from the faculty of mathematics and computer sciences at the University of Bremen and a PhD from the faculty of business administration and economics. His PhD thesis "Operational Freight Carrier Planning" has been awarded with the Dissertation award of the German Operations Research Society. Currently, Jörn Schönberger works as a research associate in the German Research Foundation funded Collaborative Research Centre "Autonomous Cooperating Logistic Processes – A Paradigm Shift and its Limitations". He is a member of the chair of logistics and works in the research project "Autonomous Adaptation of Vehicle Schedules". His research fields comprise discrete optimization problems from logistics, dynamic transportation dispatching problems, metaheuristics (evolutionary computation) and online decision making. He has published his research results in international research journals, books and proceedings.

Peter Schuster

Peter Schuster is the chairman of the Institute of Theoretical Chemistry at Vienna University since 1973. He is external faculty member of the Santa Fe Institute, Santa Fe, New Mexico, USA, since 1991. In 1992 – 1995 he was the founding director of the Institute of Molecular Biotechnology in Jena, Thüringen, Germany. He is president of the Austrian Academy of Sciences and member of several academies including the German Academy of Sciences Leopoldina, the Berlin Brandenburg Academy of Sciences and Humanities, the Saxonian Academy of Sciences and the Nordrhein Westfälische Academy of Sciences. His research interests comprise nonlinear dynamics and theory of complex evolvable systems, structures and properties of RNA molecules, sequence structure mappings and neutral networks, optimization of properties of biopolymers as well as gene regulatory and metabolic networks. He is author or coauthor of more than 300 original articles in scientific journals and nine scientific books.

Hans-Paul Schwefel

Hans-Paul Schwefel studied Aero- and Space-Technology at the Technical University of Berlin (TUB). Before and after receiving his engineer diploma in 1965 he worked at the Hermann-Foettinger-Institute of Hydrodynamics, from 1967 to 1970 at an industrial research institute, and from 1971 to 1975 again at the TUB, from where he got his Dr.-Ing. degree in 1975. Coherent during that period at Berlin was the development of a new experimental and later on also numerical optimization method called 'Evolutionsstrategie'. From 1976 to 1985 he acted as senior research fellow at the Research Centre (KFA) Jülich, where he was head of a computer aided planning tools group. Since 1985 until he was pensioned in 2006 he was holder of a Chair for Systems Analysis at the University of Dortmund, Department of Computer Science. In 1990 he was co-founder of the international conference series on Parallel Problem Solving from Nature (PPSN), which has been held biennially ever since. He acted as dean of the faculty, as spokesman of the collaborative research center on computational intelligence (SFB 531), as co-founder and president of the Informatics Centre Dortmund (ICD), and also as pro-rector for research at the university. He has been member of the editorial boards of three journals and advisory board member of two book series in the field of Evolutionary respective Natural Computation. His publication list comprises more than 160 entries. In 2002 he got an Evolutionary Computation Pioneer Award from the IEEE Neural Networks Society, later renamed IEEE Computational Intelligence Society. He was elevated to Fellow of the IEEE in 2007. In the same year The University of Birmingham admitted him to the degree of Doctor of Science, honoris causa.

Jörg Strübing

Jörg Strübing currently teaches sociology at the University of Tübingen. He holds a diploma in social assistance and a M.A. in Sociology with the University of Kassel where he acquired his doctorate in sociology. After various research and teaching positions in Kassel, Berlin, Göttingen and Urbana-Champaign (Ill.) and his habilitation with the Free University Berlin he took a position at the University of Tübingen where in 2007 he was appointed a professorship in sociology. Jörg Strübing is the speaker of the section for Methods of Qualitative Social Research within the German Sociological Association (DFG) and a member of he council of the German Society for Research on Science and Technology (GWTF e.V.). He divides his research interests between qualitative methods and methodology, interactionist social theory, and the study of science and technology issues. Among his latest books are an introduction to the life, work and theoretical achievements of Anselm Strauss (2007), a textbook on the methodological and science theoretical background of Grounded Theory (2nd Ed. 2008), and a monograph on the pragmatist approach to the study of science an technology (2005).

Philipp Süss

Philipp Süss has been a member of the research group Medical Therapy Planning of the optimization department at the Fraunhofer Insitute for Industrial Mathematics (Fraunhofer ITWM) since he was a student in 2004. He graduated from the Technical University of Kaiserslautern in 2005 with a Diploma in mathematics and has been a PhD candidate supported by the department of optimization of the ITWM since. In his Diploma thesis, he deviced an algorithm to shorten the treatment time of a specific form of radiotherapy and in his PhD thesis, he develops an optimization method based on adaptive discretizations of the problem domain to calculate efficient treatment plans reliably and fast.

Andreas Tuchscherer

Andreas Tuchscherer studied Applied Mathematics (Wirtschaftsmathematik) at the Technische Universität Berlin with focus on algorithmic discrete mathematics. Since 2003 he is research assistant and PhD student at the Konrad-Zuse-Zentrum für Informationstechnik Berlin. Moreover, he is a member of the DFG Research Center Matheon "Mathematics for Key Technologies". The main focus of his research is the optimization for combinatorial problems in the field of logistics.

Klaus Weinert

Klaus Weinert was the head of the Institute of Machining Technology (ISF, Institut für Spanende Fertigung) at the Technische Universität Dortmund from 1992 until he retired in April 2008. After his studies and his promotion to Dr.-Ing. at the TU Braunschweig, he started working in automotive industry. Before his return to university he held a leading position at VW in Kassel. The research interests of Klaus Weinert are focused on the industrial relevant machining processes as well as on their simulation and optimization. Klaus Weinert is a member of acatech — german academy of science and engineering, fellow of the International Academy for Production Engineering (CIRP) and a member of the German Academic Society for Production Engineering (WGP). During his academic research Klaus Weinert published more than 300 technical articles and other contributions to the field of machining technology.

Bibliography

[AEN79] A. Achilles, K. H. Elster, and R. Nehse. Bibliographie zur Vektoroptimierung. Math. Op. forsch. Stat., Ser. Optim. 10(2):1979.

[AH05] A. Auger and N. Hansen. A restart CMA evolution strategy with increasing population size. In B. McKay et al., editors, *Proc. 2005 Congress on Evolutionary Computation (CEC'05), Edinburgh, Scotland*, volume 2, pages 1769–1776. IEEE Press, Piscataway NJ, 2005.

[AHB89] J. A. Asenjo, L. Herrera, and B. Byrne. Development of an expert system for selection and synthesis of protein-purification processes. *Journal of Biotechnology*, 11(2–3):275–298, 1989.

[Aka73] H. Akaike. Information theory and an extension of the maximum likelihood principle. In *2nd International Symposium on information theory*, pages 267–281, Budapest, Akademiai Kiado, B. N. Petrov and F. Csaki, 1973.

[AKR00] N. Ascheuer, S. O. Krumke, and J. Rambau. Online Dial-a-Ride problems: Minimizing the completion time. In *Proceedings of the 17th Symposium on Theoretical Aspects of Computer Science*, volume 1770 of *Lecture Notes in Computer Science*, pages 639–650. Springer, Berlin 2000.

[And85] Y. Ando. *Concert Hall Acoustics*. Springer-Verlag, Berlin, 1985.

[And98] Y. Ando. *Architectural Acoustics*. Springer-Verlag, New York, Inc., 1998.

[AO97] P. Ayton and D. "Onkal. Forecasting football fixtures: Confidence and judged proportion correct. Unpublished manuscript, 1997.

[AR93] N. Ambady and R. Rosenthal. Half a minute: predicting teacher evaluations from thin slices of nonverbal behavior and physical attractiveness. *Journal of Personality and Social Psychology*, 64:431–441, 1993.

[Arn01] D. V. Arnold. Evolution strategies in noisy environments – A survey of existing work. In L. Kallel, B. Naudts, and A. Rogers, editors, *Theoretical Aspects of Evolutionary Computing, Natural Computing*, pages 239–249. Springer, Berlin, 2001.

[AS93] W. Ahnert and F. Steffen. *Beschallungstechnik*. S. Hirzel Verlag Stuttgart, Leipzig, 1993.

[Ase90] J. A. Asenjo. Selection of operations in separation processes. In J. A. Asenjo, editor, *Separation Processes in Biotechnology*, volume 9 *of Bioprocess Technology*, pages 3–16. Marcel Dekker, Inc., New York, 1990.

[ASK05] S. Ando, E. Suzuki, and S. Kobayashi. Sample-based crowding method for multimodal optimization in continuous domain. In B. McKay et al., editors, *Proc. 2005 Congress on Evolutionary Computation (CEC'05), Edinburgh, Scotland*, volume 2, pages 1867–1874. IEEE Press, Piscataway NJ, 2005.

[ASNS97] Y. Ando, S. Sato, T. Nakajima, and M. Sakurai. Acoustic design of a concert hall applying the theory of subject preference, and the acoustic measurement after construction. *Acustica – Acta Acustica*, 83:635–643, 1997.

[Bal03] M. Baltrusch. Zur objektiven Beurteilung von Musikinstrumenten. In *Fortschritte der Akustik*, Aachen, 2003. 29. Deutsche Jahrestagung für Akustik, DAGA 2003.

[BAIT03] M. Baitsch. Optimierung druckbeanspruchter Stabtragwerke unter Berücksichtigung geometrischer Imperfektionen. Dr.-Ing. Dissertation, Institut für Konstruktiven Ingenieurbau, Ruhr-Universität Bochum, 2003.

[BB03] Th. Bartz-Beielstein. Experimental analysis of evolution strategies – Overview and comprehensive introduction. Interner Bericht des Sonderforschungsbereichs 531 Computational Intelligence CI–157/03, Universität Dortmund, Germany, 2003.

[BB05] T. Bartz-Beielstein. *Experimental Research in Evolutionary Computation – The New Experimentalism*. Natural Computing Series. Springer, Berlin, 2005.

[BB06] Th. Bartz-Beielstein. *Experimental Research in Evolutionary Computation – The New Experimentalism*. Springer, Berlin, 2006.

[BBC$^+$05] L. Bianchi, M. Birattari, M. Chirandini, M. Manfrin, M. Mastrolilli, L. Paquete, O. Rossi-Doria, and Z. Schiavinotto. Hybrid Metaheuristics for the Vehicle Scheduling Problem with Stochastic Demands. Technical Report IDSIA-0605, IDSIA, 2005.

[BBLP05] T. Bartz-Beielstein, C. Lasarczyk, and M. Preuß. Sequential parameter optimization. In B. McKay et al., editors, *Proceedings 2005 Congress on Evolutionary Computation (CEC'05), Edinburgh, Scotland*, volume 1, pages 773–780, IEEE Press, Piscataway NJ, 2005.

[BBM04] Th. Bartz-Beielstein and S. Markon. Tuning search algorithms for real-world applications: A regression tree based approach. In G. W. Greenwood, editor, *Proc. 2004 Congress on Evolutionary Computation (CEC'04), Portland, OR*, volume 1, pages 1111–1118. IEEE Press, Piscataway NJ, 2004.

[BCW04] R. A. Burgelman, C. M. Christensen, and S. C. Wheelwright. *Strategic Management of Technology and Innovation*. 4th ed. McGraw-Hill, New York, 2004.

[BE03] J. Barciszewski and V. A. Erdmann. *Non-Coding RNAs Molecular Biology and Molecular Medicine*. Springer-Verlag, Berlin, 2003.

[Bec92] S. Bech. Selection and training of subjects for listening tests on sound-reproducing equipment. *Journal of the Audio Engineering Society*, 40:590–610, July/Aug 1992.

[Ber92] L. L. Beranek. Concert hall acoustics. *Journal of the Acoustical Society of America*, 92(1):1–39, 1992.

[Ber94] L. L. Beranek. The acoustical design of concert halls. *Building Acoustics*, 1(1):3–25, 1994.

[Ber96] L. L. Beranek. Concert and opera halls: how they sound. Published for the Acoustical Society of America through the American Institute of Physics, 1996.

[Ber03] L. L. Beranek. Subjective rank-orderings and acoustical measurements for fifty-eight concert halls. *Acustica – Acta acustica*, 89:494–508, 2003.

[Bey93] Hans-Georg Beyer. Towards a theory of evolution strategies: Some asymptotical results from the $(1 + /, \lambda)$-theory. *Evolutionary Computation*, 1(2):165–188, 1993.

[Bey95] H.-G. Beyer. Toward a theory of evolution strategies: On the benefit of sex – the $(\mu/\mu, \lambda)$-theory. *Evolutionary Computation*, 3(1):81–111, 1995.

[BEY98] A. Borodin and R. El-Yaniv. *Online Computation and Competitive Analysis*. Cambridge University Press, Cambridge 1998.

[Bey00] H.-G. Beyer. Evolutionary algorithms in noisy environments: Theoretical issues and guidelines for practice. *CMAME (Computer methods in applied mechanics and engineering)*, 186:239–267, 2000.

[BFM97] Th. Bäck, D. B. Fogel, and Z. Michalewicz, editors. *Handbook of Evolutionary Computation*. Oxford University Press, New York, and Institute of Physics Publ., Bristol, 1997.

[BFOS84] L. Breiman, J. H. Friedman, R. A. Olshen, and C. Stone. *Classification and regression trees.* Chapman and Hall, New York, 1984.

[BG97] C. K. Biebricher and W. C. Gardiner. Molecular evolution of RNA in vitro. *Biophysical Chemistry*, 66:179–192, 1997.

[BGAB83] L. D. Bodin, B. L. Golden, A. A. Assad, and M. Ball. Routing and scheduling of vehicles and crews, the state of the art. *Computers and Operations Research*, 10(2):63–212, 1983.

[BGL$^+$92] L. C. Bock, L. C. Griffin, J. A. Latham, E. H. Vermaas, and J. J. Toole. Selection of single-stranded dna molecules that bind and inhibit human thrombin. *Nature*, 355:564–566, 1992.

[BGS91] D. P. Bartel, M. R. Green, and J. W. Szostak. Hiv-1 rev regulation involves recognition of non- watson-crick base pairs in viral rna. *Cell*, 67:529–536, 1991.

[BH74] M. Bellmore and S. Hong. Transformation of Multisalesman Problem to the Standard Traveling Salesman Problem. *Journal of the ACM*, 21(3):500–504, 1974.

[BH02] M. Bodden and R. Heinrichs. Moderatoren der Geräuschqualität komplexer Geräusche mit tonalen Komponenten. In *Fortschritte der Akustik*, Bochum, 2002. 28. Deutsche Jahrestagung für Akustik, DAGA 2002.

[BH03] M. Bodden and R. Heinrichs. Geräuschqualität im Kontext weiterer Fahrzeugattribute: Bewertung durch Kunden in Feld und Labor. In *Fortschritte der Akustik*, Aachen, 2003. 29. Deutsche Jahrestagung für Akustik, DAGA 2003.

[BHJB92] A. Berzal-Herranz, S. Joseph, and J. M. Burke. In vitro selection of active hairpin ribozymes by sequential rna-catalyzed cleavage and ligation reactions. *Genes and Development*, 6:129–134, 1992.

[BJ94] R. R. Breaker and G. F. Joyce. A dna enzyme that cleaves rna. *Chemistry and Biology*, 1:223–229, 1994.

[BJ96] J. Blauert and U. Jekosch. Sound-quality evaluation – a multi-layered problem. *Acustica – Acta Acustica*, 83:747–753, 1996.

[BJ98] J. Blauert and U. Jekosch. Product-sound quality: A new aspect of machinery noise. *Archives of Acoustics*, 23(1):3–12, 1998.

[BMK03] D. Büche, S. Müller, and P. Koumoutsakos. Self-adaptation for multi-objective evolutionray algorithms. In C. M. Fonseca, P. J. Fleming, E. Zitzler, K. Deb, and L. Thiele, editors, *Evolutionary Multi-Criterion Optimization, Second Int.'l Conf., (EMO 2003)*, number 2632 in LNCS, pages 267–281. Springer, Berlin, 2003.

[Bol85] Béla Bollobás. *Random Graphs*. Academic Press, London, 1985.

[BOTHE93] H.-H. Bothe. *Fuzzy Logic. Einführung in Theorie und Anwendungen*, Springer-Verlag, New York, 1993.

[BR85] R. Boyd and P. J. Richerson. *Culture and the Evolutionary Process*. University of Chicago Press, Chicago, 1985.

[BR03] A. Bonaccorsi and C. Rossi. Why open source software can succeed. *Research Policy*, 32(7):1243–1258, 2003.

[Bre62] H. J. Bremermann. Optimization through evolution and recombination. In M. C. Yovits, G. T. Jacobi, and G. D. Goldstein, editors, *Self-Organizing Systems*. Spartan Books, Washington, DC, 1962.

[Bri06] H. J. Brighton. The robustness of fast and frugal heuristics. Unpublished manuscript, 2006.

[Bru55] E. Brunswik. Representative design and probabilistic theory in a functional psychology. *Psychological Review*, 62, 1955.

[BS92] T. Bäck and H.-P. Schwefel. Evolutionary algorithms: Some very old strategies for optimization and adaptation. In D. Perret-Gallix, editor, *New Computing Techniques in Physics Research II*, pages 247–254. World Scientific, Singapore, 1992.

[BS02a] V. Belton and T. J. Stewart. *Multiple Criteria Decision Analysis: An Integrated Approach*. Kluwer, Boston, 2002.

[BS02b] H.-G. Beyer and H.-P. Schwefel. Evolution strategies – A comprehensive introduction. *Natural Computing*, 1:3–52, 2002.

[BS08] K. Bauer and G. Schembecker. Synthese von Downstreamprozessen. *Chemie Ingenieur Technik*, 80(1–2):185–190, 2008.

[BSG95] R. E. Bechhofer, T. J. Santner, and D. M. Goldsman. *Design and Analysis of Experiments for Statistical Selection, Screening, and Multiple Comparisons*. Wiley, New York, 1995.

[BSL97] J. Bramel and D. Simchi-Levi. *The Logic of Logistics*. Springer, Berlin 1997.

[BSPV02] M. Birattari, T. Stützle, L. Paquete, and K. Varrentrapp. A racing algorithm for configuring metaheuristics. In W. B. Langdon et al., editors, *GECCO 2002: Proceedings of the Genetic and Evolutionary Computation Conference*, pages 11–18. Morgan Kaufmann, San Fransisco 2002.

[BSS01] J. Branke, C. Schmidt, and H. Schmeck. Efficient fitness estimation in noisy environments. In L. Spector et al., editors, *Proc. of the Genetic and Evolutionary Computation Conference (GECCO'01)*, pages 243–250. Morgan Kaufmann, San Francisco, 2001.

[BT93] J. R. Busemeyer and J. T. Townsend. Decision field theory: A dynamic-cognitive approach to decision making in an uncertain environment. *Psychological Review*, 100, 1993.

[BT05] P. A. N. Bosman and D. Thierens. The naive MIDEA: A baseline multi-objective EA. In C. A. Coello Coello, A. Hernández Aguirre, and E. Zitzler, editors, *Proc. Evolutionary Multi-Criterion Optimization: Third Int'l Conference (EMO 2005)*, volume 3410 *of LNCS*, pages 428–442. Springer, Berlin, 2005.

[BU88] K. Brockhoff and C. Urban. Die Beeinflussung der Entwicklungsdauer. *Zeitmanagement in Forschung und Entwicklung*, pages S. 1–42, 1988. Brockhoff, Klaus and Picot, Arnold and Urban, Christoph (Hrsg.).

[BW51] G. E. P. Box and K. B. Wilson. On the experimental attainment of optimum conditions. *Journal of the Royal Statistical Society*, 13(1):1–45, 1951.

[CC05] J. L. A. Coello and C. A. Coello. MRMOGA: Parallel evolutionary multiobjective optimization using multiple resolutions. In D. Corne et al., editors, *Proc. 2005 IEEE Congress on Evolutionary Computation, (CEC 2005)*, volume 3, pages 2294–2301. IEEE Press, Piscalaway, NJ, 2005.

[CGG99] J. Czerlinski, G. Gigerenzer, and D. G. Goldstein. *How good are simple heuristics?* Oxford University Press, New York, 1999.

[CH05] K. Crowston and J. Howison. The social structure of free and open source software development. http://firstmonday.org/issues/issue10_2/crowston/index.html, 10.11 2005. 10/11/2005.

[CM87] F. I. M. Craik and M. McDowd. Age differences in recall and recognition. *Journal of Experimental Psychology: Learning, Memory and Cognition*, 14, 1987.

[Coe02] C. A. Coello Coello. Theoretical and numerical constraint-handling techniques used with evolutionary algorithms: A survey of the state of the art. *Computer Methods in Applied Mechanics and Engineering*, 191(11–12):1245–1287, 2002.

[Coe06] C. A. Coello Coello. The EMOO repository: A resource for doing research in evolutionary multiobjective optimization. *IEEE Computational Intelligence Magazine*, 1(1):37–45, 2006.

[Con96] J. Conlisk. Why bounded rationality? *Journal of Economic Literature*, 34, 1996.

[Cor06] J. F. Cordeau. A Branch-and-Cut Algorithm for the Dial-a-Ride Problem. *Operations Research*, 54:573–586, 2006.

[CRL03] T. F. Cooper, D. E. Rozen, and R. E. Lenski. Parallel chemages in gene expression after 20 000 generation of evolution in *Escherichia coli*. *Proceedings of the National Academy of Sciences of the United States of America*, 100:1072–1077, 2003.

[CS95] B. Cuenoud and J. W. Szostak. A dna metalloenzyme with dna ligase activity. *Nature*, 375:611–614, 1995.

[CS03] Y. Collette and P. Siarry. *Multiobjective Optimization. Principles and Case Studies*. Cost Engineering in Practice. Springer, Berlin, 2003.

[CvVL02] C. A. Coello Coello, D. A. van Veldhuizen, and G. B. Lamont. *Evolutionary Algorithms for Solving Multi-Objective Problems*. Kluwer Academic Publishers, New York, 2002.

[CW64] G. Clarke and J. V. Wright. Scheduling of vehicles from a central depot to a number of delivery points. *Operations Research*, 12:568–581, 1964.

[CY94] G. Connell and M. Yarus. Rnas with dual specificity and dual rnas with similar specificity. *Science*, 264:1137–1141, 1994.

[DA01] M. K. Dhami and P. Ayton. Bailing and jailing the fast and frugal way. *Journal of Behavioral Decision Making*, 14, 2001.

[DAPM00a] K. Deb, S. Agarwal, A. Pratah, and T. Meyarivan. A fast elitist non-dominated sorting genetic algorithm for multi-objective optimization: NSGA-II. KanGAL report 200001, Indian Institute of Technology, Kanpur, India, 2000.

[DAPM00b] K. Deb, S. Agrawal, A. Pratap, and T. Meyarivan. A fast elitist non-dominated sorting genetic algorithm for multi-objective optimisation: NSGA-II. In M. Schoenauer, K. Deb, G. Rudolph, X. Yao, E. Lutton, J. J. Merelo, and H.-P. Schwefel, editors, *Proc. of the 6th Int'l Conf. on Parallel Problem Solving from Nature - PPSN VI*, volume 1917 of *LNCS*, pages 849–858. Springer, Berlin, 2000.

[Dar59] C. Darwin. *Of the Origin of Species by means of Natural Selection, or the Preservation of Favoured Races in the Struggle for Life*, volume 811 of *Everyman's Library (Printing 1967, J. M. Dent & Sons, London)*. Murray, London, 1859.

[Dar69] C. Darwin. *The autobiography of Charles Darwin*. Norton, New York, 1969. Original work published 1887.

[DCS95] J. A. Doudna, T. R. Cech, and B. A. Sullenger. Selection of an rna molecule that mimics a major autoantigenic epitope of human insulin receptor. *Proceedings of the National Academy of Sciences of the United States of Americas*, 92: 2355–2359, 1995.

[DDS05] Guy Desaulniers, Jacques Desrosiers, and Marius M. Solomon, editors. *Column generation*. GERAD 25th anniversary series. Springer, 2005.

[De 75] K. A. De Jong. *An analysis of the behavior of a class of genetic adaptive systems*. PhD thesis, University of Michigan, 1975.

[Dep03a] Department of Energy of the USA. Discussion of thermal comfort models, 2003.

[Dep03b] Department of Energy of the USA. EnergyPlus (e^+) Building Energy Simulation Software Home Page. *http://www.eere.energy.gov/buildings/energyplus*, (Stand Oktober 2003).
EnergyPlus is a elaborated simulation system for modeling the energetic flows and requirements in buildings, taking into account environmental conditions (solar irradiation, geographical orientation etc.) and usage dependent energetic inputs and losses. It is, in it's basic version, distributed free of charge.

[dFIfTuW03] Bericht des Fraunhofer-Instituts für Techno-und Wirtschaftsmathematik, editor. *Radiotherapy – A Large Scale Multi-Criteria Programming*, volume 43, Kaiserslautern, 2003. Intensity—Modulated.

[DH01] M. K. Dhami and C. Harries. Fast and frugal versus regression models in human judgement. *Thinking and Reasoning*, 7, 2001.

[Dha03] M. K. Dhami. Psychological models of professional decision-making. *Psychological Science*, 14, 2003.

[DJW98] S. Droste, Th. Jansen, and I. Wegener. Perhaps Not a Free Lunch But At Least a Free Appetizer. Technical Report CI-45/98, Universit"at Dortmund, Dortmund, 1998.

[DKS97] A. Drexl, R. Kolisch, and A. Sprecher. Neuere Entwicklungen in der Projektplanung. *Zeitschrift für Betriebswirtschaftliche Forschung*, H. 2:95–102, 1997.

[DMJ95] X. Dai, A. De Mesmaeker, and G. F. Joyce. Cleavage of an amide bond by a ribozyme. *Science*, 267:237–240, 1995.

[Dow01] S. Dowlatshahi. Product life cycle analysis: A goal programming approach. *Journal of the Operational Research Society*, 52(11):1201–1214, 2001.

[DP95] D. De Pay. *Informationsmanagement von Innovationen*. Gabler, Wiesbaden, 1995.

[DR59] G. B. Dantzig and J. H. Ramser. The truck dispatching problem. *Management Science*, 6:80–91, 1959.

[EBB06] V. A. Erdmann, J. Brosius, and J. Barciszewski. *RNA Towards Medicine Handbook of Experimental Pharmacology , Vol. 173, XII,*. Springer Verlag, 2006.

[ECL96] S. F. Elena, V. S. Cooper, and R. E. Lenski. Punctuated evolution caused by selection of rare beneficial mutations. *Science*, 272:1802–1804, 1996.

[EEER01] G. Elwyn, A. Edwards, M. Eccles, and D. Rovner. Decision analysis in patient care. 2001.

[EH75] H. J. Einhorn and R. M. Hogarth. Unit weighting schemes for decision making. *Organizational Behavior and Human Decision Processes*, 13, 1975.

[Ehr00] M. Ehrgott. *Multicriteria Optimization*, volume 491 *of Lecture Notes in Economics and Mathematical Systems*. Springer, Berlin, 2000.

[EL03] Santiago F. Elena and Richard E. Lenski. Evolution experiments with microorganisms: The dynamics and genetic bases of adaptation. *Nature Reviews Genetics*, 4:457–469, 2003.

[EPB08] V. A. Erdmann, W. Poller, and J. Barciszewski. Rna technologies in cardiovascular medicine and research. page approx. 410. Springer-Verlag, Berlin 2008.

[Erk98] E. Erkens. *Kostenbasierte Tourenplanung im Straßengüterverkehr*. PhD thesis, Universität Bremen, 1998.

[ES90a] A. D. Ellington and J. W. Szostak. In vitro selection of rna molecules that bind specific ligands. *Nature*, 346:818–822, 1990.

[ES90b] A. D. Ellington and J. W. Szostak. *In Vitro* selection of RNA molecules that bind specific ligands. *Nature*, 346:818–822, 1990.

[ES92] A. D. Ellington and J. W. Szostak. Selection in vitro of single-stranded dna molecules that fold into specific ligand-binding structures. *Nature*, 355:850–852, 1992.

[ES02] A. E. Eiben and M. Schoenauer. Evolutionary computing. *Information Processing Letters*, 82(1):1–6, 2002.

[ES03] A. E. Eiben and J. E. Smith. *Introduction to Evolutionary Computing*. Springer, Berlin, 2003.

[Ett04] M. Ettrich. Koordination und Kommunikation in Open-Source-Projekten. In Robert A. Gehring and Bernd Lutterbeck, editors, *Open Source Jahrbuch 2004*, pages 179–192. Lehmanns Media, Berlin, 2004.

[EU06] EU. Verordnung (EG) nr. 561/2006 des Europäischen Parlaments und des Rates vom 15. März 2006 zur Harmonisierung bestimmter Sozialvorschriften im Straßenverkehr und zur Änderung der Verordnungen (EWG) Nr. 3821/85 und (EG) nr. 2135/98 des Rates sowie zur Aufhebung der Verordnung (EWG) Nr. 3820/85 des Rates (Text von Bedeutung für den EWR) - Erklärung. Amtsblatt Nr. L 102 vom 11/04/2006 S. 0001 - 0014, 2006.

[Fam94] M. Famulok. Molecular recognition of amino acids by rna-aptamers: an l-citrulline binding motif and ist evolution into an l-arginine binder. *Journal of the American Chemistry Society*, 116:1698–1706, 1994.

[FC08] H. F. Cantu. Modelling and multicriteria optimization of the web formation in a spunbond process. *PhD Thesis, Department of Mathematics, Technische Universität Kaiserslautern*, 2008.

[FF96] C. M. Fonseca and P. J. Fleming. On the performance assessment and comparison of stochastic multiobjective optimizers. In H.-M. Voigt, W. Ebeling, I. Rechenberg, and H.-P. Schwefel, editors, *Parallel Problem Solving from Nature - PPSN IV*. Springer, Berlin, 1996.

[FG88] J. M. Fitzpatrick and J. J. Grevenstette. Genetic algorithms in noisy environments. *Machine Learning*, 3:101–120, 1988.

[FGS04] B. Fleischmann, S. Gnutzmann, and E. Sandvoß. Dynamic Vehicle Routing Based on Online Traffic Information. *Transportation Science*, 38(4):420–433, 2004.

[FHM⁺06] R. Flores, C. Hernández, A. E. Martínez de Alba, J.-A. Daròs, and F. Di Serio. Viroids and viroid-host interaction. *Annual Review Phytopathology*, 43:117–139, 2006.

[Fix78] W. Fix. *Strukturuntersuchungen und Verfahren zur Zeitplanung bei GERT-Netzplänen*. Stuttgart: Hochschulverlag, Stuttgart, 1978.

[FKS84] W. Fasold, W. Kraak, and W. Schirmer. *Taschenbuch der Akustik*. Verlag Technik, Berlin, 1984.

[Fle] H. Fleischer. Personal communication 4.11.2003.

[Fle00] H. Fleischer. Schwingungen und Schall von Glocken. In *Fortschritte der Akustik*, Oldenburg, 2000. 26. Deutsche Jahrestagung für Akustik, DAGA 2000.

[FLM02] F. Focacci, A. Lodi, and M. Milano. A Hybrid Exact Algorithm for the TSPTW. *INFORMS Journal on Computing*, 2002.

[Fog98] D. B. Fogel. *Evolutionary Computation: The Fossil Record*. Wiley–IEEE Press, New York, 1998.

[Fou91] Free Software Foundation. General public licence. version 2. http://www.fsf.org/licensing/licenses/gpl.html, 1991. 11/22/2005.

[Fou01] Free Software Foundation. Categories of free and non-free software. http://www.gnu.org/philosophy/categories.html, 11.22 2001. 11/22/2005.

[FOW65] L. J. Fogel, A. J. Owens, and M. J. Walsh. Artificial intelligence through a simulation of evolution. In A. Callahan, M. Maxfield, and L. J. Fogel, editors, *Biophysics and Cybernetic Systems*. Spartan Books, Washington, DC, 1965.

[FR06] P. Friese and J. Rambau. Online-optimization of a multi-elevator transport system with reoptimization algorithms based on set-partitioning models. *Discrete Appl. Math.*, 154(13):1908–1931, 2006. also available as ZIB Report 05-03.

[FS92] M. Famulok and J. W. Szostak. Stereospecific recognition of tryptophan agarose by in vitro selected rna. *Journal of the American Chemistry Society*, 114: 3990–3991, 1992.

[FS93] W. Fasold and U. Stephenson. Gute Akustik von Auditorien. *Bauphysik*, 15(2):40–49, 1993.

[FS94] M. Forster and E. Sober. How to tell when simpler, more unified, and less ad hoc theories will provide more accurate predictions. *British Journal of the Philosophy of Science*, 45, 1994.

[FS98a] W. Fontana and P. Schuster. Continuity in evolution. On the nature of transitions. *Science*, 280:1451–1455, 1998. Also published as: Preprint No. 98-04-030, Santa Fe Institute, Santa Fe, NM 1998.

[FS98b] W. Fontana and P. Schuster. Shaping space. The possible and the attainable in RNA genotype-phenotype mapping. *Journal of Theoretical Biology*, 194: 491–515, 1998. Also published as: Preprint No. 97-11-081, Santa Fe Institute, Santa Fe, NM 1997.

[FSS89] W. Fontana, W. Schnabl, and P. Schuster. Physical aspects of evolutionary optimization and adaptation. *Physical Review A*, 40:3301–3321, 1989.

[FSW87] W. Fasold, E. Sonntag, and H. Winkler. *Bau- und Raumakustik*. Verlag für Bauwesen, Berlin, 1987.

[Fuj87] J. H. Fujimura. Constructing "do-able" problems in cancer research: Articulating alignment. *Social Studies of Science*, 17:257–293, 1987.

[FV98] W. Fasold and E. Veres. *Schallschutz und Raumakustik in der Praxis*. Verlag für Bauwesen, Berlin, 1998.

[Ger04] A. Gerybadze. *Technologie- und Innovationsmanagement*. Vahlen, München, 2004.

[GEZS06] A. E. Grobalenya, L. Enjuanes, J. Ziebuhr, and E. J. Snijder. *Nidovirales*: Evolving the largest RNA virus genome. *Virus Research*, 117:17–37, 2006.

[GG02] D. G. Goldstein and G. Gigerenzer. Models of ecological rationality: The recognition heuristic. *Psychological Review*, 109, 2002.

[GGH+01] D. G. Goldstein, G. Gigerenzer, R. M. Hogarth, A. Kacelnik, Y. Kareev, et al. Group report: Why and when do simple heuristics work? *Bounded rationality: The adaptive toolbox*. MIT Press, Cambridge, MA, 2001.

[GGLM03] S. Ghiani, F. Guerriero, G. Laporte, and R. Musmanno. Real-time vehicle routing: Solution concepts, algorithms and parallel computing strategies. *European Journal of Operational Research*, 151(1):1–11, 2003.

[GGS+96a] W. Grüner, R. Giegerich, D. Strothmann, C. Reidys, J. Weber, I. L. Hofacker, and P. Schuster. Analysis of RNA sequence structure maps by exhaustive enumeration. II. Structures of neutral networks and shape space covering. *Mh.Chemie*, 127:375–389, 1996.

[GGS+96b] W. Grüner, R. Giegerich, D. Strothmann, C. Reidys, J. Weber, I. L. Hofacker, and P. Schuster. Analysis of RNA sequence structure maps by exhaustive enumeration. I. Neutral networks. *Mh.Chemie*, 127:355–374, 1996.

[GH97] D. Gigone and R. Hastie. The impact of information on small group choice. *Journal of Personality and Social Pscyhology*, 72, 1997.

[GHL94] M. Gendreau, A. Hertz, and G. Laporte. A tabu search heuristic for the vehicle routing problem. *Management Science*, 40:1276–1290, 1994.

[Gig04] G. Gigerenzer. *Fast and frugal heuristics: The tools of bounded rationality*. Blackwell, Oxford, 2004. Blackwell handbook of judgment and decision making.

[Gil76] D. T. Gillespie. A general method for numerically simulating the stochastic time evolution of coupled chemical reactions. *Journal of Comparative Physics*, 22:403–434, 1976.

[Gil77a] D. T. Gillespie. Concerning the validity of the stochastic approach to chemical kinetics. *Journal of Statistical Physics*, 16:311–318, 1977.

[Gil77b] D. T. Gillespie. Exact stochastic simulation of coupled chemical reactions. *The Journal of Physical Chemistry*, 81:2340–2361, 1977.

[GKR99] M. Grötschel, D. Hauptmeier, S. O. Krumke, and J. Rambau. Simulation studies for the online Dial-a-Ride problem. Technical Report SC 99–09, Zuse Institute Berlin, 1999.

[GL93] F. Glover and M. Laguna. Tabu search. In *Modern Heuristic Techniques for Combinatorial Problems*, pages 70–150. Blackwell Scientific Publications, Oxford 1993.

[GM74] B. E. Gillett and L. R. Miller. A heuristic algorithm for the vehicle dispatch problem. *Operations Research*, 22:340–349, 1974.

[GML+06] K. Grünberger, U. Mückstein, U. Langhammer, A. Svrcek-Seiler, A. Wernitznig, and P. Schuster. RNA evolution *in silico*. Technical report, Institut für Theoretische Chemie, Universität Wien, Austria 2006.

[GNV04] K. Gutenschwager, C. Niklaus, and S. Voß. Dispatching of an electric monorail system: Applying metaheuristics to an online pickup and delivery problem. *Transportation Science*, 38(4):434–446, 2004.

[Gol89a] D. Goldberg. *Genetic Algorithm in Search, Optimization & Machine Learning*. Addison Weasley Publishing Company, Prentice Hall 1989.

[Gol89b] D. E. Goldberg. *Genetic Algorithms in Search, Optimization and Machine Learning*. Addison Wesley, Reading, MA, 1989.

[GP98] M. Gendreau and J.-Y. Potvin. Dynamic Vehicle Routing and Dispatching. In T. G. Crainic and G. Laporte, editors, *Fleet Management and Logistics*, pages 115–126. Kluwer, 1998.

[GPUY95] L. Gold, B. Polisky, O. C. Uhlenbeck, and M. Yarus. Diversity of oligonucleotide functions. *Annual Review Biochemistry*, 64:763–797, 1995.

[GR87] D. E. Goldberg and J. Richardson. Genetic algorithms with sharing for multimodal function optimization. In *Proc. of the Second Int'l Conf. on Genetic Algorithms on Genetic Algorithms and Their Application*, pages 41–49. Lawrence Erlbaum Associates, Inc., Mahwah, NJ, 1987.

[GR99] M. Grötzner and P. Roosen. Thermodynamic evaluation of chemical processes as a base of structure-optimizing process synthesis. In *Europ. Symp. on Computer-Aided Process Engineering*, Hungary, 1999.

[Gra66] R. L. Graham. Bounds for certain multiprocessing anomalies. *Bell System Technical Journal*, 45:1563–1581, 1966.

[Gra02] V. Grassmuck. *Freie Software. Zwischen Privat- und Gemeineigentum*. Bundeszentrale für politische Bildung, Bonn, 2002.

[Gro05] Groundspeak Inc. Geocaching Portal. http://www.geocaching.com, September 2005. Introduction web site to the recreational use of the GPS technique for finding small hidden treasury chests. Being a relatively new occupation this idea seems to develop towards a kind of mass sports.

[GS92] G. A. Geist and V. S. Sunderam. Network based concurrent computing on the PVM system. *Journal of Concurrency: Practice and Experience*, 4(4):293–311, June 1992.

[GS93] D. K. Gode and S. Sunder. Allocative efficiency of markets with zero-intelligence traders: Market as a partial substitute for individual rationality. *Journal of Political Economy*, 101, 1993.

[GS01] G. Gigerenzer and R. Selten. Bounded rationality: The adaptive toolbox. MIT Press, Cambridge, MA, 2001.

[GS06] C. Grimme and K. Schmitt. Inside a predator-prey model for multi-objective optimization: A second study. In H.-G. Beyer et al., editors, *Proc. Genetic and Evolutionary Computation Conf. (GECCO 2006), Seattle WA*, pages 707–714. ACM Press, New York, 2006.

[GSD06] S. B. Gadewar, G. Schembecker, and M. F. Doherty. Rapid process design. *Chemical Engineering Progress*, 102(9):22–32, 2006.

[GTGM+83] C. Guerrier-Takada, K. Gardiner, T. Marsh, N. Pace, and S. Altman. The rna moiety of ribonuclease p is the catalytic subunit of the enzyme. *Cell*, 35:849–857, 1983.

[GTtARG99] G. Gigerenzer, P. M. Todd, and the ABC Research Group. Simple heuristics that make us smart. Oxford University Press, Oxford, 1999.

[Gul50] H. Gulliksen. *Theory of mental tests*. Wiley, New York, 1950.

[GW93] G. Galambos and G. J. Woeginger. Repacking helps in bounded space on-line bin-packing. *Computing*, 49:329–338, 1993.

[Gä95] B. Gätjen. Untersuchungen zur Hörsituation der Musiker bei Ihrem Spiel in Räumen mit unterschiedlicher Raumakustik. In *Fortschritte der Akustik*, Saarbrücken, 1995. 21. Deutsche Jahrestagung für Akustik, DAGA 1995.

[Gö95] S. Götze. *Die multikriterielle Entscheidungsfindung als Modell für die simultane Produktentwicklung*. Shaker, Aachen, 1995.

[Han99] T. Hanne. On the convergence of multiobjective evolutionary algorithms. *European Journal Of Operational Research*, 117(3):553–564, 1999.

[Har04] F. L. Harrison. Advanced project management. In *A Structured Approach*, chapter 4th ed. Gower, Aldershot, 2004.

[Has06] S. Hasse. *Taschenbuch der Geißerei-Praxis*. Schiele und Schön, Berlin, 2006.

[HB07] E. Heinzle and A. Biwer. *Development of Sustainable Bioprocesses - Modelling and Assessment*. John Wiley & Sons, Chichester, 2007.

[Hen91] L. Hennicke. *Wissensbasierte Erweiterung der Netzplantechnik.* Heidelberg, Berlin: Physica, Heidelberg, Berlin, 1991.

[Hen99] J. Hendry. *Other people's worlds. An introduction to social anthropology.* Macmillan, Basinstoke, 1999.

[Hen06] A. Hensel. Game of Life Homepage. *http://www.ibiblio.org/lifepatterns/*, (March 2006). Original site for Conway's Game of Life. Contains a Java applet and some links to elaborate pattern providers.

[Her97] M. Herdy. Evolutionary optimisation based on subjective selection – evolving blends of coffee. *Proceedings 5th European Congress on Intelligent Techniques and Soft Computing (EUFIT'97)*, pages 640–644, 1997.

[HFS$^+$94] I. L. Hofacker, W. Fontana, P. F. Stadler, L. S. Bonhoeffer, M. Tacker, and P. Schuster. Fast folding and comparison of RNA secondary structures. *Mh. Chemie*, 125:167–188, 1994.

[HHJ06] R. F. Hartl, G. Hasle, and G. K. Janssens. Special Issue on Rich Vehicle Routing Problems. *Central European Journal of Operations Research*, 14(2), 2006.

[HK05] R. M. Hogarth and N. Karelaia. Ignoring information in binary choice with continuous variables: When is less "more"? *Journal of Mathematical Psychology*, 49(2), 2005.

[HKR00] D. Hauptmeier, S. O. Krumbe, and J. Rambau. The online Dial-a-Ride problem under reasonable load. In *CIAC 2000*, volume 1767 of *Lecture Notes in Computer Science*, pages 125–136. Springer, 2000.

[HKR06] B. Hiller, S. O. Krumke, and J. Rambau. Reoptimization gaps versus model errors in online-dispatching of service units for ADAC. *Discrete Applied Mathematics*, 154(13):1897–1907, 2006. Traces of the Latin American Conference on Combinatorics, Graphs and Applications – A selection of papers from LACGA 2004, Santiago, Chile.

[HLHG00] U. Hoffrage, R. Lindsey, R. Hertwig, and G. Gigerenzer. Medicine—communicating statistical information. Science, 2000.

[HM79] C. L. Hwang and A. S. M. Masud. *Multiple Objective Descision Making – Methods and Applications: A State-of-the-Art Survey*, volume 186 of *Lecture Notes in Economics and mathematical Systems*. Springer, Berlin, 1979.

[Hof95] U. Hoffrage. *Zur Angemessenheit subjektiver Sicherheits-Urteile: Eine Exploration der Theorie der probabilistischen mentalen Modelle [The adequacy of subjective confidence judgments: Studies concerning the theory of probabilistic mental models]. Unpublished doctoral dissertation.* Austria, 1995.

[Hof03] I. L. Hofacker. Vienna RNA secondary structure server. *Nucleic Acids Research*, 31:3429–3431, 2003.

[Hol73] J. H. Holland. Genetic algorithms and the optimal allocation of trials. *SIAM Journal of Computing*, 2(2):88–105, 1973.

[HP97] J. Hannes and A. Plückthun. In vitro selection and evolution of functional proteins by using ribosome display. *Proceedings of the National Academy of Science*, 94:4937–4942, 1997.

[HR08] B. Helversen and J. Rieskamp. The mapping model: A cognitive theory of quantitative estimation. *Journal of Experimental Psychology: General*, 137(1):2008.

[HS95] D. E. Huizenga and J. W. Szostak. A dna aptamer that binds adenosine and atp. *Biochemistry*, 34:656–665, 1995.

[HSF96] M. A. Huynen, P. F. Stadler, and W. Fontana. Smoothness within ruggedness: The role of neutrality in adaptation. *Proceedings of the National Academy of Sciences of the United States of America*, 93:397–401, 1996.

[HT87] A. J. M. Houtsma and H. J. G. M. Tholen. A carillon of major-third bells. part ii: A perceptual evaluation. *Music Perception*, 4(3):255–266, 1987.

[HT04] R. Hertwig and P. M. Todd. *More is not always better: The benefits of cognitive limits*. Reasoning and decision making: A handbook. Wiley, Chichester, 2004.

[Hug71] E. C. Hughes. Going concerns: The study of american institutions. In E. C. Hughes, editor, *The sociological eye*, pages 52–64. Aldine, Chicago, 1971.

[IL00] S. S. Iyengar and M. R. Lepper. When choice is demotivating: Can one desire too much of a good thing? *Journal of Personality and Social Psychology*, 79, 2000.

[Jai88] P. Jaillet. A Priori Solution of a Traveling Salesman Problem in Which a Random Subset of Customers are Visited. *Operations Research*, 36(6):929–936, 1988.

[Jäs01] A. Jäschke. Artificial ribozymes and deoxyribozymes. *Current Opinion in Structural Biology*, 11:321–326, 2001.

[JB05a] U. Jekosch and J. Blauert. Einige Grundüberlegungen zur Qualität von Geräuschen. In *Fortschritte der Akustik*, München, 2005. 31. Deutsche Jahrestagung für Akustik, DAGA 2005.

[JB05b] Y. Jin and J. Branke. Evolutionary optimization in uncertain environments – a survey. *IEEE Transactions on Evolutionary Computation*, 9(3):303–318, Jun. 2005.

[Jek04] U. Jekosch. Basic concepts and terms of 'quality', reconsidered in the context of product-sound quality. *Acta Acustica*, 90:999–1006, 2004.

[Jek05] U. Jekosch. Assigning meaning to sounds – semiotics in the context of product-sound design. In J. Blauert, editor, *Communication acoustics*. Springer-Verlag, Berlin, 2005.

[Jel98] M. Jelasity. UEGO, an abstract niching technique for global optimization. In A. E. Eiben, Th. Bäck, M. Schoenauer, and H.-P. Schwefel, editors, *Proc. Parallel Problem Solving from Nature – PPSN V, Amsterdam*, pages 378–387. Springer, Berlin, 1998.

[Jen01] M. T. Jensen. *Robust and Flexible Scheduling with Evolutionary Computation.* PhD thesis, University of Aarhus, 2001.

[JGBJ94] D. Jellinek, L. S. Green, C. Bell, and N. Janjic. Inhibition of receptor binding by high-affinity rna ligands to vascular endothelial growth factor. *Biochemistry*, 33:10450–10456, 1994.

[JGPP94a] R. D. Jenison, S. C. Gill, A. Pardi, and B. Poliski. High-resolution molecular discrimination by RNA. *Science*, 263:1425–1429, 1994.

[JGPP94b] R. D. Jenison, S. C. Gill, A. Pardi, and B. Polisky. High-resolution molecular discrimination by rna. *Science*, 263:1425–1429, 1994.

[JKK06] A. Jurczyk, H. Kopfer, and M. Krajewska. Sepditionelle Auftragsdisposition eines mittelständischen Transportunternehmens. *Internationales Verkehrswesen*, 6:275–279, 2006.

[JLRJ93] D. Jellinek, C. K. Lynott, D. B. Rifkin, and N. Janjic. High-affinity rna ligands to basic fibroblast growth factor inhibit receptor binding. *Proceedings of the National Academy of Science*, 90:11227–11231, 1993.

[JR03] J. G. Johnson and M. Raab. Take the first: Option generation and resulting choices. *Organizational Behavior and Human Decision Processes*, 91, 2003.

[KC99] J. Knowles and D. Corne. The pareto archived evolution strategy: A new baseline algorithm for pareto multiobjective optimisation. In P. J. Angeline, Z. Michalewicz, M. Schoenauer, X. Yao, and A. Zalzala, editors, *Proc. Congress on Evolutionary Computation, (CEC'99)*, volume 1, pages 98–105. IEEE Press, Washington, DC, 1999.

[KdPSR01] S. O. Krumke, W. E. de Paepe, L. Stougie, and J. Rambau. Online bin coloring. In F. Meyer auf der Heide, editor, *Proceedings of the 9th Annual European Symposium on Algorithms*, volume 2161 of *Lecture Notes in Computer Science*, pages 74–84, 2001.

[Ker05] H. Kerzner. Project management: A systems approach to planning. In *Scheduling and Controlling Hoboken*. John Wiley & Sons., New Jersey 2005.

[KF00] K. V. Katsikopoulos and B. Fasolo. New tools for decision analysts. In *IEEE Transactions on Systems, Man and Cybernetics: Systems and Humans*, (in press).

[KGC96] E. J. Kleinschmidt, H. Geschka, and R. G. Cooper. *Erfolgsfaktor Markt: Kundenorientierte Produktinnovation*. Springer, Berlin et al., 1996.

[KGZ+82] K. Kruger, P. J. Grabowski, A. J. Zaug, J. Sands, D. E. Gottschling, and T. R. Cech. Self-splicing rna: autoexcision and autocyclization of the ribosomal rna intervening sequence of tetrahymena. *Cell*, 31:147–157, 1982.

[Kim83] M. Kimura. *The Neutral Theory of Molecular Evolution*. Cambridge University Press, Cambridge 1983.

[KK06] H. Kopfer and M. Krajewska. Inter- und Intraspeditionelle Auftragsdisposition. *Industrie Management*, 3:75–77, 2006.

[KK07] H. Kopfer and M. Krajweska. Approaches for Modelling and Solving the Integrated Transportation and Forwarding Problem. In H. Corsten and H. Missbauer, editors, *Produktions- und Logistikmanagement*, pages 439–455. Vahlen, Austria 2007.

[Kli87] W. Klippel. *Zusammenhang zwischen objektiven Lautsprecherparametern und subjektiver Qualitätsbeurteilung*. Dissertation, Technische Universität Dresden, 1987.

[Kli88] W. Klippel. Zusammenhang zwischen objektiven Lautsprecherparametern und subjektiver Qualitätsbeurteilung. In W. Kraak and G. Schommartz, editors, *Angewandte Akustik*, volume 1, pages 46–101. VEB Verlag Technik, Berlin, 1988.

[KLL+02] S. O. Krumke, L. Laura, M. Lipmann, A. Marchetti-Spaccamela, W. de Paepe, D. Poensgen, and L. Stougie. Non-abusiveness helps: An $O(1)$-competitive algorithm for minimizing the maximum flow time in the online traveling salesman problem. In *Proceedings of the 5th International Workshop on Approximation Algorithms for Combinatorial Optimization*, volume 2462 of *Lecture Notes in Computer Science*, pages 200–214. Springer, Berlin 2002.

[KLT03] T. G. Kolda, R. M. Lewis, and V. J. Torczon. Optimization by direct search: New perspectives on some classical and modern methods. *SIAM Review*, 45(3): 385–482, 2003.

[Klu06] S. Klussmann, editor. *The Aptamer Handbook. Functional Oligonucleotides and Their Applications*. Wiley-VCh Verlag, Weinheim, DE, 2006.

[KM00] K. V. Katsikopoulos and L. Martignon. Naïve heuristics for paired comparisons: Some results on their relative accuracy. *Journal of Mathematical Psychology*, (in press).

[KM01] B. Kogut and A. Metiu. Open-source software development and distributed innovation. *Oxford Review of Economic Policy*, 17:248–264, 2001.

[KMM02] E. Kurz-Milcke and L. Martignon. *Modelling practices and "tradition"*. Plenum, New York, 2002. Model-based Reasoning: Science, Technology, Values.

[KMRS88] A. Karlin, M. Manasse, L. Rudolph, and D. D. Sleator. Competitive snoopy caching. *Algorithmica*, 3(1):79–119, 1988.

[KMS+05] K. H. Küfer, M. Monz, A. Scherrer, P. Süss, F. V. Alonso, AS Sultan, TR Bortfeld, and C Thieke. Multicriteria optimization in intensity modulated radiotherapy planning. *Berichte des ITWM*, 77, 2005. http://www.itwm.fhg.de/zentral/download/berichte/bericht77.pdf.

[KNB+96] S. Klußmann, A. Nolte, R. Bald, V. A. Erdmann, and J. P. Fürste. Mirror-image rna that binds d-adenosine. *Nature Biotechnology*, 14:1112–1115, 1996.

[Kno02] J. Knowles. *Local-Search and Hybrid Evolutionary Algorithms for Pareto Optimization*. PhD thesis, The University of Reading, Reading, UK, 2002.

[KOCZ93] G. A. Klein, J. Orasanu, R. Calderwood, and C. E. Zsambok. *Decision Making in Action: Models and Methods*. Ablex Publishing Corporation, New Jersey, 1993.

[Kol99] P. Kollock. The economics of online cooperation: Gifts and public goods in cyberspace. In M. Smith and P. Kollock, editors, *Communities in Cyberspace*, pages 220–239. Routledge, London, 1999.

[Köl03] P. Költzsch. Festschrift zum Ehrenkolloquium „REICHARDT – KRAAK – WÖHLE". 4. Juli 2003. Technische Universität Dresden, 2003, 2003. ISBN 3-86005-370-1.

[Koz92] J. R. Koza. *Genetic Programming: On the Programming of Computers by Means of Natural Selection*. MIT Press, Cambridge, MA, 1992.

[Kra84] W. Kraak. Bericht zur raumakustischen Erprobung der wiederaufgebauten Semperoper Dresden. Technical report, Technische Universität Dresden, Bereich Akustik und Messtechnik, Dresden, 1984.

[KRT02] S. O. Krumke, J. Rambau, and L. M. Torres. Realtime-dispatching of guided and unguided automobile service units with soft time windows. In Rolf H. Möhring and Rajeev Raman, editors, *Proceedings of the 10th Annual European Symposium on Algorithms*, volume 2461 of *Lecture Notes in Computer Science*, pages 637–648. Springer, Berlin 2002.

[Kru01] S. Krumke. *Online-Optimisation – Competitive Analysis and Beyond*. Technical University of Berlin, Germany 2001.

[KS02] H. Kopfer and J. Schönberger. Interactive Solving of Vehicle Routing and Scheduling Problems: Basic Concepts and the Qualification of Tabu Search Approaches. In *Proceedings of HICCS 35, DTIST*, page 84pp, 2002.

[KS05] N. Krasnogor and J. E. Smith. A tutorial for competent memetic algorithms: Model, taxonomy and design issues. *IEEE Transactions on Evolutionary Computation*, 5(9):474–488, 2005.

[Kut00] H. Kuttruff. *Room Acoustics*. 4 edition, Spon Press, London, New York, 2000.

[LA92] E. W. Leser and J. A. Asenjo. Rational design of purification processes for recombinant proteins. *Journal of Chromatography: Biomedical Applications*, 584(1):43–57, 1992.

[LA00] M. E. Lienqueo and J. A. Asenjo. Use of expert systems for the synthesis of downstream protein processes. *Computers and Chemical Engineering*, 24:2339–2350, 2000.

[Lac04] A. Lackner. *Dynamische Tourenplanung mit ausgewählten Metaheuristiken*. Cuvillier Verlag, Germany 2004.

[Lal01] K. N. Laland. Bounded rationality: The adaptive toolbox. In *Imitation, social learning, and preparedness as mechanisms of bounded rationality*. MIT Press, Cambridge, MA, 2001.

[Lau04] A. M. St. Lautent. *Understanding Open Source & Free Software Licensing*. O' Reilly, Sebastopol, CA, 2004.

[LBPC02] J.-P. Li, M. E. Balazs, G. T. Parks, and P. J. Clarkson. A species conserving genetic algorithm for multimodal function optimization. *Evolutionary Computation*, 10(3):207–234, 2002.

[LE03] T. Lamla and V. A. Erdmann. Searching sequence space for high-affinity binding peptides using ribosome display. *J Mol Biol.*, 329:381–388, 2003.

[LE04] T. Lamla and V. A. Erdmann. The nano-tag, a streptavidin-binding peptide for the purification and detection of recombinant proteins. *Protein Expression Purification*, 33:39–47, 2004.

[Lev66] R. Levins. The strategy of model building in population biology. *American Scientist*, 54:421–431, 1966.

[LL85] C. C. Lee and D. T. Lee. A simple on-line bin-packing algorithm. *Journal of the ACM*, 32(3):562–572, 1985.

[LLA96] M. E. Lienqueo, E. W. Leser, and J. A. Asenjo. An expert system for the selection and synthesis of multistep protein separation processes. *Computers and Chemical Engineering*, 20:189–194, 1996.

[LLKS85] E. L. Lawler, J. K. Lenstra, A. H. G. Rinnooy Kan, and D. B. Shmoys, editors. *The Traveling Salesman Problem*. Wiley, New York 1985.

[LME01] T. Lamla, K. Mammeri, and V. A. Erdmann. The cell-free protein biosynthesis-applications and analysis of the system. *Acta Biochimica Polonica*, 48:453–465, 2001.

[LRS98] M. Laumanns, G. Rudolph, and H.-P. Schwefel. A spatial predator-prey approach to multi-objective optimization: A preliminary study. In A. E. Eiben, M. Schoenauer, and H.-P. Schwefel, editors, *Parallel Problem Solving From Nature – PPSN V*, pages 241–249, Springer, Berlin, Amsterdam, Holland, 1998.

[LS95] C. T. Lauhon and J. W. Szostak. Rna aptamers that bind flavin and nicotinamide cofactors. *Journal of the American Chemical Society*, 117:1246–1257, 1995.

[LSA99] M. E. Lienqueo, J. C. Salgado, and J. A. Asenjo. An expert system for selection of protein purification processes: Experimental validation. *Journal of Chemical Technology and Biotechnology*, 74(3):293–299, 1999.

[LSE02] T. Lamla, W. Stiege, and V. A. Erdmann. An improved protein bioreactor: efficient product isolation during in vitro protein biosynthesis via affinitytag. *Molecular and Cellular Proteomics.*, 6:466–471, 2002.

[Luc07] K. Lucas. *Molecular Models for Fluids*. Cambridge University Press, Cambridge, 2007.

[Lut04] B. Luthiger. Alles aus Spaß? Zur Motivation von Open-Source-Entwicklern. In R. A. Gehring and B. Lutterbeck, editors, *Open Source Jahrbuch 2004*, pages 93–106. Lehmanns Media, Berlin, 2004.

[Mah95] S. W. Mahfoud. *Niching Methods for Genetic Algorithms*. PhD thesis, University of Illinois at Urbana Champaign, 1995.

[MD05] M. G. Moehrle and E. Dönitz. *Die Notwendigkeit flexibler Mehrzielplanung von Innovationsprojekten*. J. Amelingmeyer and P. E. Harland, (Hrsg.), Wiesbaden: DUV, Technologiemanagement & marketing edition, 2005.

[Mea34] G. H. Mead. *Mind, Self & Society from the Standpoint of a Social Behaviorist*. The University of Chicago Press, Chicago, IL, 1934.

[Meh05] J. Mehnen. *Mehrkriterielle Optimierverfahren für produktionstechnische prozesse*. Schriftenreihe des ISF. Hrgs.: K. Weinert. Vulkan Verlag, Essen, 2005.

[Mey77] J. Meyer. Die Problematik der Qualitätsbestimmung bei Musikinstrumenten. *Instrumentenbau-Zeitschrift*, 2:3–8, 1977.

[MH02] L. Martignon and U. Hoffrage. lexicographic heuristics for paired comparison. theory and decision. *Fast, frugal and fit*, 52, 2002.

[Mie98] K. Miettinen. *Nonlinear Multiobjective Optimization*. Int. series in operations research and management science. Kluwer Academic Publishers, Boston, 1998.

[Mie99] K. Miettinen. *Nonlinear Multiobjective Optimization*. Kluwer, Boston, 1999.

[Mie02] K. Miettinen. Interactive Nonlinear Multiobjective Procedures. In M. Ehrgott and X. Gandibleux, editors, *Multiple Criteria Optimization*, pages 227–276. Kluwer Academic Publishers, 2002.

[MMBBH04] J. Mehnen, Th. Michelitsch, Th. Bartz-Beielstein, and N. Henkenjohann. Systematic Analyses of Multi-objective Evolutionary Algorithms Applied to Real-World Problems Using Statistical Design of Experiments. In R. Teti, editor, *Intelligent Computation in Manufacturing Engineering, 4th CIRP International Seminar on Intelligent Computation in Manufacturing Engineering, CIRP ICME '04*, pages 171–178, University of Naples, Italy, 2004.

[MMLBB05] J. Mehnen, Th. Michelitsch, Chr. Lasarczyk, and Th. Bartz-Beielstein. Multi-Objective Evolutionary Design of Mold Temperature Control using DACE for Parameter Optimization. In H. Pfützner and E. Leiss, editors, *Short Paper Proceedings of the ISEM 2005, 12th Interdisciplinary Electromagnetic, Mechanic and Biomedical Problems*, volume L11-1 of *Vienna Magnetics Group Reports*, pages 464–465, Bad Gastein, 2005.

[MMSK04] J. Mehnen, Th. Michelitsch, K. Schmitt, and T. Kohlen. pMOHypEA: Parallel evolutionary multiobjective optimization using hypergraphs. Technical Report of the Collaborative Research Centre 531 *Computational Intelligence* CI–189/04, University of Dortmund, 2004.

[MOELL04] B. Möller, M. Beer. *Fuzzy Randomness*. ISBN 3-540-40294-2, Springer Verlag, Berlin, Heidelberg, New York, 2004.

[Mon01] D. C. Montgomery. *Design and Analysis of Experiments*. 5 edition, John Wiley, New York, 2001.

[Mos89] P. Moscato. On Evolution, Search, Optimization, Genetic Algorithms and Martial Arts: Towards Memetic Algorithms. Technical Report Caltech Concurrent Computation Program, Report. 826, California Institute of Technology, Pasadena, CA, 1989.

[MPS67] D. R. Mills, R. L. Peterson, and S. Spiegelman. An extracellular Darwinian experiment with a self-duplicating nucleic acid molecule. *Proc. Natl. Acad. Sci. USA*, 58:217–224, 1967.

[MR95] R. Motwani and P. Raghavan. *Randomized algorithms*. Cambridge University Press, New York, 1995.

[MS96] Z. Michalewicz and M. Schoenauer. Evolutionary algorithms for constrained parameter optimization problems. *Evolutionary Computation*, 4(1):1–32, 1996.

[MS05] Martin G. Moehrle and Wulf-Dieter Spilgies. Quality function deployment for product service systems. *Industrie Management*, 21. Jg(H. 3), 2005.

[Nay01] C. D. Naylor. *Clinical decisions: from art to science and back again*. 358. The Lancet, 2001. Pachur, T. and Hertwig, R. (in press).

[NB00] W. Nary and J. Barnes. Solving the pickup and delivery problem with time windows using reactive tabu search. *Transportation Research Part B*, 34:107–121, 2000.

[NHZ92] H. F. Noller, V. Hoffarth, and L. Zimniak. Unusual resistance of peptidyl transferase to protein extraction procedures. *Science*, 256:1416–1419, 1992.

[NKB+96] A. Nolte, S. Klußmann, R. Bald, V. A. Erdmann, and J. P. Fürste. Mirror-design of l-oligonucleotide ligands binding to l-arginine. *Nature Biotechnology*, 14: 1116–1119, 1996.

[NM65] J. A. Nelder and R. Mead. A simplex method for function minimization. *Computer Journal*, 7(4):308–313, 1965.

[NR02] D. Naddef and G. Rinaldi. Branch-and-Cut Algorithms for the Capacitated VRP. In Toth and Vigo [TV02b], pages 53–84.

[OHMW02] T. Okuda, T. Hiroyasu, M. Miki, and S. Watanabe. DCMOGA: Distributed Cooperation model of Multi-Objective Genetic Algorithm. In *MPSN – II, The Second Workshop on Multiobjective Problem Solving from Nature*, Granada, 2002.

[Oli03] S. E. Olive. Differences in performance and preference of trained versus untrained listeners in loudspeaker tests: A case study. *Journal of the Audio Engineering Society*, 51(9):806–825, 2003.

[OW99] F. Oppacher and M. Wineberg. The shifting balance genetic algorithm: Improving the GA in a dynamic environment. In W. Banzhaf, J. Daida, A. E. Eiben, M. H. Garzon, V. Honavar, M. Jakiela, and R. E. Smith, editors, *Proc. Genetic and Evolutionary Computation Conf. (GECCO 1999), Orlando FL*, volume 1, pages 504–510. Morgan Kaufmann, San Francisco, 1999.

[PA90] G. J. Prokopakis and J. A. Asenjo. Synthesis of downstream processes. In J. A. Asenjo, editor, *Separation Processes in Biotechnology*, volume 9 of *Bioprocess Technology*, pages 571–601. Marcel Dekker, Inc., New York, 1990.

[Pan02] G. Pankratz. *Speditionelle Transportdisposition*. DUV, 2002.

[Pap04] Paper presented at the Engineering Systems Symposium. *Foundational issues in engineering systems: A framing paper*, Pages 29–31. MIT Press, Cambridge, MA, March 2004.

[Par96] V. Pareto. *Cours d'Economie Politique 1*. Lausanne, Rouge, 1896.

[Par71] V. Pareto. *Manual of Political Economy*. Augustus M. Kelley (Original in French 1927), New York, 1971.

[patICoED81] Paper presented at the 1981 International Conference on Engineering Design, editor. *Concept selection – a method that works*, Rome, Italy, 1981.

[PH06] T. Pachur and R. Hertwig. *On the psychology of the recognition heuristic: Retrieval primacy*, volume 32. 2006.

[PNR06] M. Preuss, B. Naujoks, and G. Rudolph. Pareto set and EMOA behavior for simple multimodal multiobjective functions. In Th.Ph. Runarsson et al., editors, *Parallel Problem Solving from Nature (PPSN IX)*, volume 4193 of *LNCS*, pages 513–522. Springer, Berlin, 2006.

[Pro] Proceedings of the 7th International Conference on Cognitive Modeling. *Fast and frugal trees: A theory of robust classification.*

[Psa95] H. Psaraftis. Dynamic vehicle routing: status and prospects. *Annals of Operations Research*, 61:143–164, 1995.

[PSE05] M. Preuss, L. Schönemann, and M. Emmerich. Counteracting genetic drift and disruptive recombination in $(\mu \overset{+}{,} \lambda)$-ea on multimodal fitness landscapes. In H.-G. Beyer, editor, *Proc. 2005 Conf. on Genetic and Evolutionary Computation, (GECCO 2005)*, pages 865–872. ACM Press, New York, 2005.

[PSM+99] D. Papadopoulos, D. Schneider, J. Meier-Eiss, W. Arber, R. E. Lenski, and M. Blot. Genomic evolution during a 10,000-generation experiment with bacteria. *Proceedings of the National Academy of Science*, 96:3807–3812, 1999.

[PVM] Home Page of PVM – Parallel Virtual Machine.

[Ran07] C. Rang. *Lenk- und Ruhezeiten im Straßenverkehr.* Verlag Heinrich Vogel, 2007.

[Ray97] E. Raymond. Die Kathedrale und der Basar. Das Erfolgsgeheimnis von Linux. *Linux-Magazin*, 8, 1997.

[RdD97] Donna Cooper Richard de Dear, Gail Brager. Developing an adaptive model of thermal comfort and preference. Technical Report ASHRAE RP- 884, American Society of Heating, Refrigerating and Air Conditioning Engineers, Inc., http://aws.mq.edu.au/rp-884/RP884_Final_Report.pdf, March 1997. ASHRAE Report.

[Rec71] I. Rechenberg. *Evolutionsstrategie: Optimierung technischer Systeme nach Prinzipien der biologischen Evolution.* PhD thesis, Department of Process Engineering, Technical University of Berlin, Germany, 1971.

[Rec73] I. Rechenberg. *Evolutionsstrategie: Optimierung technischer Systeme nach Prinzipien der biologischen Evolution.* Frommann-Holzboog, Stuttgart, 1973.

[Rec89] I. Rechenberg. Evolution strategy – nature's way of optimization. In H. W. Bergmann, editor, *Optimization: Methods and Applications, Possibilities and Limitations.* Springer, Berlin, 1989.

[Rec94] I. Rechenberg. *Evolutionsstrategie 94.* Stuttgart: Frommann-Holzboog, Stuttgart, 1994.

[Rei79] W. Reichardt. *Gute Akustik – aber wie*? VEB Verlag Technik, Berlin, 1 edition, 1979.

[Ric62] A. Rich. *Horizons in Biochemistry, M. Kasha and B. Puhlman.* Academic Press, New York Lieber, 1962.

[RJ90] D. L. Robertson and G. F. Joyce. Selection in vitro of an rna enzyme that specifically cleaves single-stranded dna. *Nature*, 344:467–468, 1990.

[RJS+95] S. Ringquist, T. Jones, E. E. Snyder, T. Gibson, I. Boni, and L. Gold. High-affinity rna ligands to escherichia coli ribosomes and ribosomal protein s1: comparison of natural and unnatural binding sites. *Biochemistry*, 34:3640–3648, 1995.

[RK04] T. Reimer and K. Katsikopoulos. The use of recognition in group decision-making. *Cognitive Science*, 28, 2004.

[Roo99] P. Roosen. Concept and application of the evolutionary parameter optimization toolbox EPO. *ASME-Conference "Modeling and Simulation"*, 1999.

[RPH+07] C. Richter, D. Polityko, J. Hefer, S. Guttowski, H. Reichl, M. Berger, U. Nowak, and M. Schröder. Technology aware modeling of 2.5D-SiP for automation in physical design. In *Proceedings of the 9th Electronics Packaging Technology Conference*, pages 623–630, December 2007.

[RSS97] C. Reidys, P. F. Stadler, and P. Schuster. Generic properties of combinatory maps. Neutral networks of RNA secondary structure. *Bulletin of Mathematical Biology*, 59:339–397, 1997.

[Rud98] G. Rudolph. On a multi–objective evolutionary algorithm and its convergence to the Pareto set. In D. B. Fogel, H.-P. Schwefel, Th. Bäck, and X. Yao, editors, *Proc. Fifth IEEE Conf. Evolutionary Computation (ICEC'98), Anchorage AK*, pages 511–516. IEEE Press, Piscataway NJ, 1998.

[RVP+06] W. Rypniewski, M. Vallazza, M. Perbandt, S. Klussmann, L. J. Delucas, C. Betzel, and V. A. Erdmann. The first crystal structure of an rna racemate. *Acta Crystallographica D Biological Crystallography*, 62:659–664, 2006.

[Sac00] C. Sachs. *Planung und Bewertung strategischer Investitionsprojekte auf Basis stochastischer Netzpläne*. Cottbus: Kovac, Cottbus, 2000.

[Sak00] M. Sakurai. *Computational systems for sound fields, as tools in design and diagnosis*. Dissertation, Kobe University, 2000.

[SBA02] G. Specht, C. Beckmann, and J. Amelingmeyer. *F&E-Management – Kompetenz im Innovationsmanagement*, volume 2. Aufl. Stuttgart: UTB, Stuttgart, 2002.

[Sch53] A. Schütz. Common sense and scientific interpretation of human action. *Philosophy and phenomenological research*, 14(1):1–38, 1953.

[Sch65] H.-P. Schwefel. Kybernetische Evolution als Strategie der exprimentellen Forschung in der Strömungstechnik. Master's thesis, Technical University of Berlin, Germany, 1965.

[Sch75] H.-P. Schwefel. *Evolutionsstrategie und numerische Optimierung*. PhD thesis, Department of Process Engineering, Technical University of Berlin, Germany, 1975.

[Sch77] H.-P. Schwefel. *Numerische Optimierung von Computer–Modellen mittels der Evolutionsstrategie*, volume 26 of *Interdisciplinary Systems Research*. Birkhäuser, Basle, Switzerland, 1977.

[Sch81] H.-P. Schwefel. *Numerical Optimization of Computer Models*. Wiley, Chichester, 1981.

[Sch84] J. D. Schaffer. *Multiple Objective Optimization with Vector Evaluated Genetic Algorithms*. PhD thesis, Vanderbilt University, 1984.

[Sch87] H.-P. Schwefel. Collective phenomena in evolutionary systems. In P. Checkland and I. Kiss, editors, *Problems of Constancy and Change – The Complementarity of Systems Approaches to Complexity, Proc. 31st Annual Meeting*, volume 2, pages 1025–1033. Int'l Soc. for General System Research, 1987.

[Sch95a] H.-P. Schwefel. *Evolution and Optimum Seeking*. Sixth-Generation Computer Technology. Wiley Interscience, New York, 1995.

[Sch95b] H.-P. Schwefel. *Evolution and Optimum Seeking*. Sixth-Generation Computer Technology Series. Wiley-Interscience, 1995.

[Sch99a] H. J. Schreyer. Untersuchungen zur Zeitverkürzung des Entwicklungsprozesses in der Automobilindustrie mit der Hilfe von GERT-Netzwerken. Technical report, Dissertation Technische Universität Cottbus, Cottbus, 1999.

[Sch99b] M. Schroeder. Die Akustik von Konzertsälen. *Physikalische Blätter*, 55(11): 47–50, 1999.

[Sch02] R. Schwarz. *Controlling-Systeme – Eine Einführung in Grundlagen, Komponenten und Methoden des Controlling*. Gabler, Wiesbaden, 2002.

[Sch03] P. Schuster. Molecular insight into the evolution of phenotypes. In James P. Crutchfield and Peter Schuster, editors, *Evolutionary Dynamics – Exploring the Interplay of Accident, Selection, Neutrality, and Function*, pages 163–215. Oxford University Press, New York, 2003.

[Sch05] J. Schönberger. *Operational Freight Carrier Planning*. Springer, 2005.

[Sch06a] G. Schembecker. Prozesssynthese in der Trenntechnik. In R. Goedecke, editor, *Fluidverfahrenstechnik*. Wiley-VCH, Weinheim, 2006.

[Sch06b] A. Scherrer. Adaptive approximation of nonlinear minimization problems - The adaptive clustering method in inverse radiation therapy planning. *PhD Thesis, Department of Mathematics, Technische Unversität Kaiserslautern*, 2006.

[Sch06c] P. Schuster. Prediction of RNA secondary structures: From theory to models and real molecules. *Reports on Progress in Physics*, 69:1419–1477, 2006.

[Sci96] E. Sciubba. Numerical process- and plant simulation methods. *Chemical Engineering and Technology*, 19(2):170–184, 1996.

[Sco02] A. Scott. Identifying and analysing dominant preferences in discrete choice experiments. *Journal of Economic Psychology*, 23, 2002.

[SDWS95] G. Schembecker, T. Droge, U. Westhaus, and K. H. Simmrock. Readpert – development, selection and design of chemical reactors. *Chemical Engineering and Processing*, 34(3):317–322, 1995.

[SE97] G. Strunk and T. Ederhof. Machines for automated evolution experiments in vitro based on the serial transfer concept. *Biophysical Chemistry*, 66:193–202, 1997.

[Sei00] R. Seindal. Gnu m4 development site. *http://www.seindal.dk/rene/gnu/*, (Stand Frühjahr 2000).

[SF03] P. F. Stadler and C. Flamm. Barrier trees on poset-velued landscapes. *Genet. Prog. Evolv. Mach.*, 4:7–20, 2003.

[SG02] D. Specht and M. G. Moehrle. *Lexikon Technologiemanagement*. Wiesbaden: Gabler, Wiesbaden, 2002.

[Sha92] J. Shanteau. How much information does an expert use? is it relevant? *Acta Psychologica*, 81, 1992.

[Shi05] O. M. Shir. Niching in evolution strategies. In H.-G. Beyer, editor, *Proc. 2005 Conf. on Genetic and Evolutionary Computation, (GECCO 2005)*, pages 865–872, New York, 2005. ACM Press, New York.

[SHT+04] S. Sato, T. Hayashi, A. Takizawa, A. Tani, H. Kawamura, and Y. Ando. Acoustic design of theatres applying genetic algorithms. *Journal of Temporal Design in Architecture and the Environment*, 4(1), 2004.

[Sim55] H. A. Simon. behavioral model of rational choice. *A. Quarterly Journal of Economics*, 69, 1955.

[Sim56] H. A. Simon. Rational choice and the structure of environments. *Psychological Review*, 63, 1956.

[SK00] Y. Sano and H. Kita. Optimization of Noisy Fitness Functions by Means of Genetic Algorithms using History of Search. In M. Schoenauer, K. Deb, G. Rudolph, X. Yao, E. Lutton, J. J. Merelo, and H.-P. Schwefel, editors, *Parallel Problem Solving from Nature (PPSN VI)*, volume 1917 of *LNCS*, pages 571–580. Springer, Berlin, 2000.

[SKM+04] A. Scherrer, K. H. Küfer, M. Monz, F. V. Alonso, and T. R. Bortfeld. IMRT planning on adaptive volume structures - a significant advance in computational complexity. *Berichte des ITWM*, 60, 2004. http://www.itwm.fhg.de/zentral/download/berichte/bericht60.pdf.

[SM01] W. Schlegel and A. Mahr. 3D Conformal Radiation Therapy - Multimedia Introduction to Methods and Techniques. Multimedia CD-ROM, Springer, 2001.

[SOT+02] S. Sato, K. Otori, A. Takizawa, H. Sakai, Y. Ando, and H. Kawamura. Applying genetic algorithms to the optimum design of a concert hall. *Journal of Sound and Vibration*, 258(3):517–526, 2002.

[SPG+97] R. Séguin, J.-Y. Potvin, M. Gendreau, T. G. Crainic, and P. Marcotte. Real-time decision problems: an operational research perspective. *Journal of the Operational Research Society*, 48(2):162–174, 1997.

[Spi71] S. Spiegelman. An approach to the experimental analysis of precellular evolution. *Quarterly Reviews of Biophysics*, 4:213–253, 1971.

[Spi06] W.-D. Spilgies. Realoptionen im Produktinnovationsmanagement. Technical report, Dissertation Universität Bremen, Bremen, 2006.

[SRB95] H.-P. Schwefel, G. Rudolph, and Th. Bäck. Contemporary evolution strategies. In F. Morán, A. Moreno, J. J. Merelo, and P. Chacón, editors, *Advances in Artificial Life – Proc. Third European Conf. Artificial Life (ECAL'95)*, pages 893–907. Springer, Berlin, 1995.

[SS93] M. Sassanfar and J. W. Szostak. An rna motif that binds atp. *Nature*, 364: 550–553, 1993.

[SS98] M. Savelsbergh and M. Sol. Drive: Dynamic Routing of Independent Vehicles. *Operations Research*, 46:474–490, 1998.

[SSUZ03] F. Streichert, G. Stein, H. Ulmer, and A. Zell. A clustering based niching method for evolutionary algorithms. In E. Cantú-Paz, editor, *Proc. 2003 Conf. on Genetic and Evolutionary Computation, (GECCO 2003)*, pages 644–645. Springer, Berlin, 2003.

[SSWF01] B. R. M. Stadler, P. F. Stadler, G. P. Wagner, and W. Fontana. The topology of the possible: Formal spaces underlying petterns of evolutionary change. *Journal of the Theoretical Biology*, 213:241–274, 2001.

[ST85] D. D. Sleator and R. E. Tarjan. Amortized efficiency of list update and paging rules. *Communications of the ACM*, 28(2):202–208, 1985.

[Sta95] S. L. Star. The politics of formal representation: Wizards, gurus, and organizational complexity. In S. L. Star, editor, *Ecologies of knowledge: Work and politics in science and technology*, pages 88–118. SUNY UP, Albany, NY, 1995.

[Sta96] S. L. Star. Working together: Symbolic interactionism, activity theory and information systems. In Y. Engeström and D. Middleton, editors, *Cognition and communication at work*, pages 296–318. Cambridge University Press, Cambridge, 1996.

[Sta98] P. Stagge. Averaging efficiently in the presence of noise. In A. Eiben, editor, *Parallel Problem Solving from Nature, PPSN V*, pages 188–197. Springer, Berlin, 1998.

[Ste86] R. E. Steuer. *Multiple Criteria Optimization: Theory, Computation and Application*. John Wiley, New York, 546pp, 1986.

[Ste92] T. J. Stewart. A critical survey on the status of multiple criteria decision making theory and practice. *OMEGA*, 20(5–6):569–586, 1992.

[Ste99] S. Pinker. The Seven Wonders of the World. http://pinker.wjh.harvard.edu/articles/media/1999_06_07_mcgill.html, 7 June 1999. Steven Pinker's convocation address at McGill University.

[Ste04] U. M. Stephenson. *Beugungssimulation ohne Rechenzeitexplosion: Die Methode der quantisierten Pyramidenstrahlen – ein neues Berechnungsverfahren für Raumakustik und Lärmimmissionsprognose*. Dissertation, RWTH Aachen, Aachen, 2004.

[STG92] D. Schneider, C. Tuerk, and L. Gold. Selection of high affinity rna ligands to the bacteriophage-r17 coat protein. *Journal of Molecular Biology*, 228:862–86, 1992.

[Str78] A. L. Strauss. *Negotiations: Varieties, contexts, processes and social order*. Jossey-Bass, San Francisco, 1978.

[Sue08] P. Suess. A primal-dual barrier algorithm for the imrt planning problem – an application of optimization-driven adaptive discretization. *PhD Thesis, Department of Mathematics, Technische Unversität Kaiserslautern*, 2008.

[SWJ98] M. Schonlau, W. Welch, and R. Jones. Global Versus Local Search in Constrained Optimization of Computer Models. *New Development and Applications in Experimental Design*, 34:11–25, 1998.

[SWM+02] B. Schwartz, A. Ward, J. Monterosso, S. Lyubomirsky, K. White, and D. R. Lehman. Happiness is a matter of choice. *Journal of Personality and Social Psychology*, 83, 2002.

[SWN03a] T. J. Santner, B. J. Williams, and W. I. Notz. *The Design and Analysis of Computer Experiments*. Springer, Berlin, 2003a.

[SWN03b] T. J. Santner, B. J. Williams, and W. I. Notz. *The Design and Analysis of Computer Experiments*. Springer Series in Statistics. Springer, New York, 2003b.

[TG90a] C. Tuerk and L. Gold. Systematic evolution of ligands by exponential enrichment: Rna ligands to bacteriophage t4 dna polymerase. *Science*, 249:505–510, 1990.

[TG90b] C. Tuerk and L. Gold. Systematic evolution of ligands by exponential enrichment: RNA ligands to bacteriophage T4 DNA polymerase. *Science*, 249: 505–510, 1990.

[TGtArgip19] P. M. Todd, G. Gigerenzer, and the ABC research group (in press). Ecological rationality: Intelligence in the world. Technical report, Oxford University Press, New York.

[TLWK01] D. Thirumalai, N. Lee, S. A. Woodson, and D. K. Klimov. Early events in RNA folding. *Annual Review of Physical Chemistry*, 52:751–762, 2001.

[TMG92] C. Tuerk, S. MacDougal, and L. Gold. Rna pseudoknots that inhibit human immunodeficiency virus type 1 reverse transcriptase. *Proceedings of the National Academy of Science*, 89:6988–6992, 1992.

[TMW93] C. Tuerk and S. MacDougal-Waugh. In vitro evolution of functional nucleic acids: high-affinity rna ligands of hiv-1 proteins. *Gene*, 137:33–39, 1993.

[Tom05] M. Tomassini. *Spatially Structured Evolutionary Algorithms Artificial Evolution in Space and Time*. Natural Computing Series. Springer, Berlin, 2005.

[TUA05a] S. Tanaka, Y. Uraguchi, and M. Araki. Dynamic optimization of the operation of single-car elevator systems with destination hall call registration: Part I. Formulation and simulations. *European Journal of Operational Research*, 167(2): 550–573, 2005a.

[TUA05b] S. Tanaka, Y. Uraguchi, and M. Araki. Dynamic optimization of the operation of single-car elevator systems with destination hall call registration: Part II. The solution algorithm. *European Journal of Operational Research*, 167(2):574–587, 2005b.

[TV02a] P. Toth and D. Vigo. Branch-and-bound Algorithms for the Capacitated VRP. In *The Vehicle Routing Problem* [TV02b], pages 29–51.

[TV02b] P. Toth and D. Vigo, editors. *The Vehicle Routing Problem*. SIAM, Philadelphia 2002.

[TV02c] P. Toth and D. Vigo. Vrp with Backhauls. In *The Vehicle Routing Problem* [TV02b], pages 195–224.

[Urs99] R. K. Ursem. Multinational evolutionary algorithms. In P. J. Angeline, editor, *Proc. of the Congress of Evolutionary Computation (CEC-99)*, volume 3, pages 1633–1640. IEEE Press, Piscataway, NJ, 1999.

[VALP01] E. Vásquez-Alvarez, M. E. Lienqueo, and J. M. Pinto. Optimal synthesis of protein purification processes. *Biotechnology Progress*, 17:685–696, 2001.

[vH05] E. von Hippel. *Democratizing Innovation*. PhD thesis, Cambridge, MA: MIT Press, 2005.

[Vie82] D. Viefhues. *Entscheidungskalküle zur Projektablauf – und – anpassungsplanung*. Lang, Frankfurt/Main, 1982.

[VM02] M. Vorländer and F. P. Mechel. Room acoustics. In F. P. Mechel, editor, *Formulas Acoustics*, pages 774–845. Springer-Verlag, Berlin etc., 2002.

[Vor04] M. Vorländer. Room acoustics. Lecture, RWTH Aachen, 2004.

[VRMG04] A. G. Vitreschak, D. A. Rodionov, A. A. Mironov, and M. S. Gelfand. Riboswitches: the oldest mechanism for the regulation of gene expression? *Trends in Genetics*, 20(1):44–50, 2004.

[vV96] A. van Vliet. On the asymptotic worst case behavior of harmonic fit. *Journal of Algorithms*, 20:113–136, 1996.

[vV99] D. A. van Veldhuizen. *Multiobjective Evolutionary Algorithms: Classifications, Analyses, and New Innovations*. PhD thesis, Air Force Institute of Technology, USA, 1999.

[Vö71] H. Völzgen. *Stochastische Netzwerkverfahren und deren Anwendungen*. Berlin, New York: de Gruyter, Berlin, New York, 1971.

[WA] Wikipedia-Article. Phylogenetics. http://en.wikipedia.org/wiki/Phylogenetics. Explanation of the concept of phylogenesis, i.e. the systematic order of ancestral relationships of biological species.

[Wal50] A. Wald. *Statistical decision functions*. Wiley, New York, 1950.

[WB03] W. C. Winkler and R. R. Breaker. Genetic control by metabolite-binding riboswitches. *Chembiochem*, 4(10):1024–1032, 2003.

[Web04] S. Weber. *The success of open source*. Harvard University Press, Cambridge, MA, 2004.

[WGW72] H. Wille, K. Gewald, and H. D. Weber. *Netzplantechnik - Methoden zur Planung und Überwachung von Projekten*, volume Band 1: Zeitplanung of *3. Aufl.* Muenchen: Oldenbourg, München, Oldenbourg, 1972.

[Wie03] Wieder GmbH. Werkzeugkühlung erhält Impulse. *MaschinenMarkt, MM Plastics*, 5, 2003.

[Wika] Wikipedia article. Civilization (computer game). http://en.wikipedia.org/wiki/... ...Civilization_computer_game. Short introduction in a series of computer games called 'Civilization', with a discussion of their features and game systematics.

[Wikb] Wikipedia article. Marsupial. http://en.wikipedia.org/wiki/Marsupial.

[Wikc] Wikipedia article. Teleology. http://en.wikipedia.org/wiki/Teleology. Explanation of various philosophical concepts of an assumed targeted development, mostly of nature.

[Wik06a] Wikipedia, http://en.wikipedia.org/wiki/Kriging. *Kriging*, July 2006. Wikipedia synapsis of the method and history of the Kriging interpolation technique.

[Wik06b] Wikipedia article, http://en.wikipedia.org/wiki/Prolog. *Prolog, the programming language*, July 2006. Short overview about the concepts of the computer programming language Prolog.

[Win72] A. Winfree. Experimental realisation of a belousov-zhabotinskii reaction. *Science*, 175:634–636, 1972.

[WJ99] C. Weihs and J. Jessenberger. *Statistische Methoden zur Qualitätssicherung und -optimierung in der Industrie*. Wiley-VCH, Weinheim, 1999.

[WK99] M. Wellenhofer-Klein. *Zulieferverträge im Privat- und Wirtschaftsrecht*. Münchener Universitätsschriften, 1999.

[WM97a] D. H. Wolpert and W. G. Macready. No Free Lunch Theorems for Optimization. *IEEE Transactions on Evolutionary Computation*, 1(1):67–82, 1997.

[WM97b] D. H. Wolpert and W. G. Macready. No free lunch theorems for optimization. *IEEE Transactions on Evolutionary Computation*, 1(1):67–82, 1997.

[WMM$^+$04] K. Weinert, J. Mehnen, Th. Michelitsch, K. Schmitt, and Th. Bartz-Beielstein. A Multiobjective Approach to Optimize Temperature Control Systems of Molding Tools. *Production Engineering? Research and Development, Annals of the German Academic Society for Production Engineering (WGP)*, XI(1), 2004.

[WNB02] W. Winkler, A. Nahvi, and R. R. Breaker. Thiamine derivatives bind messenger rnas directly to regulate bacterial gene expression. *Nature*, 419:952–956, 2002.

[Wri95] M. H. Wright. Direct search methods: Once scorned, now respectable. In *Proc. 1995 Dundee Biennial Conf. in Numerical Analysis)*, volume 344 *of Pitman Res. Notes Math. Ser.*, pages 191–208. CRC Press, Boca Raton, FL, 1995.

[WS99] D. S. Wilson and Jack W. Szostak. In vitro selection of functional nucleic acids. *Annual Review Biochemistry*, 68:611–647, 1999.

[WWB05] H. A. Wichman, J. Wichman, and J. J. Bull. Adaptive molecular evolution for 13,000 phage generations: A possible arms race. *Genetics*, 170:19–31, 2005.

[Yar93] M. Yarus. How many catalytic rnas? ions and the cheshire cat conjecture. *The FASEB Journal*, 7:31–39, 1993.

[ZADE65] L. A. Zadeh. Fuzzy sets. *Information and Control*, 8:338–353, 1965.

[ZÖ0] G. Zäpfel. *Strategisches Produktionsmanagement*. Oldenbourg, Russia 2000.

[Zie95] G. Ziegenhals. Vergleich von Hörtests aus Spieler- und Hörerperspektive. In *Fortschritte der Akustik*, Saarbrücken, 1995. 21. Deutsche Jahrestagung für Akustik, DAGA 1995.

[Zie96] G. Ziegenhals. Zur Beurteilung von Gitarren aus Spieler- und Zuschauerperspektive. In *Fortschritte der Akustik*, Bonn, 1996. 22. Deutsche Jahrestagung für Akustik, DAGA 1996.

[Zie00] G. Ziegenhals. Beurteilung objektiver Merkmale von Musikinstrumenten. In *Fortschritte der Akustik*, Oldenburg, 2000. 26. Deutsche Jahrestagung für Akustik, DAGA 2000.

[Zie02] G. Ziegenhals. Zur objektiven Beurteilung von Klavieren. In *Fortschritte der Akustik*, Bochum, 2002. 28. Deutsche Jahrestagung für Akustik, DAGA 2002.

[ZLB03] E. Zitzler, M. Laumanns, and S. Bleuler. A tutorial on evolutionary multiobjective optimization. In *Workshop on Multiple Objective Metaheuristics (MOMH 2002)*. Springer, Berlin, 2003.

[ZLT01a] E. Zitzler, M. Laumanns, and L. Thiele. SPEA2: Improving the Strength Pareto Evolutionary Algorithm. Technical Report 103, Computer Engineering and Communication Networks Lab (TIK), Swiss Federal Institute of Technology (ETH), Gloriastrasse 35, CH-8092 Zürich, 2001.

[ZLT01b] E. Zitzler, M. Laumanns, and L. Thiele. SPEA2: Improving the strength pareto evolutionary algorithm for multiobjective optimization. In K. C. Giannakoglou, D. T. Tsahalis, J. Periaux, K. D. Papailiou, and T. Fogarty, editors, *Evolutionary Methods for Design, Optimization and Control with Applications to Industrial Problems*, pages 1–6. International Center for Numerical Methods in Engineering(CIMNE), Barcelona, 2001.

[ZS81] M. Zuker and P. Stiegler. Optimal computer folding of large RNA sequences using thermodynamics and auxiliary information. *Nucleic Acids Research*, 9:133–148, 1981.

[Zuk89] Michael Zuker. On finding all suboptimal foldings of an RNA molecule. *Science*, 244:48–52, 1989.

[Zö97] O. Zöllner. Optimierte Werkzeugtemperierung. Technical Report ATI 1104, Bayer, Plastics Business Group, 1997.

GPSR Compliance

The European Union's (EU) General Product Safety Regulation (GPSR) is a set of rules that requires consumer products to be safe and our obligations to ensure this.

If you have any concerns about our products, you can contact us on ProductSafety@springernature.com

In case Publisher is established outside the EU, the EU authorized representative is:

Springer Nature Customer Service Center GmbH
Europaplatz 3
69115 Heidelberg, Germany

Batch number: 10755442

Printed by Printforce, the Netherlands